Kamal Kumar, Shahid Ahmad Bhat, Paul Augustine Ejegwa, Reeta Bhardwaj (Eds.)
Fuzzy Decision-Making Algorithms

De Gruyter Series on the Applications of Mathematics in Engineering and Information Sciences

Edited by
Mangey Ram

Volume 26

Fuzzy Decision-Making Algorithms

Mathematical Optimization, Metaheuristic Optimization, Supply Chain Management

Edited by
Kamal Kumar, Shahid Ahmad Bhat, Paul Augustine Ejegwa, and Reeta Bhardwaj

DE GRUYTER

Editors

Dr Kamal Kumar
Department of Mathematics
Amity School of Applied Sciences
Amity University Haryana
Pachgaon, Manesar
Gurugram 122413, Haryana
India
kamalkumarrajput92@gmail.com

Dr Shahid Ahmad Bhat
LUT – University School of Business
and Management
Yliopistonkatu 34
FI-53850 Lappeenranta
Finland
shahid.bhat@lut.fi

Dr Paul Augustine Ejegwa
Department of Mathematics
Joseph Sarwuan Tarka University
P.M.B. 2373
970101 Makurdi
Nigeria
ejegwa.augustine@uam.edu.ng

Dr Reeta Bhardwaj
Department of Mathematics
Amity School of Applied Sciences
Amity University Haryana
Pachgaon, Manesar
Gurugram 122413, Haryana
India
bhardwajreeta84@gmail.com

ISSN 2626-5427
ISBN 978-3-11-914266-3
ISBN 978-3-11-222224-9 (PDF)
ISBN 978-3-11-222254-6 (EPUB)
DOI https://doi.org/10.1515/9783112222249

Library of Congress Control Number: 2026932444

Bibliographic information published by the Deutsche Nationalbibliothek
The Deutsche Nationalbibliothek lists this publication in the Deutsche Nationalbibliografie; detailed bibliographic data are available on the internet at http://dnb.dnb.de.

De Gruyter and Walter de Gruyter GmbH are part of De Gruyter Brill.
www.degruyterbrill.com

Questions about General Product Safety Regulation:
productsafety@degruyterbrill.com

Cover illustration: MF3d/E+/Getty Images
Printing: CPI books GmbH, Leck

MIX
Papier | Fördert gute Waldnutzung
FSC® C083411

Preface

Contemporary supply chains function within a highly dynamic context shaped by globalization, sustainability demands, and significant uncertainty. Decision-makers must simultaneously consider economic performance, environmental stewardship, and social responsibility, often while working with vague, incomplete, or conflicting information. This book responds to these challenges by presenting an integrated approach that combines metaheuristic optimization techniques with sophisticated fuzzy decision-making models.

Metaheuristic algorithms, which are inspired by natural, biological, and physical phenomena, have proven effective in addressing complex and large-scale optimization problems. When these algorithms are coupled with fuzzy decision-making approaches, they offer robust mechanisms for representing uncertainty and managing the inherent complexity of modern supply chain systems. This book delivers a thorough exploration of such hybrid methodologies, encompassing theoretical foundations, methodological innovations, and practical implementations in supply chain management.

The volume discusses a broad spectrum of decision-making tools, including entropy-based measures, distance and similarity metrics, correlation coefficients, possibility measures, aggregation operators, and well-established multi-criteria decision-making techniques such as TOPSIS and EDAS. These methodological developments are reinforced with real-world case studies that demonstrate the practical value of metaheuristic fuzzy decision-making algorithms in improving supply chain sustainability and resilience. Illustrative applications, such as green supplier selection, show how environmental policies, product quality, and regulatory constraints can be effectively incorporated under uncertain conditions.

Several significant contributions are advanced in this book. These include the development of modified Pythagorean fuzzy distance measures embedded within a metaheuristic fuzzy framework for circular economy decision-making, as well as novel q-rung orthopair fuzzy correlation coefficient-based methods for multi-criteria decision-making under high uncertainty. The book further addresses food waste management using p, q-quasirung orthopair fuzzy models and extends Fermatean fuzzy theory through new measures and correlation coefficients for sustainable and organic food supply chains. Moreover, multi-criteria group decision-making models under probabilistic dual hesitant fuzzy environments are proposed for healthcare supply chain management, alongside orthopair fuzzy composite relations for waste management optimization.

By integrating methods of methodology with fuzzy decision-making paradigms, this book offers practical and implementable frameworks for addressing the complexity and uncertainty of contemporary supply chains. It is intended for researchers, practitioners, sustainability professionals, and policymakers seeking advanced computational approaches to support informed, adaptive, and sustainable decision-making.

Editors

 | https://doi.org/10.1515/9783112222249-202

Contents

List of Contributing Authors

Chimele Orji Abraham
Department of Mathematics
Joseph Sarwuan Tarka University
Makurdi
Nigeria

Paul Augustine Ejegwa
Department of Mathematics
Joseph Sarwuan Tarka University
Makurdi
Nigeria

Abolape Deborah Akwu
Department of Mathematics
Joseph Sarwuan Tarka University
Makurdi
Nigeria

Dristi Rajbonshi
Department of Mathematics
Dibrugarh University
Dibrugarh, Assam
India

Palash Dutta
Department of Mathematics
Dibrugarh University
Dibrugarh 786004, Assam
India

Raina Ahuja
Department of Mathematics
Thapar Institute of Engineering & Technology
(Deemed to be University)
Patiala, Punjab
India

Ajendra Kumar
Department of Mathematics and Statistics
Gurukula Kangri (Deemed to be University)
Haridwar 249404, Uttarakhand
India

Vishal Kumar
Department of Mathematics and Statistics
Gurukula Kangri (Deemed to be University)
Haridwar 249404, Uttarakhand
India

Reeta Bhardwaj
Department of Mathematics
Amity School of Applied Sciences
Amity University Haryana Gurugram 122413
India

Ashu Redhu
Department of Mathematics
Amity School of Applied Sciences
Amity University Haryana Gurugram 122413
India

Kamal Kumar
Department of Mathematics
Amity School of Applied Sciences
Amity University Haryana Gurugram 122413
Haryana
India

Rishu Arora
Department of Mathematics
Amity Institute of Applied Sciences
Amity University Noida, Uttar Pradesh
India

Dr Divya Mol K. T.
Department of Basic Science
Amal Jyothi College (Autonomous) of
Engineering
Kanjirappally, Kerala
India

Abhilash Kangsha Banik
Department of Mathematics
Dibrugarh University
Dibrugarh 786004, Assam
India

 | https://doi.org/10.1515/9783112222249-204

Alakananda Konwar
Department of Mathematics
Dibrugarh University
Dibrugarh 786004, Assam
India

Chukwudi Obinna Nwokoro
Department of Computer Science
Nigerian British University
Asa, Abia State
Nigeria

Benard Alechenu
Department of Mathematics and Statistics
Federal University
Kashere, Gombe
Nigeria

Fathima Banu
Department of Science and Humanities (Mathematics)
University College of Engineering
Konam, Nagercoil
Tamil Nadu
India

Subramanian Petchimuthu
Department of Science and Humanities (Mathematics)
University College of Engineering
Konam, Nagercoil
Tamil Nadu
India

Pooja Yadav
Department of Mathematics
Amity School of Applied Sciences
Amity University Haryana
Gurugram, Haryana
India

Gagandeep Kaur
Department of Mathematics
Amity School of Applied Sciences
Amity University Haryana
Gurugram, Haryana
India

Rohit
Department of Mathematics
Amity School of Applied Sciences
Amity University Haryana
Gurugram, Haryana
India

Grace Ataguba
Department of Computer Science
Faculty of Computer Science
Dalhousie University
Halifax, NS
Canada

Rita Orji
Department of Computer Science
Faculty of Computer Science
Dalhousie University
Halifax, NS
Canada

Parul Tomar
Department of Applied Sciences and Humanities
Dronacharya Group of Institutions
Greater Noida, Uttar Pradesh
India

Kirti
Department of Mathematics
National Institute of Technology Kurukshetra
136119 Kurukshetra, Haryana
India

Chirag Dhankhar
Amity College of Commerce
Amity University Haryana Gurugram, Haryana
India

Anju Kumari
Amity College of Commerce
Amity University Haryana Gurugram, Haryana
India

Amit Sharma
Department of Mathematics
Amity School of Applied Sciences
Amity University Haryana Gurugram, Haryana
India

Chimele Orji Abraham, Paul Augustine Ejegwa and
Abolape Deborah Akwu

A Novel Distance Metric Between Fermatean Fuzzy Sets with Application in Vehicle Selection Process Through the TOPSIS Technique

Abstract: This study introduces a novel Fermatean fuzzy distance metric (FFDM), which is integrated with the Technique for Order Preference by Similarity to Ideal Solution in response to the growing need for effective decision-making under uncertainty and sustainable transportation solutions. The main objective is to improve multi-criteria decision-making by providing a more mathematically consistent and reliable tool for tackling higher degrees of hesitancy common in Fermatean fuzzy sets (FFSs). The proposed FFDM, unlike existing distance measures that usually overlook the hesitation degree or lack precision, includes all three parameters of FFSs together with parametric weights, thereby improving stability, interpretability, and discrimination power. The method is applied to the problem of eco-friendly vehicle selection, where alternatives are assessed based on criteria such as pollutant emission, cost, reliability, technical requirements, maintenance, and social acceptance. The comparative analysis shows that the proposed FFDM gives more consistent results and is in line with domain expectations, and offers improved support in decision-making. The findings have significant implications for consumers, policy makers, and industry stakeholders, as well as other potential applications in areas beyond transportation, which may include energy resource management, healthcare, and supply chain.

Keywords: Fermatean fuzzy distance metric, decision-making, MCDM, TOPSIS, vehicle selection, Fermatean fuzzy sets

1 Introduction

One of the most interesting topics for world leaders is the emission of greenhouse gas (GHG) into the atmosphere, thereby resulting in climate change. The quest to reduce

Chimele Orji Abraham, Department of Mathematics, Joseph Sarwuan Tarka University, Makurdi, Nigeria, e-mail: aborji03@gmail.com
Paul Augustine Ejegwa, Department of Mathematics, Joseph Sarwuan Tarka University, Makurdi, Nigeria, e-mail: ejegwa.augustine@uam.edu.ng
Abolape Deborah Akwu, Department of Mathematics, Joseph Sarwuan Tarka University, Makurdi, Nigeria, e-mail: abolaopeyemi@yahoo.co.uk

 | https://doi.org/10.1515/9783112222249-001

the emission of GHG (like carbon dioxide [CO_2]) has raised concerns about the use of fossil-fuel-powered vehicles. The global interest is in the use of eco-friendly vehicles to minimize the emission of GHG. Vehicles, the most common means of transport, are powered by petrol (gasoline), diesel, gas, electricity, biofuel, and solar. These vehicles can be classified based on their power sources, which are: petrol (gasoline)-powered vehicles (PPVs), diesel-powered vehicles (DPVs), gas-powered vehicles (GPVs) (comprising of liquefied petroleum gas and compressed natural gas), electric vehicles (EVs) (including fuel cell EVs, or hydrogen vehicles, hybrid electric vehicles (HEVs), battery electric vehicles (BEVs), and plug-in HEVs), biofuel-powered vehicles (BPVs), and solar-powered vehicles (SPVs), respectively. However, PPVs and DPVs are the most commonly used and are categorized as internal combustion engine (ICE) vehicles [1–4].

To optimize the vehicle selection process, criteria like eco-friendly, cost of purchase, maintenance cost, risk and social acceptance, pollutant emission, reliability, and technical know-how are considered to arrive at the selection of optimum eco-friendly and efficient vehicles due to their power-based sources. The vehicle selection process is a decision-making problem, bedeviled by several uncertainties and vagueness. Decision-making is an essential aspect of human activities, encompassing a vast range of domains, from personal choices to complex industrial systems. Recently, the development of advanced tools of mathematics that can handle nonclear and imprecise situations has been necessitated by the intricacy and unpredictability in the decision-making process [5–7]. The fuzzy sets theory, which was invented by Zadeh [8], is one of such developments, which provides a structure for handling unpredictability in decision-making.

There have been many suggestions on extensions of fuzzy set from Zadeh's first work aimed at enhancing the capacity to capture more difficult forms of uncertainty. Some of these extensions are IFSs [9] and PFSs [10] and although they have been effective in various applications, there exist some limitations especially in their ability to capture higher hesitancy degrees in decision-making. For dealing with hesitancy and unpredictability in decision making, the idea of Fermatean fuzzy sets (FFS) [11] was seen as a more adaptable and general structure so that these constraints can be addressed. The Fermatean fuzzy (FF) structure has been seen to have considerable effect in areas such as social network analysis, supply chain management, security analysis, and medical diagnostics [12–16], although its usage in multi-criteria decision-making (MCDM) problems, such as the optimization of vehicle selection, continues to be a growing research area. In this context, the Technique for Order Preference by Similarity to Ideal Solution (TOPSIS) has proven to be a robust and effective method for solving MCDM problems. The aim of this study is to develop a novel approach for addressing vehicle selection optimization problems by integrating distance measure under FFSs with TOPSIS under conditions of uncertainty. Lots of research have been carried out on the real-life applications of the fuzzy set theory (FST), and Chen et al. [17] championed methods for making fuzzy rules to address real-world data analysis challenges while also enhancing classification accuracy in ambiguous scenarios. Enormous advancements in fuzzy set applications have come to be as a result of the foundation laid by their work. Over time, fuzzy set has grown to handle limitations faced in the handling of higher degrees of uncertainties. The IFSs, proposed by Atanassov [9], improved on the FST by introducing a degree of nonmembership (DNM) alongside degree of membership (DM), thereby providing a more inclusive and detailed representation of uncertainty. The IFSs expanded the potentials of FST in various fields. For example, the IFS has been generally adopted in public sector management [21], medical diagnostics [18], supply chain management [20], social network analysis [19], among others.

It is essential to handle uncertainty and imprecision when dealing with real-world decision-making. The PFSs, pioneered by Yager [10], improved on the IFS by allowing the sum of the square of the DM and DNM to be less than or equal to 1. With this, there is additional flexibility that allows for better representation of uncertainty as compared to IFS and FST. This additional relaxation of the constraint of IFS enhances a more nuanced uncertainty representation [10]. PFSs enhance a greater modeling flexibility, which allows decision-makers to address complex situations with higher precision, unlike the IFS that restricts the sum of the DM and DNM to at most 1. With this increased degree of freedom, PFSs are more effective in representing hesitation, which is a key factor in decision-making under uncertainty. PFSs are definitely useful and have been applied in MCDM problems where there is a need to address conflicting criteria and factors simultaneously. Because of the ability of PFSs to model uncertainty effectively, they have become an ideal tool in areas such as supplier selection [22], decision-making [23, 24], medical image diagnosis [25, 26], and pattern recognition [27]. It should, however, be noted that despite these advantages, PFSs face certain challenges, one of which is that they are less practical for large-scale applications due to an increased computational overhead resulting from the additional degree of freedom.

Surpassing the capabilities of FST, IFSs, and PFSs are the FFSs, which have now become a great instrument for handling uncertainties in decision-making processes [28]. FFSs can be used in a wide range of applications, especially due to their ability to handle uncertain complex information, and are particularly effective in depicting expert reasoning information [29]. FFSs have been found to be useful in different fields, including MCDM [30], pattern analysis [31], medical diagnosis [32, 33], among others. MCDM under fuzzy structures has recently attracted so much attention as a result of its ability to tackle complex decision problems with uncertainty. In MCDM techniques like the TOPSIS, information measures play an essential role in fuzzy systems. These measures help to effectively rank alternatives and assess how alternatives are close to the ideal solutions. Over time, different information measures under FFSs have been proposed by researchers, which include similarity measures, aggregation operators, correlation coefficients, etc. The TOPSIS technique, pioneered by Hwang and Yoon [34], is a widely used MCDM method. It ranks alternatives based on how far they are from a negative-ideal solution and their closeness to the positive-ideal solution. The approach has been expanded to accommodate FF environment. In FFS-based TOPSIS, distance metrics are essential in determining how close alternatives are to the ideal solution. Applications of FFS-TOPSIS have been explored in various domains. Poongavanam et al. [35] applied TOPSIS to select refrigerants that have low global warming potential for automobile air conditioning, demonstrating the versatility of the TOPSIS method in various automotive applications. Gul et al. [36] introduced a fuzzy risk assessment model by integrating FFSs with the TOPSIS method to rank potential hazards in manufacturing settings. Aydoğan and Çetinsaya [37] employed FFSs to overcome limitations in membership functions and the expression of hesitation in MCDM methods. They extended the TOPSIS method with the technique of step-wise weight assessment ratio analysis. The suggested methodology was tried on two actual case studies: ranking for performance evaluation, the Turkish research universities and selecting for a company in the beverage industry, the optimal facility location. Sindhu et al. [38] conducted research that provided an alternative distance measure to the approach of TOPSIS for solving MCDM uncertain problems.

Transportation is a fundamental component of contemporary society, making the movement of people and goods across various distances easier. Vehicles, which are a primary means of transportation, contribute to social connectivity, technological advancements, and economic growth. Efficient transportation systems improve industrial productivity, boost overall quality of life, and enhance trade. Moreover, the choice of vehicles affects urban planning, energy consumption, and environmental sustainability, thereby making vehicle selection a decision essential in modern day transportation planning [39]. Vehicles are grouped based on their energy sources, which include, BPV, DPV, EV, GPV, PPV, and SPV. The application of fuzzy systems in vehicle selection has drawn reasonable attention in recent years. For instance, Dwivedi and Sharma [40] did a work on the ranking and evaluation of battery EVs based on TOPSIS methods and Shannon's entropy. The vehicle selection process, due to the different choices available in the automobile industry, is a complex decision-making task that has become increasingly challenging [41].

In addition, as sustainability becomes increasingly crucial, factors such as environmental impact and corporate social responsibility (CSR) are being integrated into the selection process, particularly in industries with significant environmental concerns [42]. Interestingly, there has been a focus on specific aspects of vehicle selection by some research, such as wheel hub design. The study, which examined the perceptual matching quality between car types and wheel hubs, established that matching quality has a good relationship with the similarity of styles of design between components and the overall car image [43]. This brings to light the essence of looking at aesthetic harmony together with functional criteria during the selection process. Governmental as well as regulatory policies also impact decision-making, as many countries provide incentives for green vehicles that influence consumer choices while introducing variability that is policy-dependent. The selection of an optimal vehicle involves uncertain and subjective criteria, making traditional decision-making methods deficient [44].

Distance measures are important in precisely characterizing the variations between FFSs, with applications in MCDM, clustering, pattern analysis, and medical diagnosis [45]. Researchers have suggested numerous distance metrics for FFSs, including Euclidean distance, hybrid distance, and Hamming distance measures. The work of Senapati and Yager [11], who initiated a distance metric to measure the difference between FFSs, laid the foundation for applying FFSs in numerous MCDM problems based on distance metric, especially where higher degrees of uncertainty are available. However, this method neglected the cardinality of the set in its calculation, which could have an impact on the computed distance result. Ejegwa et al. [14] instituted a distance metric for FFSs, which aimed at enhancing the evaluation of security crises in Nigeria, precisely the North-Central Region. The instituted distance metric was applied to evaluate numerous forms of insecurity, such as kidnappings, banditry, and terrorism, by gathering data through FF linguistic variables (LVs) from security experts. Its performance in other application domains, however, remains to be validated, and the rise in complications of the metric may introduce computational challenges, eventually having an impact in its practical applicability.

Ganie [29] instituted Fermatean fuzzy distance metrics (FFDMs), aiming to enhance the discrimination power in MCDM applications. Ganie's metrics were used in numerous decision-making situations, including resource allocation and risk assessment, where capturing nuanced differences between options is essential. Ganie's FFDMs did not consider the hesitation degree (HD), although the cardinality was considered. Liu [45] suggested an advanced distance metric for FFSs, aimed at increasing the discrimination power in decision-making processes involving strategic planning and medical diagnosis. The method that failed to include the HD, considered the cardinality of the sets involved. Two FFDMs based on triangular divergence and Hellinger metric distance were introduced by Deng and Wang [46], and applied to medical diagnosis and pattern recognition problems. One method did not include the HD, and the other failed to include parametric differences. In Onyeke and Ejegwa [47], the FFDM in Senapati and Yager [11] was modified by the addition of HD and applied in tertiary institutions for the placement of courses. The application in a particular area may limit the generalizability of the metric, and its performance in other contexts remains to be validated. Kirisci [48] introduced FFDMs and used them in risk assessment and supplier selection. Ganie et al. [49] proposed some comprehensive FFDMs, which were focused on improving accuracy in decision-making, as well as structural representation.

While reviewing these existing FFDMs, it was revealed that many distance measures failed to include the HD value in their models, which is crucial for capturing the hesitancy between sets in a more complete manner. Moreover, only a limited number of the developed methods satisfy the fundamental distance metric properties, such as the triangle inequality, identity, symmetry, and nonnegativity. These limitations raise concerns about the reliability and effectiveness of such distance techniques in real-world applications. Addressing these challenges requires the creation of a more reliable and mathematically sound distance metrics for FF environments. By modifying the distance method in Deng and Wang [30] under PFSs and extending it to FFSs, a novel FFDM is constructed in this work based on 3D approach, which integrates all the three parameters of FFSs. In addition, the weights of the feature space are included, leading to precise outcomes and reliable interpretations. The novel FFDM is applied to vehicle selection process via TOPSIS technique, to enhance the selection of an eco-friendly vehicle.

This study aims to propose a novel distance metric between FFSs with application in vehicle selection process through the TOPSIS technique. The main objectives of this study include:

1. Presenting the novel FFDM and analyzing its properties in line with the classical distance metric axioms.
2. Integrating the novel FFDM with the TOPSIS technique to enhance its utility in practical domains.
3. Exploring the optimization of the vehicle selection problem based on the novel distance metric via the TOPSIS approach.
4. Evaluating the impact of the feature space weights on the selection process and comparing the results of the novel distance method with existing FFDMs.

The remaining chapter is arranged as follows: Section 2 presents the preliminaries of FFSs and reiterates the existing FFDMs. Section 3 discusses the new FFDM and presents its theoretical properties. Section 4 presents some illustrative examples. Section 5 presents the use of the new FFDM in the discussion of the selection problem of eco-friendly vehicles based on powered sources. The section also discusses the sensitivity analysis of the new FFDM with respect to the influence of weight. In addition, a comparison is presented to showcase the superiority of the new FFDM over other FFDMs. Section 6 discusses the conclusion, findings of the work, and recommendations for further investigations.

2 Preliminaries

In this section, we review some of the fundamentals of FFSs and distance metrics between FFSs. Throughout this work, we take $\mathcal{M} = \{m_1, m_2, \ldots, m_q\}$ as a nonempty set, where q is the number of elements in $\mathcal{M}$, and FFS($\mathcal{M}$) is the collection of all FFSs in $\mathcal{M}$.

Definition 2.1 ([8]). A fuzzy set represented by $\wp$ in $\mathcal{M}$ is of the form: $\wp = \{m_i, \wp_\mu(m_i) | m_i \in \mathcal{M}\}$, where $\wp_\mu \colon \mathcal{M} \to [0,1]$ is a function representing the DM.

Definition 2.2 ([9]). An IFS depicted by $\bar{\wp}$ in $\mathcal{M}$ is of the form:

$$\bar{\wp} = \left\{ \left(m_i, \bar{\wp}_\mu(m_i), \bar{\wp}_a(m_i) \right) | m_i \in \mathcal{M} \right\}$$

where $\bar{\wp}_{\mu}:\mathcal{M} \longrightarrow [0, 1]$ and $\bar{\wp}_{a}:\mathcal{M} \longrightarrow [0, 1]$ define the DM and DNM, respectively, and for each $m_i \in \mathcal{M}$, $0 \le \bar{\wp}_{\mu}(m_i) + \bar{\wp}_{a}(m_i) \le 1$. In addition, the function, $\bar{\wp}_{\pi}(m_i) = 1 - (\bar{\wp}_{\mu}(m_i) + \bar{\wp}_{a}(m_i))$ is called the HD of $\bar{\wp}$ and $0 \le \bar{\wp}_{\pi}(m_i) \le 1$, for every $m_i \in \mathcal{M}$.

Definition 2.3 ([10]). A PFS represented by $\hat{\wp}$ in $\mathcal{M}$ is described as:

$$\hat{\wp} = \left\{ \left(m_i, \hat{\wp}_{\mu}(m_i), \hat{\wp}_{a}(m_i) \right) | m_i \in \mathcal{M} \right\}$$

where $\hat{\wp}_{\mu}:\mathcal{M} \to [0,1]$, $\hat{\wp}_{a}:\mathcal{M} \to [0,1]$, such that $0 \le \hat{\wp}_{\mu}^2(m_i) + \hat{\wp}_{a}^2(m_i) \le 1$ for all $m_i \in \mathcal{M}$, where $\hat{\wp}_{\mu}(m_i)$ is the DM and $\hat{\wp}_{a}(m_i)$ is the DNM of $m_i \in \mathcal{M}$. The HD denoted as $\hat{\wp}_{\pi}(m_i)$ defined by $\hat{\wp}_{\pi}(m_i) = (1 - (\hat{\wp}_{\mu}^2(m_i) + \hat{\wp}_{a}^2(m_i)))^{1/2}$ is the HD of $m_i \in \mathcal{M}$, to the set $\hat{\wp}$ and $\hat{\wp}_{\pi}(m_i) \in [0,1]$.

Definition 2.4 ([11]). An FFS represented by $\Re$ in $\mathcal{M}$ can be described as:

$$\Re = \{ (m_i, \Re_{\mu}(m_i), \Re_{a}(m_i)) | m_i \in \mathcal{M} \}$$

where $\Re_{\mu}:\mathcal{M} \to [0,1]$, $\Re_{a}:\mathcal{M} \to [0,1]$, such that $0 \le \Re_{\mu}^3(m_i) + \Re_{a}^3(m_i) \le 1$ for all $m_i \in \mathcal{M}$, where $\Re_{\mu}(m_i)$ is the DM and $\Re_{a}(m_i)$ is the DNM of m_i in $\mathcal{M}$. The HD denoted as $\Re_{\pi}(m_i)$ defined by $\Re_{\pi}(m_i) = (1 - \Re_{\mu}^3(m_i) - \Re_{a}^3(m_i))^{1/3}$ is the HD of $m_i \in \mathcal{M}$, to the set $\Re$ and $\Re_{\pi}(m_i) \in [0,1]$, for $i = 1, 2, \ldots, q$.

Definition 2.5 ([11]). Suppose we have two FFSs, $\hat{\Re}$ and $\tilde{\Re}$ in $\mathcal{M}$, then we have the following operations:

1. Complement: $\hat{\Re}^c = \left\{ \left(m_i, \hat{\Re}_{a}(m_i), \hat{\Re}_{\mu}(m_i) \right) | m_i \in \mathcal{M} \right\}$
2. Intersection:
$\hat{\Re} \cap \tilde{\Re} = \left\{ \left(m_i, \min\left\{ \hat{\Re}_{\mu}(m_i), \tilde{\Re}_{\mu}(m_i) \right\}, \max\left\{ \hat{\Re}_{a}(m_i), \tilde{\Re}_{a}(m_i) \right\} \right) | m_i \in \mathcal{M} \right\}$
3. Union: $\hat{\Re} \cup \tilde{\Re} = \left\{ \left(m_i, \max\left\{ \hat{\Re}_{\mu}(m_i), \tilde{\Re}_{\mu}(m_i) \right\}, \min\left\{ \hat{\Re}_{a}(m_i), \tilde{\Re}_{a}(m_i) \right\} \right) | m_i \in \mathcal{M} \right\}$
4. Inclusion relation: $\hat{\Re} \subseteq \tilde{\Re} \Leftrightarrow \hat{\Re}_{\mu}(m_i) \le \tilde{\Re}_{\mu}(m_i)$ and $\hat{\Re}_{a}(m_i) \ge \tilde{\Re}_{a}(m_i)$ for all $m_i \in \mathcal{M}$
5. Equality: $\hat{\Re} = \tilde{\Re} \Leftrightarrow \hat{\Re}_{\mu}(m_i) = \tilde{\Re}_{\mu}(m_i)$ and $\hat{\Re}_{a}(m_i) = \tilde{\Re}_{a}(m_i)$ for all $m_i \in \mathcal{M}$
6. Sum:
$$\hat{\Re} \oplus \tilde{\Re} = \left\{ \left(m_i, \left(\hat{\Re}_{\mu}^3(m_i) + \tilde{\Re}_{\mu}^3(m_i) - \hat{\Re}_{\mu}^3(m_i)\tilde{\Re}_{\mu}^3(m_i) \right)^{\frac{1}{3}}, \hat{\Re}_{a}(m_i), \tilde{\Re}_{a}(m_i) \right) | m_i \in \mathcal{M} \right\}$$
7. Product:
$$\hat{\Re} \otimes \tilde{\Re} = \left\{ \left(m_i, \hat{\Re}_{\mu}(m_i), \tilde{\Re}_{\mu}(m_i), \left(\hat{\Re}_{a}^3(m_i) + \tilde{\Re}_{a}^3(m_i) - \hat{\Re}_{a}^3(m_i)\tilde{\Re}_{a}^3(m_i) \right)^{\frac{1}{3}} \right) | m_i \in \mathcal{M} \right\}$$

Definition 2.6 ([49]). Let $\eth : \mathrm{FFS}(\mathcal{M}) \times \mathrm{FFS}(\mathcal{M}) \to [0,1]$ be a function, and let $\hat{\Re}, \tilde{\Re}$, and $\acute{\Re}$ be three FFSs in $\mathcal{M}$, then $\eth\left(\hat{\Re}, \tilde{\Re}\right)$ is called a FFDM between $\hat{\Re}$ and $\tilde{\Re}$ if it satisfies the following properties:

1. $0 \leq \eth(\hat{\Re}, \tilde{\Re}) \leq 1$
2. $\eth(\hat{\Re}, \tilde{\Re}) = 0$ if and only if $\hat{\Re} = \tilde{\Re}$
3. $\eth(\tilde{\Re}, \hat{\Re}) = \eth(\hat{\Re}, \tilde{\Re})$
4. $\eth(\hat{\Re}, \hat{\Re}) = 1$, if and only if $\hat{\Re}$ is a crisp set
5. If $\hat{\Re}, \tilde{\Re}, \acute{\Re} \in PFS(\mathcal{M})$ and $\hat{\Re} \subseteq \tilde{\Re} \subseteq \acute{\Re}$, then $\eth(\hat{\Re}, \acute{\Re}) \geq d(\hat{\Re}, \tilde{\Re})$ and $\eth(\hat{\Re}, \acute{\Re}) \geq \eth(\tilde{\Re}, \acute{\Re})$
6. $\eth(\hat{\Re}, \tilde{\Re}) \geq \min\{\eth(\hat{\Re}, \acute{\Re}), (\acute{\Re}, \tilde{\Re})\}$
7. $\eth(\hat{\Re}, \tilde{\Re}) \geq \max\{\eth(\hat{\Re}, \acute{\Re}), (\acute{\Re}, \tilde{\Re})\}$

Definition 2.7 ([14]). Suppose $\hat{\Re} = \left\{ \left(m_i, \hat{\Re}_\mu(m_i), \hat{\Re}_\alpha(m_i) \right) | m_i \in \mathcal{M} \right\}$ and $\tilde{\Re} = \left\{ \left(m_i, \tilde{\Re}_\mu(m_i), \tilde{\Re}_\alpha(m_i) \right) | m_i \in \mathcal{M} \right\}$ are FFSs in $\mathcal{M}$. The weights of the elements of $\mathcal{M}$ in the FFSs denoted as ω_i can be computed as eq. (1):

$$\omega_i = \frac{\frac{3\hat{\Re}^3_\mu(m_i) + \hat{\Re}^3_\alpha(m_i) + \hat{\Re}^3_\pi(m_i)}{3} + \frac{3\tilde{\Re}^3_\mu(m_i) + \tilde{\Re}^3_\alpha(m_i) + \tilde{\Re}^3_\pi(m_i)}{3}}{\sum_{i=1}^{q} \left(\frac{3\hat{\Re}^3_\mu(m_i) + \hat{\Re}^3_\alpha(m_i) + \hat{\Re}^3_\pi(m_i)}{3} + \frac{3\tilde{\Re}^3_\mu(m_i) + \tilde{\Re}^3_\alpha(m_i) + \tilde{\Re}^3_\pi(m_i)}{3} \right)} \tag{1}$$

for $\omega_i \in [0, 1]$ and $\sum_{i=1}^{q} \omega_i = 1$.

2.1 Existing Distance Functions on FFSs

Here, we outline some existing distance metrics for FFSs and evaluate the efficiency of each method numerically. Given two FFSs as follows:

$$\hat{\Re} = \left\{ \left(m_i, \hat{\Re}_\mu(m_i), \hat{\Re}_\alpha(m_i) \right) | m_i \in \mathcal{M} \right\} \text{ and } \tilde{\Re} = \left\{ \left(m_i, \tilde{\Re}_\mu(m_i), \tilde{\Re}_\alpha(m_i) \right) | m_i \in \mathcal{M} \right\}$$

in $\mathcal{M} = \{m_1, m_2, \ldots, m_q\}$. Then, the existing distance metrics for FFSs are as follows:

– FFDM in Senapati and Yager [11]:

$$\eth_{SY}\left(\hat{\Re}, \tilde{\Re}\right) = \sqrt{\frac{1}{2} \sum_{i=1}^{q} \left(\begin{array}{c} \left(\hat{\Re}^3_\mu(m_i) - \tilde{\Re}^3_\mu(m_i)\right)^2 + \left(\hat{\Re}^3_\alpha(m_i) - \tilde{\Re}^3_\alpha(m_i)\right)^2 \\ + \left(\hat{\Re}^3_\pi(m_i) - \tilde{\Re}^3_\pi(m_i)\right)^2 \end{array} \right)} \tag{2}$$

This method (2) ignored the cardinality of $\mathcal{M}$ in its computation, which could affect the computed distance output.

– FFDMs in Ganie [29]:

$$\eth_{G1}\left(\hat{\Re}, \tilde{\Re}\right) = \frac{1}{q} \sum_{i=1}^{q} \left[\frac{\left|\hat{\Re}^3_\mu(m_i) - \tilde{\Re}^3_\mu(m_i)\right| + \left|\hat{\Re}^3_\alpha(m_i) - \tilde{\Re}^3_\alpha(m_i)\right|}{1 + \left|\hat{\Re}^3_\mu(m_i) - \tilde{\Re}^3_\mu(m_i)\right| \left|\hat{\Re}^3_\alpha(m_i) - \tilde{\Re}^3_\alpha(m_i)\right|} \right] \tag{3}$$

$$\eth_{G2}\left(\hat{\Re},\tilde{\Re}\right)=$$

$$\frac{1}{q}\sum_{i=1}^{q}\left(\frac{\left|\hat{\Re}_{\mu}^{3}(m_i)-\tilde{\Re}_{\mu}^{3}(m_i)\right|+\left|\hat{\Re}_{a}^{3}(m_i)-\tilde{\Re}_{a}^{3}(m_i)\right|-2\left|\hat{\Re}_{\mu}^{3}(m_i)-\tilde{\Re}_{\mu}^{3}(m_i)\right|\left|\hat{\Re}_{a}^{3}(m_i)-\tilde{\Re}_{a}^{3}(m_i)\right|}{1-\left|\hat{\Re}_{\mu}^{3}(m_i)-\tilde{\Re}_{\mu}^{3}(m_i)\right|\left|\hat{\Re}_{a}^{3}(m_i)-\tilde{\Re}_{a}^{3}(m_i)\right|}\right) \tag{4}$$

$$\eth_{G3}\left(\hat{\Re},\tilde{\Re}\right)=\frac{1}{q}\sum_{j=1}^{q}\left(\begin{array}{c}\left|\hat{\Re}_{\mu}^{3}(m_i)-\tilde{\Re}_{\mu}^{3}(m_i)\right|+\left|\hat{\Re}_{a}^{3}(m_i)-\tilde{\Re}_{a}^{3}(m_i)\right|\\-\left|\hat{\Re}_{\mu}^{3}(m_i)-\tilde{\Re}_{\mu}^{3}(m_i)\right|\left|\hat{\Re}_{a}^{3}(m_i)-\tilde{\Re}_{a}^{3}(m_i)\right|\end{array}\right) \tag{5}$$

$$\eth_{G4}\left(\hat{\Re},\tilde{\Re}\right)=\frac{1}{q}\sum_{i=1}^{q}\min\left\{1,\left|\hat{\Re}_{\mu}^{3}(m_i)-\tilde{\Re}_{\mu}^{3}(m_i)\right|+\left|\hat{\Re}_{a}^{3}(m_i)-\tilde{\Re}_{a}^{3}(m_i)\right|\right\} \tag{6}$$

The methods (eq. (3) and eq. (4)) considered the cardinality of $\mathcal{M}$, but ignored the HD. In eq. (5) and eq. (6), the values of the HDs are not included and eq. (6) uses other minimum values, which may lead to error of exclusion.

– FFDMs in Deng and Wang [46]:

$$\eth_{DW1}\left(\hat{\Re},\tilde{\Re}\right)=\sqrt{\frac{1}{2q}\sum_{i=1}^{q}\left(\frac{\left(\hat{\Re}_{\mu}^{3}(m_i)-\tilde{\Re}_{\mu}^{3}(m_i)\right)^2}{\hat{\Re}_{\mu}^{3}(m_i)+\tilde{\Re}_{\mu}^{3}(m_i)}+\frac{\left(\hat{\Re}_{a}^{3}(m_i)-\tilde{\Re}_{a}^{3}(m_i)\right)^2}{\hat{\Re}_{a}^{3}(m_i)+\tilde{\Re}_{a}^{3}(m_i)}\right)} \tag{7}$$

$$\eth_{DW2}\left(\hat{\Re},\tilde{\Re}\right)=\sqrt{\frac{1}{2q}\sum_{i=1}^{q}\left(\begin{array}{c}\left(\hat{\Re}_{\mu}^{3}(m_i)-\tilde{\Re}_{\mu}^{3}(m_i)\right)^2+\left(\hat{\Re}_{a}^{3}(m_i)-\tilde{\Re}_{a}^{3}(m_i)\right)^2\\+\left(\hat{\Re}_{\pi}^{3}(m_i)-\tilde{\Re}_{\pi}^{3}(m_i)\right)^2\end{array}\right)} \tag{8}$$

$$\eth_{DW3}\left(\hat{\Re},\tilde{\Re}\right)=\sqrt{\frac{1}{2q}\sum_{i=1}^{q}\left(\left(\sqrt{\hat{\Re}_{\mu}^{3}(m_i)}-\sqrt{\tilde{\Re}_{\mu}^{3}(m_i)}\right)^2+\left(\sqrt{\hat{\Re}_{a}^{3}(m_i)}-\sqrt{\tilde{\Re}_{a}^{3}(m_i)}\right)^2\right)} \tag{9}$$

The method in eq. (7) does not include the HD in its model. In eq. (8), the number of parametric differences is not incorporated in the computation and in eq. (9), the value of the HD is not included, which may alter the outcome of the distance method.

– FFDM in Onyeke and Ejegwa [47]:

$$\eth_{OE}\left(\hat{\Re},\tilde{\Re}\right)=\sqrt{\frac{1}{3q}\sum_{j=1}^{q}\left(\begin{array}{c}\left(\hat{\Re}_{\mu}^{3}(m_j)-\tilde{\Re}_{\mu}^{3}(m_j)\right)^2+\left(\hat{\Re}_{a}^{3}(m_j)-\tilde{\Re}_{a}^{3}(m_j)\right)^2\\+\left(\hat{\Re}_{\pi}^{3}(m_j)-\tilde{\Re}_{\pi}^{3}(m_j)\right)^2\end{array}\right)} \tag{10}$$

The FFDM in eq. (10) incorporated all the functions of the FFSs but lacks precision.

- FFDM in Kirisci [48]:

$$\eth_{K}\left(\hat{\Re},\tilde{\Re}\right)=\frac{1-d_1-d_2}{2} \tag{11}$$

where

$$d_1=\frac{1}{q}\sum_{i=1}^{q}\left(\frac{\left(\hat{\Re}_{\mu}^{3}(m_i)\tilde{\Re}_{\mu}^{3}(m_i)\right)^3+\left(\hat{\Re}_{a}^{3}(m_i)\tilde{\Re}_{a}^{3}(m_i)\right)^3+\left(\hat{\Re}_{\pi}^{3}(m_i)\tilde{\Re}_{\pi}^{3}(m_i)\right)^3}{\sqrt[3]{\hat{\Re}_{\mu}^{6}(m_i)+\hat{\Re}_{a}^{6}(m_i)+\hat{\Re}_{\pi}^{6}(m_i)}\sqrt[3]{\tilde{\Re}_{\mu}^{6}(m_i)+\tilde{\Re}_{a}^{6}(m_i)+\tilde{\Re}_{\pi}^{6}(m_i)}}\right)$$

$$d_2=\sqrt{\frac{1}{2q}\sum_{i=1}^{q}\left(\left|\hat{\Re}_{\mu}^{3}(m_i)-\tilde{\Re}_{\mu}^{3}(m_i)\right|^2+\left|\hat{\Re}_{a}^{3}(m_i)-\tilde{\Re}_{a}^{3}(m_i)\right|^2+\left|\hat{\Re}_{\pi}^{3}(m_i)-\tilde{\Re}_{\pi}^{3}(m_i)\right|^2\right)}$$

The FFDM (11) incorporated all the functions of the FFSs but lacks precision.

- FFDM in Liu [45]:

$$\eth_{L}\left(\hat{\Re},\tilde{\Re}\right)=\sqrt{\frac{3}{2q}\sum_{i=1}^{q}\left(\frac{\left(\hat{\Re}_{\mu}^{3}(m_i)-\tilde{\Re}_{\mu}^{3}(m_i)\right)^2}{\hat{\Re}_{\mu}^{3}(m_i)+\tilde{\Re}_{\mu}^{3}(m_i)+2}+\frac{\left(\hat{\Re}_{a}^{3}(m_i)-\tilde{\Re}_{a}^{3}(m_i)\right)^2}{\hat{\Re}_{a}^{3}(m_i)+\tilde{\Re}_{a}^{3}(m_i)+2}\right)} \tag{12}$$

The FFDM (12) considered the cardinality of $\mathcal{M}$, but it does not consider the HDs in its model.

- FFDM in Ganie et al. [49]:

$$\eth_{Ge}\left(\hat{\Re},\tilde{\Re}\right)=\frac{1}{2k}\sum_{j=1}^{k}\left[\begin{array}{c}\left|\hat{\Re}_{\mu}^{3}(m_i)-\tilde{\Re}_{\mu}^{3}(m_i)\right|+\left|\hat{\Re}_{a}^{3}(m_i)-\tilde{\Re}_{a}^{3}(m_i)\right|\\+\left|\hat{\Re}_{\pi}^{3}(m_i)-\tilde{\Re}_{\pi}^{3}(m_i)\right|\end{array}\right] \tag{13}$$

The FFDM (13) considered all three parameters of the FFSs but lacks precision.

3 The New FFDM

Here, we propose a new enhanced FFDM, discuss the properties of the new FFDM, and compare the results of the new method to the existing distance metrics between FFSs [11, 14, 29, 45–49]. The new FFDM is an extension of a distance metric under PFSs [30], which is based on 2D framework, neglecting the hesitation component. The new FFDM is built on the FF setting and 3D framework, including the HD to enhance reliability, unlike the approach in Deng and Wang [30]. This enhancement strengthens the accuracy of distance-based analyses, particularly in applications requiring precise ranking and alternatives classification. As a result, the proposed method improves decision-making reliability, and makes it a more effective tool used for handling uncertainty in practical applications.

Definition 3.1. Let $\hat{\mathfrak{R}}$ and $\tilde{\mathfrak{R}}$ be two FFSs in $\mathscr{M} = \{m_1, m_2, \ldots, m_q\}$ defined as follows:

$$\hat{\mathfrak{R}} = \left\{ \left(m_i, \hat{\mathfrak{R}}_\mu(m_i), \hat{\mathfrak{R}}_\alpha(m_i) \right) | m_i \in \mathscr{M} \right\} \text{ and } \hat{\mathfrak{R}} = \left\{ \left(m_i, \hat{\mathfrak{R}}_\mu(m_i), \hat{\mathfrak{R}}_\alpha(m_i) \right) | m_i \in \mathscr{M} \right\}$$

where $\hat{\mathfrak{R}}_\pi(m_i) = \left(1 - \left(\hat{\mathfrak{R}}^3_\mu(m_i) + \hat{\mathfrak{R}}^3_\alpha(m_i)\right)\right)^{1/3}$ and $\tilde{\mathfrak{R}}_\pi(m_i) = \left(1 - \left(\tilde{\mathfrak{R}}^3_\mu(m_i) + \tilde{\mathfrak{R}}^3_\alpha(m_i)\right)\right)^{1/3}$
Then, the new method is presented as follows:

$$\eth_*\left(\hat{\mathfrak{R}}, \tilde{\mathfrak{R}}\right) = \frac{1}{q}\sum_{i=1}^{q} \left(\begin{array}{l} \left(\frac{\hat{\mathfrak{R}}^3_\mu(m_i) - \hat{\mathfrak{R}}^3_\alpha(m_i) - \hat{\mathfrak{R}}^3_\pi(m_i)}{3} - \frac{\tilde{\mathfrak{R}}^3_\mu(m_i) - \tilde{\mathfrak{R}}^3_\alpha(m_i) - \tilde{\mathfrak{R}}^3_\pi(m_i)}{3} \right)^2 + \\ \frac{1}{3}\left(\frac{\hat{\mathfrak{R}}^3_\mu(m_i) + \hat{\mathfrak{R}}^3_\alpha(m_i) + \hat{\mathfrak{R}}^3_\pi(m_i)}{3} - \frac{\tilde{\mathfrak{R}}^3_\mu(m_i) + \tilde{\mathfrak{R}}^3_\alpha(m_i) + \tilde{\mathfrak{R}}^3_\pi(m_i)}{3} \right)^2 \end{array} \right)^{1/3} \tag{14}$$

Due to the significance of the weights of FFSs parameters under the defining elements, the approach (14) is modified as follows:

$$\eth^\omega_*\left(\hat{\mathfrak{R}}, \tilde{\mathfrak{R}}\right) = \sum_{i=1}^{q} \omega_i \left(\begin{array}{l} \left(\frac{\hat{\mathfrak{R}}^3_\mu(m_i) - \hat{\mathfrak{R}}^3_\alpha(m_i) - \hat{\mathfrak{R}}^3_\pi(m_i)}{3} - \frac{\tilde{\mathfrak{R}}^3_\mu(m_i) - \tilde{\mathfrak{R}}^3_\alpha(m_i) - \tilde{\mathfrak{R}}^3_\pi(m_i)}{3} \right)^2 + \\ \frac{1}{3}\left(\frac{\hat{\mathfrak{R}}^3_\mu(m_i) + \hat{\mathfrak{R}}^3_\alpha(m_i) + \hat{\mathfrak{R}}^3_\pi(m_i)}{3} - \frac{\tilde{\mathfrak{R}}^3_\mu(m_i) + \tilde{\mathfrak{R}}^3_\alpha(m_i) + \tilde{\mathfrak{R}}^3_\pi(m_i)}{3} \right)^2 \end{array} \right)^{1/3} \tag{15}$$

where ω_i is defined in eq. (1).

Next, we show that the new enhanced FFDM satisfies all the axioms of a distance metric with the aid of a theorem and validate its efficiency and effectiveness in distinguishing between closely related FFSs with the aid of a numerical example.

Theorem 3.1. The newly constructed FFDM is a satisfied distance metric.
Proof. To establish, we verify the conditions of Definition 2.6. We first prove eq. (1)., i.e., $0 \le \eth_*(\hat{\mathfrak{R}}, \tilde{\mathfrak{R}}) \le 1$ for two FFSs $\hat{\mathfrak{R}}$ and $\ddot{\mathfrak{R}}$ defined in $\mathscr{M}$. From Definition 2.4, we get $0 \le \hat{\mathfrak{R}}^3_\mu(m_i) + \hat{\mathfrak{R}}^3_\alpha(m_i) \le 1$ and $0 \le \tilde{\mathfrak{R}}^3_\mu(m_i) + \tilde{\mathfrak{R}}^3_\alpha(m_i) \le 1$ such that

$$\hat{\mathfrak{R}}_\pi(m_i) = \left(1 - \hat{\mathfrak{R}}^3_\mu(m_i) - \hat{\mathfrak{R}}^3_\alpha(m_i)\right)^{1/3} \text{ and } \tilde{\mathfrak{R}}_\pi(m_i) = \left(1 - \tilde{\mathfrak{R}}^3_\mu(m_i) - \tilde{\mathfrak{R}}^3_\alpha(m_i)\right)^{1/3}$$

Since $\hat{\mathfrak{R}}_\mu(m_i), \hat{\mathfrak{R}}_\alpha(m_i), \hat{\mathfrak{R}}_\pi(m_i) \in [0,1]$ and $\tilde{\mathfrak{R}}_\mu(m_i), \tilde{\mathfrak{R}}_\alpha(m_i), \tilde{\mathfrak{R}}_\pi(m_i) \in [0,1]$, the maximum possible difference between the corresponding components is 1. Hence, from eq. (14), we have

$$\left(\frac{\hat{\mathfrak{R}}^3_\mu(m_i) - \hat{\mathfrak{R}}^3_\alpha(m_i) - \hat{\mathfrak{R}}^3_\pi(m_i)}{3} - \frac{\tilde{\mathfrak{R}}^3_\mu(m_i) - \tilde{\mathfrak{R}}^3_\alpha(m_i) - \tilde{\mathfrak{R}}^3_\pi(m_i)}{3} \right)^2 \ge 0$$

and

$$\left(\frac{\hat{\Re}^3_\mu(m_i)+\hat{\Re}^3_\alpha(m_i)+\hat{\Re}^3_\pi(m_i)}{3}-\frac{\tilde{\Re}^3_\mu(m_i)+\tilde{\Re}^3_\alpha(m_i)+\tilde{\Re}^3_\pi(m_i)}{3}\right)^2\geq 0$$

Thus, $\eth_*(\hat{\Re},\tilde{\Re})\geq 0$, for all $i\in\{1,2,3,\ldots,q\}$. Since $\eth_*(\hat{\Re},\tilde{\Re})\geq 0$ has been proved, we need to show that $\eth_*(\hat{\Re},\tilde{\Re})\leq 1$.

From eq. (15), we have:

$$\eth_*\left(\hat{\Re},\tilde{\Re}\right)=\frac{1}{q}\sum_{i=1}^{q}\left(\begin{array}{c}\left(\frac{\hat{\Re}^3_\mu(m_i)-\hat{\Re}^3_\alpha(m_i)-\hat{\Re}^3_\pi(m_i)}{3}-\frac{\tilde{\Re}^3_\mu(m_i)-\tilde{\Re}^3_\alpha(m_i)-\tilde{\Re}^3_\pi(m_i)}{3}\right)^2+\\ \frac{1}{3}\left(\frac{\hat{\Re}^3_\mu(m_i)+\hat{\Re}^3_\alpha(m_i)+\hat{\Re}^3_\pi(m_i)}{3}-\frac{\tilde{\Re}^3_\mu(m_i)+\tilde{\Re}^3_\alpha(m_i)+\tilde{\Re}^3_\pi(m_i)}{3}\right)^2\end{array}\right)^{1/3}$$

$$=\frac{1}{q}\left[\begin{array}{c}\frac{\sum_{i=1}^{q}\left(\hat{\Re}^3_\mu(m_i)-\hat{\Re}^3_\alpha(m_i)-\hat{\Re}^3_\pi(m_i)-\tilde{\Re}^3_\mu(m_i)+\tilde{\Re}^3_\alpha(m_i)+\tilde{\Re}^3_\pi(m_i)\right)^2}{9}\\ +\frac{\sum_{i=1}^{q}\left(\hat{\Re}^3_\mu(m_i)+\hat{\Re}^3_\alpha(m_i)+\hat{\Re}^3_\pi(m_i)-\tilde{\Re}^3_\mu(m_i)-\tilde{\Re}^3_\alpha(m_i)-\tilde{\Re}^3_\pi(m_i)\right)^2}{27}\end{array}\right]^{\frac{1}{3}}$$

$$=\frac{1}{9q}\left[\begin{array}{c}\left(\sum_{i=1}^{q}\left(\hat{\Re}^3_\mu(m_i)-\hat{\Re}^3_\alpha(m_i)-\hat{\Re}^3_\pi(m_i)\right)-\sum_{i=1}^{q}\left(\tilde{\Re}^3_\mu(m_i)-\tilde{\Re}^3_\alpha(m_i)-\tilde{\Re}^3_\pi(m_i)\right)\right)^2\\ +\frac{\left(\sum_{i=1}^{q}\left(\hat{\Re}^3_\mu(m_i)+\hat{\Re}^3_\alpha(m_i)+\hat{\Re}^3_\pi(m_i)\right)-\sum_{i=1}^{q}\left(\tilde{\Re}^3_\mu(m_i)+\tilde{\Re}^3_\alpha(m_i)+\tilde{\Re}^3_\pi(m_i)\right)\right)^2}{3}\end{array}\right]^{\frac{1}{3}}$$

Hence,

$$\eth_*\left(\hat{\Re},\tilde{\Re}\right)=\frac{1}{9q}\left[\begin{array}{c}\left(\sum_{i=1}^{q}\left(\hat{\Re}^3_\mu(m_i)-\hat{\Re}^3_\alpha(m_i)-\hat{\Re}^3_\pi(m_i)\right)-\sum_{i=1}^{q}\left(\tilde{\Re}^3_\mu(m_i)-\tilde{\Re}^3_\alpha(m_i)-\tilde{\Re}^3_\pi(m_i)\right)\right)^2\\ +\frac{\left(\sum_{i=1}^{q}\left(\hat{\Re}^3_\mu(m_i)+\hat{\Re}^3_\alpha(m_i)+\hat{\Re}^3_\pi(m_i)\right)-\sum_{i=1}^{q}\left(\tilde{\Re}^3_\mu(m_i)+\tilde{\Re}^3_\alpha(m_i)+\tilde{\Re}^3_\pi(m_i)\right)\right)^2}{3}\end{array}\right]^{\frac{1}{3}}$$

To simplify the proof, we assume that:

$$\sum_{i=1}^{q}\left(\hat{\Re}^3_\mu(m_i)-\hat{\Re}^3_\alpha(m_i)-\hat{\Re}^3_\pi(m_i)\right)=a,\ \sum_{i=1}^{q}\left(\tilde{\Re}^3_\mu(m_i)-\tilde{\Re}^3_\alpha(m_i)-\tilde{\Re}^3_\pi(m_i)\right)=b$$

$$\sum_{i=1}^{q}\left(\hat{\Re}^3_\mu(m_i)+\hat{\Re}^3_\alpha(m_i)+\hat{\Re}^3_\pi(m_i)\right)=\bar{a},\ \sum_{i=1}^{q}\left(\tilde{\Re}^3_\mu(m_i)+\tilde{\Re}^3_\alpha(m_i)+\tilde{\Re}^3_\pi(m_i)\right)=\bar{b}$$

Then,

$$\partial_*\left(\hat{\mathfrak{R}}, \tilde{\mathfrak{R}}\right) = \frac{1}{9q}\left[(a-b)^2 + \frac{\left(\bar{a}-\bar{b}\right)^2}{3}\right]^{\frac{1}{3}} \Rightarrow$$

$$\partial_*\left(\hat{\mathfrak{R}}, \tilde{\mathfrak{R}}\right)^3 = \frac{1}{729q^3}\left[(a-b)^2 + \frac{\left(\bar{a}-\bar{b}\right)^2}{3}\right]$$

$$= \frac{1}{729q^3}\left[\frac{3(a-b)^2 + \left(\bar{a}-\bar{b}\right)^2}{3}\right] = \frac{3(a-b)^2 + \left(\bar{a}-\bar{b}\right)^2}{2,187q^3}$$

By subtracting -1 from both sides, we have:

$$\partial_*\left(\hat{\mathfrak{R}}, \tilde{\mathfrak{R}}\right)^3 - 1 = \frac{3(a-b)^2 + \left(\bar{a}-\bar{b}\right)^2}{2187q^3} - 1$$

$$= \frac{3(a-b)^2 + \left(\bar{a}-\bar{b}\right)^2 - 2187q^3}{2187q^3}$$

$$= -\frac{\left(2187q^3 - 3(a-b)^2 - \left(\bar{a}-\bar{b}\right)^2\right)}{2187q^3} \leq 0$$

which means $\partial_*\left(\hat{\mathfrak{R}}, \tilde{\mathfrak{R}}\right)^3 - 1 \leq 0$, and so $\partial_*\left(\hat{\mathfrak{R}}, \tilde{\mathfrak{R}}\right)^3 \leq 1$. Thus, $\partial_*\left(\hat{\mathfrak{R}}, \tilde{\mathfrak{R}}\right) \leq 1$. Hence, the property (1) is satisfied.

Next, we prove eq. (2). Suppose $\partial_*\left(\hat{\mathfrak{R}}, \tilde{\mathfrak{R}}\right) = 0$, we can get $\frac{\hat{\mathfrak{R}}^3_\mu(m_i) - \hat{\mathfrak{R}}^3_\alpha(m_i) - \hat{\mathfrak{R}}^3_\pi(m_i)}{3} - \frac{\tilde{\mathfrak{R}}^3_\mu(m_i) - \tilde{\mathfrak{R}}^3_\alpha(m_i) - \tilde{\mathfrak{R}}^3_\pi(m_i)}{3} = 0$ and $\frac{\hat{\mathfrak{R}}^3_\mu(m_i) + \hat{\mathfrak{R}}^3_\alpha(m_i) + \hat{\mathfrak{R}}^3_\pi(m_i)}{3} - \frac{\tilde{\mathfrak{R}}^3_\mu(m_i) + \tilde{\mathfrak{R}}^3_\alpha(m_i) \mp \tilde{\mathfrak{R}}^3_\pi(m_i)}{3}$ $= 0$, for all $i \in \{1, 2, 3, \ldots, q\}$.

Hence,

$$\frac{\hat{\mathfrak{R}}^3_\mu(m_i) - \hat{\mathfrak{R}}^3_\alpha(m_i) - \hat{\mathfrak{R}}^3_\pi(m_i)}{3} = \frac{\tilde{\mathfrak{R}}^3_\mu(m_i) - \tilde{\mathfrak{R}}^3_\alpha(m_i) - \tilde{\mathfrak{R}}^3_\pi(m_i)}{3}$$

and

$$\frac{\hat{\mathfrak{R}}^3_\mu(m_i) + \hat{\mathfrak{R}}^3_\alpha(m_i) + \hat{\mathfrak{R}}^3_\pi(m_i)}{3} = \frac{\tilde{\mathfrak{R}}^3_\mu(m_i) + \tilde{\mathfrak{R}}^3_\alpha(m_i) + \tilde{\mathfrak{R}}^3_\pi(m_i)}{3}$$

Thus, $\hat{\mathfrak{R}}^3_\mu(m) - \hat{\mathfrak{R}}^3_\alpha(m_i) - \hat{\mathfrak{R}}^3_\pi(m_i) = \tilde{\mathfrak{R}}^3_\mu(m_i) - \tilde{\mathfrak{R}}^3_\alpha(m_i) - \tilde{\mathfrak{R}}^3_\pi(m_i)$, $\hat{\mathfrak{R}}^3_\mu(m_i) + \hat{\mathfrak{R}}^3_\alpha(m_i) + \hat{\mathfrak{R}}^3_\pi(m_i)$ $= \tilde{\mathfrak{R}}^3_\mu(m_i) + \tilde{\mathfrak{R}}^3_\alpha(m_i) + \tilde{\mathfrak{R}}^3_\pi(m_i)$, which imply that $\tilde{\mathfrak{R}}^3_\mu(m_i) = \hat{\mathfrak{R}}^3_\mu(m_i)$, $\tilde{\mathfrak{R}}^3_\alpha(m_i) = \hat{\mathfrak{R}}^3_\alpha(m_i)$ and $\tilde{\mathfrak{R}}^3_\pi(m_i) = \hat{\mathfrak{R}}^3_\pi(m_i)$. Hence, $\tilde{\mathfrak{R}} = \hat{\mathfrak{R}}$.

Conversely, assume $\tilde{\Re} = \hat{\Re}$, then $\frac{\hat{\Re}^3_{\mu}(m_i) - \hat{\Re}^3_{\alpha}(m_i) - \hat{\Re}^3_{\pi}(m_i)}{3} - \frac{\tilde{\Re}^3_{\mu}(m_i) - \tilde{\Re}^3_{\alpha}(m_i) - \tilde{\Re}^3_{\pi}(m_i)}{3} = 0$ and $\frac{\hat{\Re}^3_{\mu}(m_i) + \hat{\Re}^3_{\alpha}(m_i) + \hat{\Re}^3_{\pi}(m_i)}{3} - \frac{\tilde{\Re}^3_{\mu}(m_i) + \tilde{\Re}^3_{\alpha}(m_i) \mp \tilde{\Re}^3_{\pi}(m_i)}{3} = 0$, which imply that $\partial_{*}\left(\hat{\Re}, \tilde{\Re}\right) = 0$. Hence, eq. (2) is proved.

For the proof of eq. (3), we have

$$\partial_{*}\left(\hat{\Re}, \tilde{\Re}\right) = \frac{1}{q}\sum_{i=1}^{q}\left(\begin{array}{l}\left(\frac{\hat{\Re}^3_{\mu}(m_i) - \hat{\Re}^3_{\alpha}(m_i) - \hat{\Re}^3_{\pi}(m_i)}{3} - \frac{\tilde{\Re}^3_{\mu}(m_i) - \tilde{\Re}^3_{\alpha}(m_i) - \tilde{\Re}^3_{\pi}(m_i)}{3}\right)^2 + \\ \frac{1}{3}\left(\frac{\hat{\Re}^3_{\mu}(m_i) + \hat{\Re}^3_{\alpha}(m_i) + \hat{\Re}^3_{\pi}(m_i)}{3} - \frac{\tilde{\Re}^3_{\mu}(m_i) + \tilde{\Re}^3_{\alpha}(m_i) + \tilde{\Re}^3_{\pi}(m_i)}{3}\right)^2\end{array}\right)^{1/3}$$

$$= \frac{1}{q}\sum_{i=1}^{q}\left(\begin{array}{l}\left(\frac{\tilde{\Re}^3_{\mu}(m_i) - \tilde{\Re}^3_{\alpha}(m_i) - \tilde{\Re}^3_{\pi}(m_i)}{3} - \frac{\hat{\Re}^3_{\mu}(m_i) - \hat{\Re}^3_{\alpha}(m_i) - \hat{\Re}^3_{\pi}(m_i)}{3}\right)^2 + \\ \frac{1}{3}\left(\frac{\tilde{\Re}^3_{\mu}(m_i) + \tilde{\Re}^3_{\alpha}(m_i) + \tilde{\Re}^3_{\pi}(m_i)}{3} - \frac{\hat{\Re}^3_{\mu}(m_i) + \hat{\Re}^3_{\alpha}(m_i) + \hat{\Re}^3_{\pi}(m_i)}{3}\right)^2\end{array}\right)^{1/3}$$

$$= \partial_{*}\left(\tilde{\Re}, \hat{\Re}\right)$$

Therefore, condition (3) holds.

For the proof of eq. (4), suppose $\hat{\Re}$ and $\tilde{\Re}$ are two FFSs in $\mathscr{M}$. Since $\hat{\Re}$ is crisp then either $\hat{\Re} = 0$ or $\hat{\Re} = 1$. In both cases, $\partial_{*}\left(\hat{\Re}, \hat{\Re}^{c}\right) = 1$. Thus, property (4) is satisfied.

For the proof of eq. (5), considering the condition, $\hat{\Re} \subseteq \tilde{\Re} \subseteq \Re'$ we have the following inequalities:

$$\hat{\Re}^3_{\mu}(m_i) \leq \tilde{\Re}^3_{\mu}(m_i) \leq \Re'^{3}_{\mu}(m_i),\ \hat{\Re}^3_{\alpha}(m_i) \geq \tilde{\Re}^3_{\alpha}(m_i) \geq \Re'^{3}_{\alpha}(m_i),$$

$$\hat{\Re}^3_{\pi}(m_i) \geq \tilde{\Re}^3_{\pi}(m_i) \geq \Re'^{3}_{\pi}(m_i)$$

From eq. (14), we derive a function $f(r, s, t)$, which is defined as:

$$f(r, s, t) = ((r - s - t) - (x - y - z))^2 + \frac{1}{3}((r + s + t) - (x + y + z))^2$$

where the three variables r, s, and t represent the FF parameters and $r, s, t, x, y, z \in [0, 1]$.

Now, we find the partial derivatives of $f(r, s, t)$ with respect to r, s, and t:

$$\frac{\partial_f}{\partial_r} = 2((r-s-t)-(x-y-z)) + \frac{2}{3}((r+s+t)-(x+y+z))$$

$$= 2(r-s-t-x+y+z) + \frac{2}{3}(r+s+t-x-y-z)$$

$$= \frac{8}{3}r - \frac{4}{3}s - \frac{4}{3}t - \frac{8}{3}x + \frac{4}{3}y + \frac{4}{3}z$$

$$= \frac{8}{3}(r-x) - \frac{4}{3}(s+t-y-z)$$

$$= \frac{4}{3}(2(r-x)-(s+t-y-z))$$

$$\frac{\partial_f}{\partial_r} = -2((r-s-t)-(x-y-z)) + \frac{2}{3}((r+s+t)-(x+y+z))$$

$$= -2(r-s-t-x+y+z) + \frac{2}{3}(r+s+t-x-y-z)$$

$$= -\frac{4}{3}r + \frac{8}{3}s + \frac{8}{3}t + \frac{4}{3}x - \frac{8}{3}y - \frac{8}{3}z$$

$$= -\frac{4}{3}(r-x) + \frac{8}{3}(s+t-y-z)$$

$$= \frac{4}{3}(2(s+t-y-z)-(r-x))$$

Similarly, we have:

$$\frac{\partial_f}{\partial_t} = \frac{4}{3}(2(s+t-y-z)-(r-x))$$

Thus,

$$\frac{\partial_f}{\partial_r} = \frac{4}{3}(2(r-x)-(s+t-y-z)), \quad \frac{\partial_f}{\partial_s} = \frac{4}{3}(2(s+t-y-z)-(r-x))$$

$$\frac{\partial_f}{\partial_t} = \frac{4}{3}(2(s+t-y-z)-(r-x))$$

To conclude the verification, we consider the following cases:

Suppose $0 \le x \le r \le 1$, $0 \le s \le y \le 1$, and $0 \le t \le z \le 1$. Then, we have $\frac{\partial_f}{\partial_r} \ge 0$, $\frac{\partial_f}{\partial_s} \le 0$ and $\frac{\partial_f}{\partial_t} \le 0$

Thus, $f(r,s,t)$ is an increasing and decreasing function for r, s, and t, respectively. Let $x = \hat{\Re}^3_\mu(m_i)$, $y = \hat{\Re}^3_\alpha(m_i)$, and $z = \hat{\Re}^3_\pi(m_i)$, then $x = \hat{\Re}^3_\mu(m_i) \le \tilde{\Re}^3_\mu(m_i) \le \acute{\Re}'^{3}_\mu(m_i)$, $y = \hat{\Re}^3_\alpha(m_i) \ge \tilde{\Re}^3_\alpha(m_i) \ge \acute{\Re}'^{3}_\alpha(m_i)$ and $z = \hat{\Re}^3_\pi(m_i) \ge \tilde{\Re}^3_\pi(m_i) \ge \acute{\Re}'^{3}_\pi(m_i)$ for $i = 1, 2, 3, \ldots, q$. From the monotonicity of $f(r,s,t)$, we observe that $f\left(\tilde{\Re}^3_\mu(m_i), \tilde{\Re}^3_\alpha(m_i), \tilde{\Re}^3_\pi(m_i)\right) \le$

$\left(\acute{\Re}^{3}_{\mu}(m_i), \acute{\Re}^{3}_{\alpha}(m_i), \acute{\Re}^{3}_{\pi}(m_i)\right)$ for all $i=1,2,3,\ldots,q$. Using, $x=\hat{\Re}^3_\mu(m_i)$, $y=\hat{\Re}^3_\mu(m_i)$, and $z=\hat{\Re}^3_\pi(m_i)$ for all $i=1,2,3,\ldots,q$, the specified statements are satisfied:

$$\eth_*\left(\hat{\Re},\tilde{\Re}\right)=\frac{1}{2q}\sum_{i=1}^{q}\sqrt{f\left(\tilde{\Re}^3_\mu(m_i),\tilde{\Re}^3_\alpha(m_i),\tilde{\Re}^3_\pi(m_i)\right)}$$

and

$$\eth_*\left(\hat{\Re},\acute{\Re}\right)=\frac{1}{2q}\sum_{i=1}^{q}\sqrt{f\left(\acute{\Re}^{3}_\mu(m_i),\acute{\Re}^{3}_\alpha(m_i),\acute{\Re}^{3}_\pi(m_i)\right)}$$

So, we have $\eth_*(\hat{\Re},\tilde{\Re})\geq\eth_*(\hat{\Re},\acute{\Re})$.

Suppose $0\leq x\leq r\leq 1$, $0\leq y\leq s\leq 1$, and $0\leq t\leq z\leq 1$. Then, we have $\frac{\partial f}{\partial r}\leq 0$, $\frac{\partial f}{\partial s}\geq 0$ and $\frac{\partial f}{\partial t}\geq 0$.

So $f(r,s,t)$ is an increasing and decreasing function for variables r,s, and t. If $x=\acute{\Re}^{3}_\mu(m_i), y=\acute{\Re}^{3}_\alpha(m_i), z=\acute{\Re}^{3}_\pi(m_i)$, then $x=\hat{\Re}^3_\mu(m_i)\leq\tilde{\Re}^3_\mu(m_i)\leq\acute{\Re}^{3}_\mu(m_i)$, $y=\hat{\Re}^3_\alpha(m_i)\geq\tilde{\Re}^3_\alpha(m_i)\geq\acute{\Re}^{3}_\alpha(m_i)$ and $z=\hat{\Re}^3_\pi(m_i)\geq\tilde{\Re}^3_\pi(m_i)\geq\acute{\Re}^{3}_\pi(m_i)$. From the monotonicity of $f(r,s,t)$, we have $\left(\tilde{\Re}^3_\mu(m_i),\tilde{\Re}^3_\alpha(m_i),\tilde{\Re}^3_\pi(m_i)\right)\leq\left(\acute{\Re}^{3}_\mu(m_i),\acute{\Re}^{3}_\alpha(m_i),\acute{\Re}^{3}_\pi(m_i)\right)$ for all $i=1,2,3,\ldots,q$. For $x=\tilde{\Re}^3_\mu(m_i), y=\tilde{\Re}^3_\alpha(m_i), z=\tilde{\Re}^3_\pi(m_i)$, for all $i=1,2,3,\ldots,q$, the specified statements are satisfied:

$$\eth_*\left(\tilde{\Re},\acute{\Re}\right)=\frac{1}{2q}\sum_{i=1}^{q}\sqrt{f\left(\hat{\Re}^3_\mu(m_i),\hat{\Re}^3_\alpha(m_i),\hat{\Re}^3_\pi(m_i)\right)}$$

and

$$\eth_*\left(\tilde{\Re},\acute{\Re}\right)=\frac{1}{2q}\sum_{i=1}^{q}\sqrt{f\left(\tilde{\Re}^3_\mu(m_i),\tilde{\Re}^3_\alpha(m_i),\tilde{\Re}^3_\pi(m_i)\right)}$$

Thus, we obtain $\eth_*(\hat{\Re},\acute{\Re})\geq{}_*(\tilde{\Re},\acute{\Re})$.

From (I) and (II), we conclude that $\eth_*(\hat{\Re},\tilde{\Re})\geq\eth_*(\hat{\Re},\acute{\Re})$ and $\eth_*(\hat{\Re},\acute{\Re})\geq\eth_*(\tilde{\Re},\acute{\Re})$.

Hence, the condition of axiom (5) is satisfied.

Since $\eth_*(\hat{\Re},\tilde{\Re})\geq\eth_*(\hat{\Re},\acute{\Re})$ and $\eth_*(\hat{\Re},\tilde{\Re})\geq\eth_*(\acute{\Re},\tilde{\Re})$, it follows that $\eth_*(\hat{\Re},\tilde{\Re})\geq\min\{\eth_*(\hat{\Re},\acute{\Re}),\eth_*(\acute{\Re},\tilde{\Re})\}$ and $\eth_*(\hat{\Re},\tilde{\Re})\geq\max\{\eth_*(\hat{\Re},\acute{\Re}),\eth_*(\acute{\Re},\tilde{\Re})\}$. Hence, eqs. (6) and (7) are established.

Theorem 3.2. The weighted version of the new FFDM is a satisfied distance metric.
Proof.
The verification is similar to the proof of Theorem 3.1.

3.1 Illustrative Examples

Here, we use the example below to evaluate the distance between the three FFSs $\hat{\mathfrak{R}}$, $\tilde{\mathfrak{R}}$, and $\acute{\mathfrak{R}}$ using the new FFDM.

Example 3.1 Given $\hat{\mathfrak{R}}, \tilde{\mathfrak{R}}, \acute{\mathfrak{R}} \in FFS(\mathcal{M})$, where $\mathcal{M} = \{m_1, m_2\}$, are defined as follows:

$$\hat{\mathfrak{R}} = \{(m_1, 0.6, 0.5), (m_2, 0.4, 0.8)\}, \ \tilde{\mathfrak{R}} = \{(m_1, 0.7, 0.7), (m_2, 0.6, 0.4)\}$$

$$\acute{\mathfrak{R}} = \{(m_1, 0.5, 0.4), (m_2, 0.8, 0.5)\}$$

Now, we apply the proposed FFDM to compute the distance between the three FFSs and get:

$$\eth_*\left(\hat{\mathfrak{R}}, \tilde{\mathfrak{R}}\right) = 0.1297, \eth_*\left(\hat{\mathfrak{R}}, \acute{\mathfrak{R}}\right) = 0.2264, \eth_*\left(\tilde{\mathfrak{R}}, \acute{\mathfrak{R}}\right) = 0.1958.$$

The numerical evaluation using the newly proposed FFDM shows that, between the three FFSs, $\hat{\mathfrak{R}}$ and $\tilde{\mathfrak{R}}$ are more related with a distance value of 0.1297.

Next, we apply the new weighted technique to compute the distances between the three FFSs. First, we compute the weights using eq. (1) and get: $\omega = \{(\omega_1 0.4883), (\omega_2 0.5117)\}$. Incorporating these into the new FFDM, we have:

$$\eth_*^{\omega}\left(\hat{\mathfrak{R}}, \tilde{\mathfrak{R}}\right) = 0.1031, \eth_*^{\omega}\left(\hat{\mathfrak{R}}, \acute{\mathfrak{R}}\right) = 0.1810, \eth_*^{\omega}\left(\tilde{\mathfrak{R}}, \acute{\mathfrak{R}}\right) = 0.1558.$$

The numerical evaluation using the newly proposed weighted FFDM in eq. (15) shows that, between the three FFSs, $\hat{\mathfrak{R}}$ and $\tilde{\mathfrak{R}}$are more related with a distance value of 0.1031. The weighted version of the new FFDM yields the same interpretation as the nonweighted, but the weighted version provides more efficient distance outputs.

Here, we perform a comparative analysis for the new and the existing FFDMs with the aid of numerical examples, so as to illustrate the efficiency and effectiveness of the new technique over the already existing techniques.

Example 3.2 Given a finite set $\mathcal{M} = \{m_1, m_2, m_3\}$. Assume there are two FFSs $\hat{\mathfrak{R}}$ and $\tilde{\mathfrak{R}}$ in $\mathcal{M}$ as:

$$\hat{\mathfrak{R}} = \{(m_1, 0.60, 0.65), (m_2, 0.54, 0.68), (m_3, 0.42, 0.78)\}$$
$$\tilde{\mathfrak{R}} = \{(m_1, 0.70, 0.35), (m_2, 0.63, 0.40), (m_3, 0.54, 0.28)\}$$

Suppose we have a similar FFS to $\hat{\mathfrak{R}}$ and $\tilde{\mathfrak{R}}$, defined as:

$$\acute{\mathfrak{R}} = \{(m_1, 0.55, 0.50), (m_2, 0.70, 0.35), (m_2, 0.20, 0.80)\}$$

We use the new and the existing FFDMs to determine which of the given sets is more similar to $\tilde{\mathfrak{R}}$. The computed distance outputs between the FFSs are presented in Table 1.

Table 1: Distance outputs for Example 3.

FFDMs	$(\hat{\mathfrak{R}}, \acute{\mathfrak{R}})$	$(\tilde{\mathfrak{R}}, \acute{\mathfrak{R}})$	Ordering	Most similar pair
$\eth_{SY}$ [11]	0.3055	0.4688	$(\hat{\mathfrak{R}}, \acute{\mathfrak{R}}) \prec (\tilde{\mathfrak{R}}, \acute{\mathfrak{R}})$	$(\hat{\mathfrak{R}}, \acute{\mathfrak{R}})$
$\eth_{G1}$[29]	0.2225	0.2703	$(\hat{\mathfrak{R}}, \acute{\mathfrak{R}}) \prec (\tilde{\mathfrak{R}}, \acute{\mathfrak{R}})$	$(\hat{\mathfrak{R}}, \acute{\mathfrak{R}})$
$\eth_{G2}$ [29]	0.1870	0.2285	$(\hat{\mathfrak{R}}, \acute{\mathfrak{R}}) \prec (\tilde{\mathfrak{R}}, \acute{\mathfrak{R}})$	$(\hat{\mathfrak{R}}, \acute{\mathfrak{R}})$
$\eth_{DW1}$[46]	0.2557	0.3428	$(\hat{\mathfrak{R}}, \acute{\mathfrak{R}}) \prec (\tilde{\mathfrak{R}}, \acute{\mathfrak{R}})$	$(\hat{\mathfrak{R}}, \acute{\mathfrak{R}})$
$\eth_{DW2}$[46]	0.1764	0.2707	$(\hat{\mathfrak{R}}, \acute{\mathfrak{R}}) \prec (\tilde{\mathfrak{R}}, \acute{\mathfrak{R}})$	$(\hat{\mathfrak{R}}, \acute{\mathfrak{R}})$
$\eth_{DW3}$[46]	0.1946	0.2825	$(\hat{\mathfrak{R}}, \acute{\mathfrak{R}}) \prec (\tilde{\mathfrak{R}}, \acute{\mathfrak{R}})$	$(\hat{\mathfrak{R}}, \acute{\mathfrak{R}})$
$\eth_{G3}$ [29]	0.2072	0.2546	$(\hat{\mathfrak{R}}, \acute{\mathfrak{R}}) \prec (\tilde{\mathfrak{R}}, \acute{\mathfrak{R}})$	$(\hat{\mathfrak{R}}, \acute{\mathfrak{R}})$
$\eth_{G4}$ [29]	0.2533	0.3333	$(\hat{\mathfrak{R}}, \acute{\mathfrak{R}}) \prec (\tilde{\mathfrak{R}}, \acute{\mathfrak{R}})$	$(\hat{\mathfrak{R}}, \acute{\mathfrak{R}})$
$\eth_{OE}$ [47]	0.1440	0.2210	$(\hat{\mathfrak{R}}, \acute{\mathfrak{R}}) \prec (\tilde{\mathfrak{R}}, \acute{\mathfrak{R}})$	$(\hat{\mathfrak{R}}, \acute{\mathfrak{R}})$
$\eth_{K}$ [48]	0.2707	0.0788	$(\tilde{\mathfrak{R}}, \acute{\mathfrak{R}}) \prec (\hat{\mathfrak{R}}, \acute{\mathfrak{R}})$	$(\hat{\mathfrak{R}}, \acute{\mathfrak{R}})$
$\eth_{L}$[45]	0.1703	0.2492	$(\hat{\mathfrak{R}}, \acute{\mathfrak{R}}) \prec (\tilde{\mathfrak{R}}, \acute{\mathfrak{R}})$	$(\hat{\mathfrak{R}}, \acute{\mathfrak{R}})$
$\eth_{Ge}$ [49]	0.1790	0.2532	$(\hat{\mathfrak{R}}, \acute{\mathfrak{R}}) \prec (\tilde{\mathfrak{R}}, \acute{\mathfrak{R}})$	$(\hat{\mathfrak{R}}, \acute{\mathfrak{R}})$
$\eth_{*}$	0.0879	0.1008	$(\hat{\mathfrak{R}}, \acute{\mathfrak{R}}) \prec (\tilde{\mathfrak{R}}, \acute{\mathfrak{R}})$	$(\hat{\mathfrak{R}}, \acute{\mathfrak{R}})$
$\eth_{*}^{\omega}$	0.0598	0.0708	$(\hat{\mathfrak{R}}, \acute{\mathfrak{R}}) \prec (\tilde{\mathfrak{R}}, \acute{\mathfrak{R}})$	$(\hat{\mathfrak{R}}, \acute{\mathfrak{R}})$

Example 3.3 For a finite set $\mathscr{M} = \{m_1, m_2, m_3\}$, given three known models represented as FFSs $\hat{\mathfrak{R}}$, $\tilde{\mathfrak{R}}$, and $\acute{\mathfrak{R}}$ in $\mathscr{M}$ as:

$$\hat{\mathfrak{R}} = \{(m_1, 0.40, 0.65), (m_2, 0.74, 0.35), (m_3, 0.45, 0.68)\}$$
$$\tilde{\mathfrak{R}} = \{(m_1, 0.50, 0.30), (m_2, 0.30, 0.40), (m_3, 0.80, 0.20)\}$$
$$\acute{\mathfrak{R}} = \{(m_1, 0.65, 0.45), (m_2, 0.40, 0.35), (m_3, 0.90, 0.15)\}$$

An unknown model to be classified is also given in terms of FFS represented by:

$$\mathfrak{B} = \{(m_1, 0.80, 0.25), (m_2, 0.10, 0.65), (m_3, 0.40, 0.70)\}$$

We classify the unknown model $\mathfrak{B}$ with the known models using both the new and the existing techniques for a comparative analysis. The computed outputs and classification are presented in Table 2.

In Table 1, the newly developed FFDM, $\eth_{*}$, and its weighted version, $\eth_{*}^{\omega}$, both yield the lowest distance values among all compared methods. The smaller values indicate closer similarity between $\hat{\mathfrak{R}}$ and $\acute{\mathfrak{R}}$, which is in line with the findings from most existing methods. The newly developed method is more precise and preserves the correct ordering and classification, confirming its effectiveness and robustness in identifying similar FFSs.

Table 2: Distance outputs for Example 4.

FFDMs	$\left(\hat{\mathfrak{R}}, \mathfrak{B}\right)$	$\left(\tilde{\mathfrak{R}}, \mathfrak{B}\right)$	$\left(\acute{\mathfrak{R}}, \mathfrak{B}\right)$	Ordering	Similar model
$\mathfrak{d}_{SY}$ [11]	0.5253	0.5898	0.6472	$\left(\acute{\mathfrak{R}}, \mathfrak{B}\right) \succ \left(\tilde{\mathfrak{R}}, \mathfrak{B}\right) \succ \left(\hat{\mathfrak{R}}, \mathfrak{B}\right)$	$\left(\hat{\mathfrak{R}}, \mathfrak{B}\right)$
$\mathfrak{d}_{G1}$[29]	0.3201	0.3195	0.3308	$\left(\acute{\mathfrak{R}}, \mathfrak{B}\right) \succ \left(\hat{\mathfrak{R}}, \mathfrak{B}\right) \succ \left(\tilde{\mathfrak{R}}, \mathfrak{B}\right)$	$\left(\tilde{\mathfrak{R}}, \mathfrak{B}\right)$
$\mathfrak{d}_{G2}$ [29]	0.2978	0.2939	0.3225	$\left(\acute{\mathfrak{R}}, \mathfrak{B}\right) \succ \left(\hat{\mathfrak{R}}, \mathfrak{B}\right) \succ \left(\tilde{\mathfrak{R}}, \mathfrak{B}\right)$	$\left(\tilde{\mathfrak{R}}, \mathfrak{B}\right)$
$\mathfrak{d}_{DW1}$[46]	0.4391	0.4206	0.4557	$\left(\acute{\mathfrak{R}}, \mathfrak{B}\right) \succ \left(\hat{\mathfrak{R}}, \mathfrak{B}\right) \succ \left(\tilde{\mathfrak{R}}, \mathfrak{B}\right)$	$\left(\tilde{\mathfrak{R}}, \mathfrak{B}\right)$
$\mathfrak{d}_{DW2}$[46]	0.3033	0.3405	0.3737	$\left(\acute{\mathfrak{R}}, \mathfrak{B}\right) \succ \left(\tilde{\mathfrak{R}}, \mathfrak{B}\right) \succ \left(\hat{\mathfrak{R}}, \mathfrak{B}\right)$	$\left(\hat{\mathfrak{R}}, \mathfrak{B}\right)$
$\mathfrak{d}_{DW3}$[46]	0.3748	0.3376	0.3779	$\left(\acute{\mathfrak{R}}, \mathfrak{B}\right) \succ \left(\hat{\mathfrak{R}}, \mathfrak{B}\right) \succ \left(\tilde{\mathfrak{R}}, \mathfrak{B}\right)$	$\left(\tilde{\mathfrak{R}}, \mathfrak{B}\right)$
$\mathfrak{d}_{G3}$ [29]	0.3140	0.3128	0.3293	$\left(\acute{\mathfrak{R}}, \mathfrak{B}\right) \succ \left(\hat{\mathfrak{R}}, \mathfrak{B}\right) \succ \left(\tilde{\mathfrak{R}}, \mathfrak{B}\right)$	$\left(\tilde{\mathfrak{R}}, \mathfrak{B}\right)$
$\mathfrak{d}_{G4}$ [29]	0.3333	0.3333	0.3333	$\left(\acute{\mathfrak{R}}, \mathfrak{B}\right) \succeq \left(\hat{\mathfrak{R}}, \mathfrak{B}\right) \succeq \left(\tilde{\mathfrak{R}}, \mathfrak{B}\right)$	$\left(\tilde{\mathfrak{R}}, \mathfrak{B}\right)$
$\mathfrak{d}_{OE}$ [47]	0.2476	0.2781	0.3051	$\left(\acute{\mathfrak{R}}, \mathfrak{B}\right) \succ \left(\tilde{\mathfrak{R}}, \mathfrak{B}\right) \succ \left(\hat{\mathfrak{R}}, \mathfrak{B}\right)$	$\left(\hat{\mathfrak{R}}, \mathfrak{B}\right)$
$\mathfrak{d}_{K}$ [48]	0.3405	0.0768	0.3405	$\left(\acute{\mathfrak{R}}, \mathfrak{B}\right) \succ \left(\hat{\mathfrak{R}}, \mathfrak{B}\right) \succ \left(\tilde{\mathfrak{R}}, \mathfrak{B}\right)$	$\left(\tilde{\mathfrak{R}}, \mathfrak{B}\right)$
$\mathfrak{d}_{L}$[45]	0.3154	0.3177	0.3574	$\left(\acute{\mathfrak{R}}, \mathfrak{B}\right) \succ \left(\tilde{\mathfrak{R}}, \mathfrak{B}\right) \succ \left(\hat{\mathfrak{R}}, \mathfrak{B}\right)$	$\left(\hat{\mathfrak{R}}, \mathfrak{B}\right)$
$\mathfrak{d}_{Ge}$ [49]	0.2936	0.3485	0.3780	$\left(\acute{\mathfrak{R}}, \mathfrak{B}\right) \succ \left(\hat{\mathfrak{R}}, \mathfrak{B}\right) \succ \left(\tilde{\mathfrak{R}}, \mathfrak{B}\right)$	$\left(\tilde{\mathfrak{R}}, \mathfrak{B}\right)$
$\mathfrak{d}_{*}$	0.1818	0.1795	0.2022	$\left(\acute{\mathfrak{R}}, \mathfrak{B}\right) \succ \left(\hat{\mathfrak{R}}, \mathfrak{B}\right) \succ \left(\tilde{\mathfrak{R}}, \mathfrak{B}\right)$	$\left(\tilde{\mathfrak{R}}, \mathfrak{B}\right)$
$\mathfrak{d}_{*}^{\omega}$	0.1285	0.1250	0.1370	$\left(\acute{\mathfrak{R}}, \mathfrak{B}\right) \succ \left(\hat{\mathfrak{R}}, \mathfrak{B}\right) \succ \left(\tilde{\mathfrak{R}}, \mathfrak{B}\right)$	$\left(\tilde{\mathfrak{R}}, \mathfrak{B}\right)$

In Table 2, the newly developed FFDM, $\mathfrak{d}_*$, and its weighted version, both yield minimal distance values, with both identifying $\tilde{\mathfrak{R}}$ as the most similar FFS to $\mathfrak{B}$, which aligns with majority of the other existing and effective methods [29, 46, 48, 49]. Moreover, the weighted version shows even more discrimination in the similarity ranking, affirming its sensitivity and precision, while the newly developed method provides accurate and consistent classification results, reinforcing its suitability for real-world decision-making and classification tasks involving fuzzy information.

4 Application to Vehicle Selection Process Based on TOPSIS Approach

In this section, we apply the new FFDM in a vehicle selection process. Given the complexity of choosing a suitable vehicle among different vehicles based on petrol-powered, diesel-powered, solar-powered, among others, decision-makers must assess multiple criteria, including the cost of maintenance, cost of purchase, fuel efficiency, environmental impact, and the technical know-how required for operation. FFSs, with their enhanced capabilities to model uncertainty, enable a more precise evaluation of alternatives. The new FFDM is deployed to carry out the selection exercise based on the TOPSIS approach.

4.1 Case Study

We considered six vehicle models based on their power sources for the selection exercise, which are represented in the set, $C = \{C_1, C_2, C_3, C_4, C_5, C_6\}$, where C_1 represents PPV, C_2 represents DPV, C_3 represents GPV, C_4 represents EV, C_5 represents BPV, and C_6 represents SPV. These vehicles are described as follows:

PPV (C_1): This vehicle model uses gasoline as fuel, an ICE, smooth operation, and widely available refueling stations. The merits of PPVs are as follows: lower upfront cost, quick refueling, lightweight engines, and good performance in terms of speed and acceleration.

DPV (C_2): This vehicle model runs on diesel fuel, has a more efficient ICE, higher torque, and a longer engine lifespan. The merits include better fuel efficiency than petrol cars, durable engines, cost-effective for long-distance travel, and ideal for heavy-duty applications.

GPV (C_3): This vehicle model uses liquefied petroleum gas or compressed natural gas, has a dual-fuel option in some models, and emits lower pollutants. The advantages of this model over the others include longer engine life due to cleaner combustion, lower fuel costs compared to petrol and diesel, and reduced GHG emissions.

EV (C_4): The features of this vehicle model include plug-in HEVs, BEVs, HEVs, and fuel cell electric vehicles (FCEVs). These vehicles use electric motors instead of conventional ICEs. The advantages include zero tailpipe emissions (for BEVs and FCEVs), lower maintenance costs, higher energy efficiency, and government incentives in many countries.

BPV (C_5): This vehicle model runs on biofuels such as ethanol, biodiesel, or biogas, which are gotten from plants and animal waste that are renewable sources. The advantages include the fact that it reduces reliance on lower carbon footprint, biodegradable and renewable fuel sources, fossil fuels, and supports agricultural economies.

SPV (C_6): This vehicle model uses photovoltaic (solar) panels to generate electricity for propulsion, often supplemented with batteries for energy storage. Its advantages include a new renewable and sustainable energy source, zero emissions, low operational costs, and minimal dependence on external fuel sources.

In order to make the decision on the best car model based on power source, seven criteria were considered as presented in the set, $\mathfrak{H} = \{\mathfrak{H}_1, \mathfrak{H}_2, \mathfrak{H}_3, \mathfrak{H}_4, \mathfrak{H}_5, \mathfrak{H}_6, \mathfrak{H}_7\}$, where, $\mathfrak{H}_1$ represents eco-friendly, $\mathfrak{H}_2$ represents cost of purchase, $\mathfrak{H}_3$ represents maintenance cost,

$\mathfrak{H}_4$ represents risk and social acceptance, $\mathfrak{H}_5$ represents pollutant emission, $\mathfrak{H}_6$ represents reliability, and $\mathfrak{H}_7$ represents the technical know-how.

Eco-friendliness ($\mathfrak{H}_1$): Refers to the environmental impact of a vehicle, including carbon footprint, fuel efficiency, and emissions.

Cost of purchase ($\mathfrak{H}_2$): The initial price of acquiring a vehicle.

Maintenance cost ($\mathfrak{H}_3$): Covers expenses related to servicing, repairs, and part replacements.

Risk and social acceptance ($\mathfrak{H}_4$): Include factors such as safety, fuel availability, and public perception.

Pollutant emission ($\mathfrak{H}_5$): Measures the level of harmful gases released into the environment.

Reliability ($\mathfrak{H}_6$): Assesses the durability and consistent performance of the vehicle.
Technical know-how ($\mathfrak{H}_7$): Refers to the level of expertise needed for vehicle's operation and maintenance.

A knowledge-based system of data collection was employed where data are collected from three vehicle experts. The vehicle experts were selected because of their broad knowledge regarding different vehicles' power sources. These experts used the LVs and FF numbers (FFNs) to present their insights. The LVs and FFNs adopted for this work are given in Table 3. The experts' opinions are presented in Table 4, and then converted to FFNs, as presented in Table 5.

Table 3: Vehicle selection linguistic variables.

LVs	FFNs
Extremely high (EH)	(1.00, 0.00)
Very very high (VVH)	(0.90, 0.10)
Very high (VH)	(0.80, 0.15)
High (H)	(0.70, 0.20)
Moderately high (MH)	(0.60, 0.30)
Fair (F)	(0.50, 0.40)
Moderately low (ML)	(0.40, 0.50)
Low (L)	(0.25, 0.60)
Very low (VL)	(0.10, 0.75)
Very very low (VVL)	(0.15, 0.90)
Extremely low (EL)	(0.00, 1.00)

To identify the best vehicle model based on power sources using the collected data, we utilize the TOPSIS technique.

Table 4: LVs representing vehicle experts' opinions.

Vehicle models	**Vehicle expert I**						
	$\mathfrak{h}_1$	$\mathfrak{h}_2$	$\mathfrak{h}_3$	$\mathfrak{h}_4$	$\mathfrak{h}_5$	$\mathfrak{h}_6$	$\mathfrak{h}_7$
$\mathcal{C}_1$	L	H	H	VH	VH	VH	VH
$\mathcal{C}_2$	L	VVH	VVH	H	VH	H	F
$\mathcal{C}_3$	H	H	H	H	H	H	H
$\mathcal{C}_4$	H	EH	EH	EH	ML	EH	H
$\mathcal{C}_5$	H	EH	EH	H	H	H	H
$\mathcal{C}_6$	EH	EH	EH	ML	EL	EH	H
	Vehicle expert II						
$\mathcal{C}_1$	H	VVH	H	MH	F	H	F
$\mathcal{C}_2$	EL	VH	H	H	F	MH	H
$\mathcal{C}_3$	F	F	MH	F	H	F	MH
$\mathcal{C}_4$	H	VH	H	MH	F	H	H
$\mathcal{C}_5$	F	MH	MH	F	ML	ML	F
$\mathcal{C}_6$	VH	VH	MH	H	MH	H	MH
	Vehicle expert III						
$\mathcal{C}_1$	EL	EH	VH	VVH	VL	EH	L
$\mathcal{C}_2$	VL	VVH	EH	VH	EH	EH	EH
$\mathcal{C}_3$	VH	VH	VH	VVH	L	H	MH
$\mathcal{C}_4$	VH	VVH	VH	VVH	EL	MH	MH
$\mathcal{C}_5$	H	MH	MH	F	VVL	MH	F
$\mathcal{C}_6$	VVH	VH	VH	MH	L	F	F

Table 5: FFNs representing vehicle experts' opinions.

Vehicle models	**Vehicle expert I**						
	$\mathfrak{h}_1$	$\mathfrak{h}_2$	$\mathfrak{h}_3$	$\mathfrak{h}_4$	$\mathfrak{h}_5$	$\mathfrak{h}_6$	$\mathfrak{h}_7$
$\mathcal{C}_1$	$\begin{pmatrix}0.25\\0.60\end{pmatrix}$	$\begin{pmatrix}0.70\\0.20\end{pmatrix}$	$\begin{pmatrix}0.70\\0.20\end{pmatrix}$	$\begin{pmatrix}0.80\\0.15\end{pmatrix}$	$\begin{pmatrix}0.80\\0.15\end{pmatrix}$	$\begin{pmatrix}0.80\\0.15\end{pmatrix}$	$\begin{pmatrix}0.80\\0.15\end{pmatrix}$
$\mathcal{C}_2$	$\begin{pmatrix}0.25\\0.60\end{pmatrix}$	$\begin{pmatrix}0.90\\0.10\end{pmatrix}$	$\begin{pmatrix}0.90\\0.10\end{pmatrix}$	$\begin{pmatrix}0.70\\0.20\end{pmatrix}$	$\begin{pmatrix}0.80\\0.15\end{pmatrix}$	$\begin{pmatrix}0.70\\0.20\end{pmatrix}$	$\begin{pmatrix}0.50\\0.40\end{pmatrix}$
$\mathcal{C}_3$	$\begin{pmatrix}0.70\\0.20\end{pmatrix}$	$\begin{pmatrix}0.70\\0.20\end{pmatrix}$	$\begin{pmatrix}0.70\\0.20\end{pmatrix}$	$\begin{pmatrix}0.70\\0.20\end{pmatrix}$	$\begin{pmatrix}0.70\\0.20\end{pmatrix}$	$\begin{pmatrix}0.70\\0.20\end{pmatrix}$	$\begin{pmatrix}0.70\\0.20\end{pmatrix}$
$\mathcal{C}_4$	$\begin{pmatrix}0.70\\0.20\end{pmatrix}$	$\begin{pmatrix}1.00\\0.00\end{pmatrix}$	$\begin{pmatrix}1.00\\0.00\end{pmatrix}$	$\begin{pmatrix}1.00\\0.00\end{pmatrix}$	$\begin{pmatrix}0.40\\0.50\end{pmatrix}$	$\begin{pmatrix}1.00\\0.00\end{pmatrix}$	$\begin{pmatrix}0.70\\0.20\end{pmatrix}$
$\mathcal{C}_5$	$\begin{pmatrix}0.70\\0.20\end{pmatrix}$	$\begin{pmatrix}1.00\\0.00\end{pmatrix}$	$\begin{pmatrix}1.00\\0.00\end{pmatrix}$	$\begin{pmatrix}0.70\\0.20\end{pmatrix}$	$\begin{pmatrix}0.70\\0.20\end{pmatrix}$	$\begin{pmatrix}0.70\\0.20\end{pmatrix}$	$\begin{pmatrix}0.70\\0.20\end{pmatrix}$
$\mathcal{C}_6$	$\begin{pmatrix}1.00\\0.00\end{pmatrix}$	$\begin{pmatrix}1.00\\0.00\end{pmatrix}$	$\begin{pmatrix}1.00\\0.00\end{pmatrix}$	$\begin{pmatrix}0.40\\0.50\end{pmatrix}$	$\begin{pmatrix}0.00\\1.00\end{pmatrix}$	$\begin{pmatrix}1.00\\0.00\end{pmatrix}$	$\begin{pmatrix}0.70\\0.20\end{pmatrix}$

Table 5 (continued)

Vehicle models	$\mathfrak{H}_1$	$\mathfrak{H}_2$	$\mathfrak{H}_3$	$\mathfrak{H}_4$	$\mathfrak{H}_5$	$\mathfrak{H}_6$	$\mathfrak{H}_7$
	Vehicle expert I						
	Vehicle expert II						
$\mathcal{C}_1$	$\binom{0.70}{0.20}$	$\binom{0.90}{0.10}$	$\binom{0.70}{0.20}$	$\binom{0.60}{0.30}$	$\binom{0.50}{0.40}$	$\binom{0.70}{0.20}$	$\binom{0.50}{0.40}$
$\mathcal{C}_2$	$\binom{0.00}{1.00}$	$\binom{0.80}{0.15}$	$\binom{0.70}{0.20}$	$\binom{0.70}{0.20}$	$\binom{0.50}{0.40}$	$\binom{0.60}{0.30}$	$\binom{0.70}{0.20}$
	$\binom{0.50}{0.40}$	$\binom{0.50}{0.40}$	$\binom{0.60}{0.30}$	$\binom{0.50}{0.40}$	$\binom{0.70}{0.20}$	$\binom{0.50}{0.40}$	$\binom{0.60}{0.30}$
$\mathcal{C}_4$	$\binom{0.70}{0.20}$	$\binom{0.80}{0.15}$	$\binom{0.70}{0.20}$	$\binom{0.60}{0.30}$	$\binom{0.50}{0.40}$	$\binom{0.70}{0.20}$	$\binom{0.70}{0.20}$
$\mathcal{C}_5$	$\binom{0.50}{0.40}$	$\binom{0.60}{0.30}$	$\binom{0.60}{0.30}$	$\binom{0.50}{0.40}$	$\binom{0.40}{0.50}$	$\binom{0.40}{0.50}$	$\binom{0.50}{0.40}$
$\mathcal{C}_6$	$\binom{0.80}{0.15}$	$\binom{0.80}{0.15}$	$\binom{0.60}{0.30}$	$\binom{0.70}{0.20}$	$\binom{0.60}{0.30}$	$\binom{0.70}{0.20}$	$\binom{0.60}{0.30}$
	Vehicle expert III						
$\mathcal{C}_1$	$\binom{0.00}{1.00}$	$\binom{1.00}{0.00}$	$\binom{0.80}{0.15}$	$\binom{0.90}{0.10}$	$\binom{0.10}{0.75}$	$\binom{1.00}{0.00}$	$\binom{0.25}{0.60}$
$\mathcal{C}_2$	$\binom{0.10}{0.75}$	$\binom{0.90}{0.10}$	$\binom{1.00}{0.00}$	$\binom{0.80}{0.15}$	$\binom{1.00}{0.00}$	$\binom{1.00}{0.00}$	$\binom{1.00}{0.00}$
$\mathcal{C}_3$	$\binom{0.80}{0.15}$	$\binom{0.80}{0.15}$	$\binom{0.80}{0.15}$	$\binom{0.90}{0.10}$	$\binom{0.25}{0.60}$	$\binom{0.70}{0.20}$	$\binom{0.60}{0.30}$
$\mathcal{C}_4$	$\binom{0.80}{0.15}$	$\binom{0.90}{0.10}$	$\binom{0.80}{0.15}$	$\binom{0.90}{0.10}$	$\binom{0.00}{1.00}$	$\binom{0.60}{0.30}$	$\binom{0.60}{0.30}$
$\mathcal{C}_5$	$\binom{0.70}{0.20}$	$\binom{0.60}{0.30}$	$\binom{0.60}{0.30}$	$\binom{0.50}{0.40}$	$\binom{0.15}{0.90}$	$\binom{0.60}{0.30}$	$\binom{0.50}{0.40}$
$\mathcal{C}_6$	$\binom{0.90}{0.10}$	$\binom{0.80}{0.15}$	$\binom{0.80}{0.15}$	$\binom{0.80}{0.15}$	$\binom{0.25}{0.60}$	$\binom{0.50}{0.40}$	$\binom{0.50}{0.40}$

Table 6: Average FFNs of the vehicle experts.

Vehicle models	$\mathfrak{H}_1$	$\mathfrak{H}_2$	$\mathfrak{H}_3$	$\mathfrak{H}_4$	$\mathfrak{H}_5$	$\mathfrak{H}_6$	$\mathfrak{H}_7$
$\mathcal{C}_1$	$\binom{0.3167}{0.6000}$	$\binom{0.8667}{0.1000}$	$\binom{0.7333}{0.1833}$	$\binom{0.7667}{0.1833}$	$\binom{0.4667}{0.4333}$	$\binom{0.8333}{0.1167}$	$\binom{0.5167}{0.3833}$
$\mathcal{C}_2$	$\binom{0.1167}{0.7833}$	$\binom{0.8667}{0.1067}$	$\binom{0.8667}{0.1000}$	$\binom{0.7333}{0.1833}$	$\binom{0.7667}{0.1833}$	$\binom{0.7667}{0.1667}$	$\binom{0.7333}{0.2000}$
$\mathcal{C}_3$	$\binom{0.6667}{0.2500}$	$\binom{0.6667}{0.2500}$	$\binom{0.7000}{0.2167}$	$\binom{0.7000}{0.2333}$	$\binom{0.5500}{0.3333}$	$\binom{0.6333}{0.2667}$	$\binom{0.6333}{0.2667}$
$\mathcal{C}_4$	$\binom{0.7333}{0.1833}$	$\binom{0.9000}{0.0833}$	$\binom{0.8333}{0.1167}$	$\binom{0.8333}{0.1333}$	$\binom{0.3000}{0.6333}$	$\binom{0.7667}{0.1667}$	$\binom{0.6667}{0.2333}$

Table 6 (continued)

Vehicle models	$\mathfrak{H}_1$	$\mathfrak{H}_2$	$\mathfrak{H}_3$	$\mathfrak{H}_4$	$\mathfrak{H}_5$	$\mathfrak{H}_6$	$\mathfrak{H}_7$
C_5	$\begin{pmatrix}0.6333\\0.2667\end{pmatrix}$	$\begin{pmatrix}0.7333\\0.2000\end{pmatrix}$	$\begin{pmatrix}0.7333\\0.2000\end{pmatrix}$	$\begin{pmatrix}0.5667\\0.3333\end{pmatrix}$	$\begin{pmatrix}0.4167\\0.5333\end{pmatrix}$	$\begin{pmatrix}0.5667\\0.3333\end{pmatrix}$	$\begin{pmatrix}0.5667\\0.3333\end{pmatrix}$
C_6	$\begin{pmatrix}0.9000\\0.0833\end{pmatrix}$	$\begin{pmatrix}0.8667\\0.1000\end{pmatrix}$	$\begin{pmatrix}0.8000\\0.1500\end{pmatrix}$	$\begin{pmatrix}0.6333\\0.2833\end{pmatrix}$	$\begin{pmatrix}0.2833\\0.6333\end{pmatrix}$	$\begin{pmatrix}0.7333\\0.2000\end{pmatrix}$	$\begin{pmatrix}0.6000\\0.3000\end{pmatrix}$

4.1.1 Algorithm of TOPSIS

Step I: Secure the opinions of vehicle experts according to the LVs in Table 3.
Step II: Transform the LVs to FFNs to obtain the FF decision matrices denoted by $\mathfrak{M}_q$, which is defined by eq. (16):

$$\mathfrak{M}_q = \left(C_j(\mathfrak{H}_i)\right)_{6\times 7} = \begin{array}{c} \\ C_1 \\ \vdots \\ C_6 \end{array} \begin{array}{c} \begin{array}{ccc} \mathfrak{H}_1 & \cdots & \mathfrak{H}_7 \end{array} \\ \begin{pmatrix} (\mu,\alpha)_{1,1} & \cdots & (\mu,\alpha)_{1,7} \\ \vdots & \ddots & \vdots \\ (\mu,\alpha)_{6,1} & \cdots & (\mu,\alpha)_{6,7} \end{pmatrix} \end{array} \tag{16}$$

where C_j indicates the vehicle models available $(j = 1, 2, \ldots, 6)$, $\mathfrak{H}_i$ indicates the criteria $(i = 1, 2, \ldots, 7)$, and e denotes the number of experts $(e = 1, 2, 3)$.

Step III: Compute the average FF decision matrix, $\widehat{\mathfrak{M}}$ using the following equation:

$$\widehat{\mathfrak{M}} = \frac{\left(C_j(\mathfrak{H}_i)\right)_{6\times 7}}{e} \tag{17}$$

Step IV: Take note the cost criterion (CC), which is the least $\mathfrak{H}_i$, and the benefit criteria, which is the rest of $\mathfrak{H}_i$. Then obtain the CC using the following equation:

$$\bigwedge_{1\le j\le 6}\left\{\sum_{i=1}^{7} C_{j_\mu}(\mathfrak{H}_i)\right\} \tag{18}$$

where $\wedge$ is a minimum operator.

Step V: Obtain the normalized FF decision matrix using $\widetilde{\mathfrak{M}}$, which is defined as follows:

$$\widetilde{\mathfrak{M}} = \begin{cases} \left(C_{j_\mu}(\mathfrak{H}_i), C_{j_\alpha}(\mathfrak{H}_i)\right), & \text{for BC of } C_j \\ \left(C_{j_\alpha}(\mathfrak{H}_i), C_{j_\mu}(\mathfrak{H}_i)\right), & \text{for CC of } C_j \end{cases} \tag{19}$$

Step VI: Formulate the positive ideal solution (PIS) represented by $\widetilde{\mathfrak{M}}^+ = \{\widetilde{\mathfrak{M}}_1^+ \ldots, \widetilde{\mathfrak{M}}_q^+\}$ and the negative ideal solution (NIS) represented by $\widetilde{\mathfrak{M}}^- = \{\widetilde{\mathfrak{M}}_1^- \ldots, \widetilde{\mathfrak{M}}_q^-\}$ defined as follows:

$$\widetilde{\mathfrak{M}}^+ = \begin{Bmatrix} \left(\vee_{1\le j\le 6}\left\{\mathcal{C}_{j_\mu}(\mathfrak{H}_i)\right\}, \wedge_{1\le j\le 6}\left\{\mathcal{C}_{j_a}(\mathfrak{H}_i)\right\}\right), \text{if } \mathfrak{H}_i \text{ is the BC} \\ \left(\wedge_{1\le j\le 6}\left\{\mathcal{C}_{j_\mu}(\mathfrak{H}_i)\right\}, \vee_{1\le j\le 6}\left\{\mathcal{C}_{j_a}(\mathfrak{H}_i)\right\}\right), \text{if } \mathfrak{H}_i \text{ is the CC} \end{Bmatrix}$$
$$\widetilde{\mathfrak{M}}^- = \begin{Bmatrix} \left(\wedge_{1\le j\le 6}\left\{\mathcal{C}_{j_\mu}(\mathfrak{H}_i)\right\}, \vee_{1\le j\le 6}\left\{\mathcal{C}_{j_a}(\mathfrak{H}_i)\right\}\right), \text{if } \mathfrak{H}_i \text{ is the } BC \\ \left(\vee_{1\le j\le 6}\left\{\mathcal{C}_{j_\mu}(\mathfrak{H}_i)\right\}, \wedge_{1\le j\le 6}\left\{\mathcal{C}_{j_a}(\mathfrak{H}_i)\right\}\right), \text{if } \mathfrak{H}_i \text{ is the } CC \end{Bmatrix} \tag{20}$$

where $\vee$ is a maximum operator.

Step VII: Compute the distances indexes, $\eth_*(\widetilde{\mathfrak{M}}^+, \mathcal{C}_j)$ and $\eth_*(\widetilde{\mathfrak{M}}^-, \mathcal{C}_j)$.

Step VIII: Calculate for each alternative $\mathcal{C}_j$ using eq. (21), the closeness coefficient $\Delta(\mathcal{C}_j)$:

$$\Delta(\mathcal{C}_j) = \frac{\eth_*\left(\widetilde{\mathfrak{M}}^+, \mathcal{C}_j\right)}{\eth_*\left(\widetilde{\mathfrak{M}}^+, \mathcal{C}_j\right) + \eth_*\left(\widetilde{\mathfrak{M}}^-, \mathcal{C}_j\right)} \tag{21}$$

Step IX: Obtain the ranking of the vehicle models in an ascending order of $\Delta(\mathcal{C}_j)$ to ascertain the vehicle that is most suitable based on eco-friendliness.

Step X: Select the vehicle model with the least closeness coefficient.

4.1.2 TOPSIS Algorithm's Implementation

In order to determine the best vehicle type among the different types in $\mathcal{C}$ based on the criteria in $\mathfrak{H}$, we implement the TOPSIS algorithm as follows:

Using Step I, the opinions of three car maintenance experts are presented in Table 4 as LVs.

The LVs in Step I, by Step II, are converted to FFNs and presented as the FF decision matrices in Table 5. From Step III, the average FF decision matrix, $(\widehat{\mathfrak{M}})$ is presented in Table 6. By Step IV, the CC, which is the least criteria, is $\mathfrak{H}_5$, while the benefit criteria, which are the other criteria, are computed using eq. (18).

The normalized FF decision matrix $(\widetilde{\mathfrak{M}})$ is presented in Table 7 using Step V. By Step VI, The PIS $(\widetilde{\mathfrak{M}}^+)$ and NIS $(\widetilde{\mathfrak{M}}^-)$ are also presented in Table 7. The distance indexes $\eth_*\left(\widetilde{\mathfrak{M}}^+, \mathcal{C}_j\right)$ and $\eth_*\left(\widetilde{\mathfrak{M}}^-, \mathcal{C}_j\right)$ were calculated with the aid of JavaScript and presented in Table 8 using Step VII.

By Step VIII, $\Delta(\mathcal{C}_j)$ were computed and presented in Table 8. Using Step IX, the vehicle types were ranked in the order $\mathcal{C}_4 \prec \mathcal{C}_6 \prec \mathcal{C}_1 \prec \mathcal{C}_2 \prec \mathcal{C}_3 \prec \mathcal{C}_5$. Hence, the best vehicle type to select amongst the six types evaluated based on the TOPSIS approach via the proposed FFDM is $\mathcal{C}_4$ (EV).

Table 7: Normalized FF decision matrices.

Vehicle models	$\mathfrak{H}_1$	$\mathfrak{H}_2$	$\mathfrak{H}_3$	$\mathfrak{H}_4$	$\mathfrak{H}_5$	$\mathfrak{H}_6$	$\mathfrak{H}_7$
C_1	$\begin{pmatrix}0.3167\\0.6000\end{pmatrix}$	$\begin{pmatrix}0.8667\\0.1000\end{pmatrix}$	$\begin{pmatrix}0.7333\\0.1833\end{pmatrix}$	$\begin{pmatrix}0.7667\\0.1833\end{pmatrix}$	$\begin{pmatrix}0.4333\\0.4667\end{pmatrix}$	$\begin{pmatrix}0.8333\\0.1167\end{pmatrix}$	$\begin{pmatrix}0.5167\\0.3833\end{pmatrix}$
C_2	$\begin{pmatrix}0.1167\\0.7833\end{pmatrix}$	$\begin{pmatrix}0.8667\\0.1067\end{pmatrix}$	$\begin{pmatrix}0.8667\\0.1000\end{pmatrix}$	$\begin{pmatrix}0.8667\\0.1000\end{pmatrix}$	$\begin{pmatrix}0.1833\\0.7667\end{pmatrix}$	$\begin{pmatrix}0.7667\\0.1667\end{pmatrix}$	$\begin{pmatrix}0.7333\\0.2000\end{pmatrix}$
C_3	$\begin{pmatrix}0.6667\\0.2500\end{pmatrix}$	$\begin{pmatrix}0.6667\\0.2500\end{pmatrix}$	$\begin{pmatrix}0.7000\\0.2167\end{pmatrix}$	$\begin{pmatrix}0.7000\\0.2333\end{pmatrix}$	$\begin{pmatrix}0.3333\\0.5500\end{pmatrix}$	$\begin{pmatrix}0.6333\\0.2667\end{pmatrix}$	$\begin{pmatrix}0.6333\\0.2667\end{pmatrix}$
C_4	$\begin{pmatrix}0.7333\\0.1833\end{pmatrix}$	$\begin{pmatrix}0.9000\\0.0833\end{pmatrix}$	$\begin{pmatrix}0.8333\\0.1167\end{pmatrix}$	$\begin{pmatrix}0.8333\\0.1333\end{pmatrix}$	$\begin{pmatrix}0.6333\\0.3000\end{pmatrix}$	$\begin{pmatrix}0.7667\\0.1667\end{pmatrix}$	$\begin{pmatrix}0.6667\\0.2333\end{pmatrix}$
C_5	$\begin{pmatrix}0.6333\\0.2667\end{pmatrix}$	$\begin{pmatrix}0.7333\\0.2000\end{pmatrix}$	$\begin{pmatrix}0.7333\\0.2000\end{pmatrix}$	$\begin{pmatrix}0.5667\\0.3333\end{pmatrix}$	$\begin{pmatrix}0.5333\\0.4167\end{pmatrix}$	$\begin{pmatrix}0.5667\\0.3333\end{pmatrix}$	$\begin{pmatrix}0.5667\\0.3333\end{pmatrix}$
C_6	$\begin{pmatrix}0.9000\\0.0833\end{pmatrix}$	$\begin{pmatrix}0.8667\\0.1000\end{pmatrix}$	$\begin{pmatrix}0.8000\\0.1500\end{pmatrix}$	$\begin{pmatrix}0.6333\\0.2833\end{pmatrix}$	$\begin{pmatrix}0.6333\\0.2833\end{pmatrix}$	$\begin{pmatrix}0.7333\\0.2000\end{pmatrix}$	$\begin{pmatrix}0.6000\\0.3000\end{pmatrix}$
				PIS/NIS			
$\widetilde{\mathfrak{M}}^{+}$	$\begin{pmatrix}0.9000\\0.0833\end{pmatrix}$	$\begin{pmatrix}0.9000\\0.0833\end{pmatrix}$	$\begin{pmatrix}0.8667\\0.1000\end{pmatrix}$	$\begin{pmatrix}0.8333\\0.1333\end{pmatrix}$	$\begin{pmatrix}0.1833\\0.7667\end{pmatrix}$	$\begin{pmatrix}0.8333\\0.1167\end{pmatrix}$	$\begin{pmatrix}0.7333\\0.2000\end{pmatrix}$
$\widetilde{\mathfrak{M}}^{-}$	$\begin{pmatrix}0.1167\\0.7833\end{pmatrix}$	$\begin{pmatrix}0.6667\\0.2500\end{pmatrix}$	$\begin{pmatrix}0.7000\\0.2167\end{pmatrix}$	$\begin{pmatrix}0.5667\\0.3333\end{pmatrix}$	$\begin{pmatrix}0.6333\\0.2833\end{pmatrix}$	$\begin{pmatrix}0.5667\\0.3333\end{pmatrix}$	$\begin{pmatrix}0.5167\\0.3833\end{pmatrix}$

Table 8: Distance indexes and closeness coefficients.

New FFDM	$(\widetilde{\mathfrak{M}}^{+}, C_1)$	$(\widetilde{\mathfrak{M}}^{+}, C_2)$	$(\widetilde{\mathfrak{M}}^{+}, C_3)$	$(\widetilde{\mathfrak{M}}^{+}, C_4)$	$(\widetilde{\mathfrak{M}}^{+}, C_5)$	$(\widetilde{\mathfrak{M}}^{+}, C_6)$
∂_*	0.0944	0.1005	0.0957	0.0571	0.0998	0.0634
	$(\widetilde{\mathfrak{M}}^{-}, C_1)$	$(\widetilde{\mathfrak{M}}^{-}, C_2)$	$(\widetilde{\mathfrak{M}}^{-}, C_3)$	$(\widetilde{\mathfrak{M}}^{-}, C_4)$	$(\widetilde{\mathfrak{M}}^{-}, C_5)$	$(\widetilde{\mathfrak{M}}^{-}, C_6)$
∂_*	0.0791	0.0831	0.0565	0.0969	0.0524	0.1009
	$\Delta(C_1)$	$\Delta(C_2)$	$\Delta(C_3)$	$\Delta(C_4)$	$\Delta(C_5)$	$\Delta(C_6)$
∂_*	0.5441	0.5474	0.6288	0.3708	0.6557	0.3859

4.2 Sensitivity Analysis

Here, we evaluate the impact of the computed weights on the proposed FFDM by comparing the results and interpretations of the distance method without the weights and the distance method with the weights, based on the TOPSIS approach. Using eq. (1), we get the weight values as

$$\omega_i = \{0.1357, 0.1848, 0.1677, 0.1357, 0.1065, 0.1489, 0.1207\}$$

The results for the methods are presented in Table 9 and Figures 1–3.

Table 9: Sensitivity analysis for distance indexes and closeness coefficients.

New FFDM	$\left(\widetilde{\mathfrak{M}}^{+}, \mathcal{C}_1\right)$	$\left(\widetilde{\mathfrak{M}}^{+}, \mathcal{C}_2\right)$	$\left(\widetilde{\mathfrak{M}}^{+}, \mathcal{C}_3\right)$	$\left(\widetilde{\mathfrak{M}}^{+}, \mathcal{C}_4\right)$	$\left(\widetilde{\mathfrak{M}}^{+}, \mathcal{C}_5\right)$	$\left(\widetilde{\mathfrak{M}}^{+}, \mathcal{C}_6\right)$
$\eth_*$	0.0944	0.1005	0.0957	0.0571	0.0998	0.0634
$\eth_*^{\omega}$	0.3411	0.3554	0.3592	0.2062	0.3690	0.2306
	$\left(\widetilde{\mathfrak{M}}^{-}, \mathcal{C}_1\right)$	$\left(\widetilde{\mathfrak{M}}^{-}, \mathcal{C}_2\right)$	$\left(\widetilde{\mathfrak{M}}^{-}, \mathcal{C}_3\right)$	$\left(\widetilde{\mathfrak{M}}^{-}, \mathcal{C}_4\right)$	$\left(\widetilde{\mathfrak{M}}^{-}, \mathcal{C}_5\right)$	$\left(\widetilde{\mathfrak{M}}^{-}, \mathcal{C}_6\right)$
$\eth_*$	0.0791	0.0831	0.0565	0.0969	0.0524	0.1009
$\eth_*^{\omega}$	0.2979	0.3116	0.2019	0.3605	0.1866	0.3697
	$\Delta(\mathcal{C}_1)$	$\Delta(\mathcal{C}_2)$	$\Delta(\mathcal{C}_3)$	$\Delta(\mathcal{C}_4)$	$\Delta(\mathcal{C}_3)$	$\Delta(\mathcal{C}_6)$
$\eth_*$	0.5441	0.5474	0.6288	0.3708	0.6557	0.3859
$\eth_*^{\omega}$	0.5338	0.5328	0.6402	0.3639	0.6641	0.3841

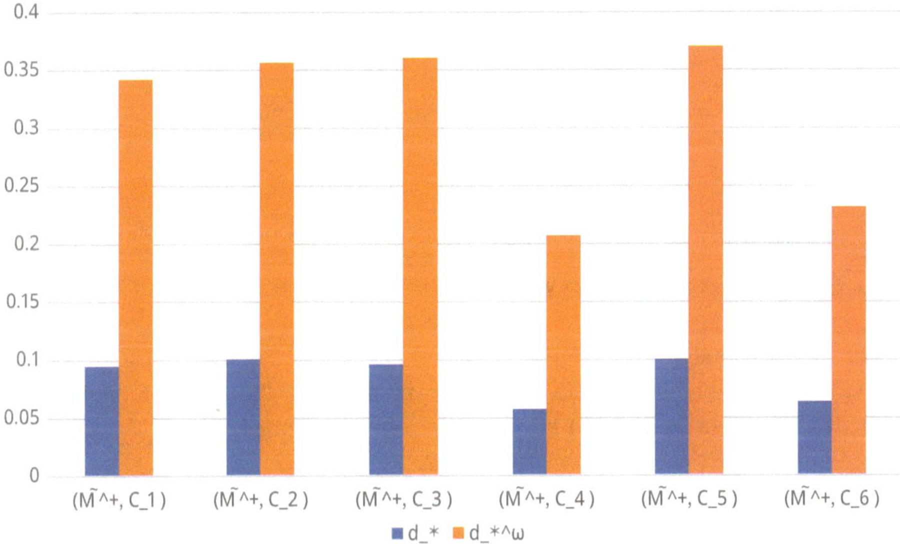

Figure 1: Sensitivity analysis of distance indexes with PIS.

Table 9 and Figures 1–3 show that incorporating the computed weights of the criteria improve distance results. However, the ranking order remains relatively stable in most cases, i.e., $\mathcal{C}_4 \prec \mathcal{C}_6 \prec \mathcal{C}_1 \prec \mathcal{C}_2 \prec \mathcal{C}_3 \prec \mathcal{C}_5$ and $\mathcal{C}_4 \prec \mathcal{C}_6 \prec \mathcal{C}_2 \prec \mathcal{C}_1 \prec \mathcal{C}_3 \prec \mathcal{C}_5$ for $\eth_*$and $\eth_*^{\omega}$, respectively. The closeness coefficients derived using the weighted distance measures show the proposed method's stability and ability to reflect the influence of criterion importance in a more precise manner.

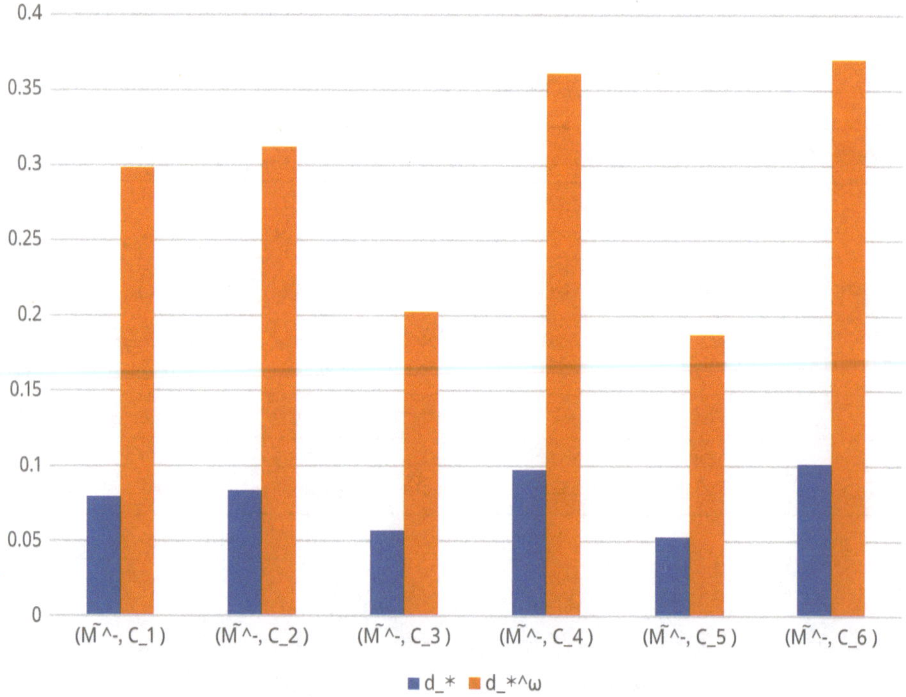

Figure 2: Sensitivity analysis of distance indexes with NIS.

5 Comparative Analysis

A comparative analysis of the novel and existing FFDMs using the vehicle models' selection data based on the TOPSIS algorithm is presented in this section to indicate the effectiveness of the new FFDM in comparison to the existing FFDMs. The results of the comparison are presented in Tables 10–12.

Table 10 presents the computed distance values for each FFDM across the six vehicle alternatives, with respect to both the PIS and NIS. The proposed FFDM ($\eth_*$) consistently yields lower distances to the PIS and NIS, which aligns with the desirable behavior of a robust distance method in decision-making scenarios. On the other hand, methods such as $\eth_{G1}$and $\eth_{G2}$ [29] show unstable or even extreme values, which is against the condition of FFDM. Additionally, $\eth_{G4}$ [29] showed constant values, indicating the inability to discriminate among alternatives. These outcomes underline the discriminative, stability, and superior sensitivity capacity of the proposed FFDM.

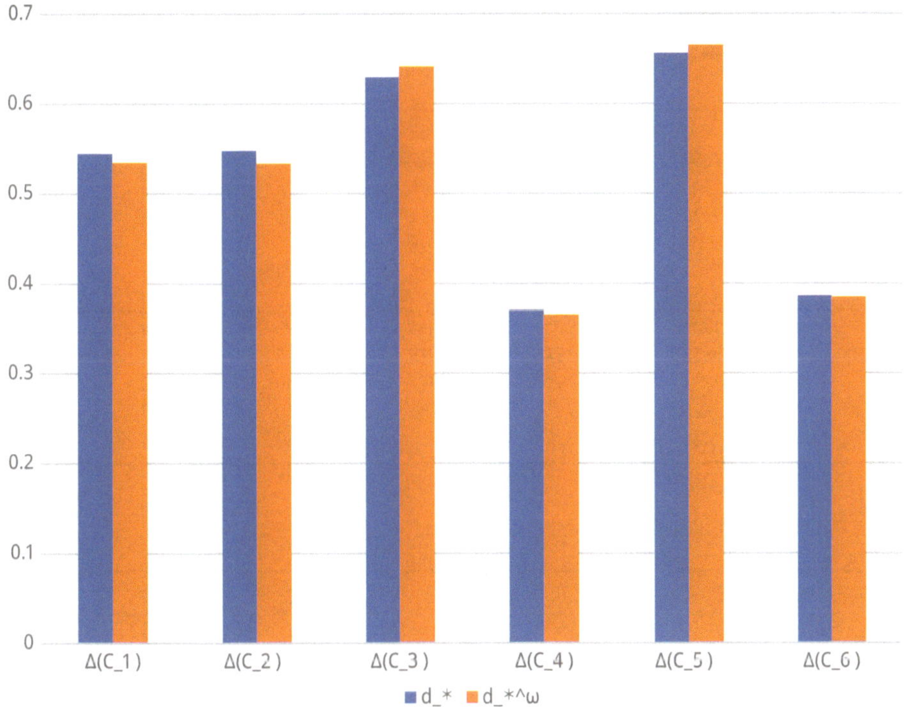

Figure 3: Sensitivity analysis for closeness coefficients.

Table 10: Comparative distance indexes.

FFDMs	$(\widetilde{\mathfrak{M}}^{+}, \mathcal{C}_1)$	$(\widetilde{\mathfrak{M}}^{+}, \mathcal{C}_2)$	$(\widetilde{\mathfrak{M}}^{+}, \mathcal{C}_3)$	$(\widetilde{\mathfrak{M}}^{+}, \mathcal{C}_4)$	$(\widetilde{\mathfrak{M}}^{+}, \mathcal{C}_5)$	$(\widetilde{\mathfrak{M}}^{+}, \mathcal{C}_6)$
∂_{SY}[11]	0.7968	0.8144	0.8695	0.4190	0.8911	0.4707
∂_{G1}[29]	0.1561	0.1451	0.1804	0.1085	0.1910	0.1359
∂_{G2} [29]	1.0504	0.1309	−3, 501.79	0.0976	1.3209	0.1318
∂_{DW1} [46]	0.3210	0.3898	0.2975	0.1222	0.2927	0.1539
∂_{DW2}[46]	0.3011	0.3011	0.3286	0.3286	0.3368	0.1779
∂_{DW3}[46]	0.3210	0.3898	0.2975	0.1222	0.2927	0.1539
∂_{G3} [29]	1.1830	1.0382	1.5258	0.7268	1.6477	0.9403
∂_{G4} [29]	0.1429	0.1429	0.1429	0.1235	0.1429	0.1429
∂_{OE} [47]	0.2459	0.2513	0.2683	0.1293	0.2750	0.1453
∂_{K} [48]	0.3565	0.3529	0.3425	0.4288	0.3384	0.4190
∂_{L}[45]	0.2582	0.3161	0.2536	0.1143	0.2548	0.1338
∂_{Ge} [49]	0.2551	0.2232	0.3268	0.1185	0.3387	0.1573
∂_{*}	0.0944	0.1005	0.0957	0.0571	0.0998	0.0634
	$(\widetilde{\mathfrak{M}}^{-}, \mathcal{C}_1)$	$(\widetilde{\mathfrak{M}}^{-}, \mathcal{C}_2)$	$(\widetilde{\mathfrak{M}}^{-}, \mathcal{C}_3)$	$(\widetilde{\mathfrak{M}}^{-}, \mathcal{C}_4)$	$(\widetilde{\mathfrak{M}}^{-}, \mathcal{C}_5)$	$(\widetilde{\mathfrak{M}}^{-}, \mathcal{C}_6)$
∂_{SY} [11]	0.6429	0.6391	0.4583	0.8423	0.4465	0.8130

Table 10 (continued)

FFDMs	$(\widetilde{\mathfrak{M}}^{+}, C_1)$	$(\widetilde{\mathfrak{M}}^{+}, C_2)$	$(\widetilde{\mathfrak{M}}^{+}, C_3)$	$(\widetilde{\mathfrak{M}}^{+}, C_4)$	$(\widetilde{\mathfrak{M}}^{+}, C_5)$	$(\widetilde{\mathfrak{M}}^{+}, C_6)$
∂_{G1}[29]	0.1571	0.2011	0.1308	0.1520	0.1283	0.1523
∂_{G2} [29]	0.1868	0.2383	0.1142	0.1100	0.1097	0.0945
∂_{DW1}[46]	0.2317	0.2255	0.2508	0.3736	0.2539	0.3695
∂_{DW2}[46]	0.2430	0.2416	0.1732	0.3183	0.1688	0.3073
∂_{DW3}[46]	0.2317	0.2255	0.2508	0.3736	0.2539	0.3695
∂_{G3} [29]	1.1502	1.5063	0.8816	1.1779	0.8583	1.1644
∂_{G4} [29]	0.1429	0.1429	0.1429	0.1429	0.1429	0.1429
∂_{OE} [47]	0.1984	0.1972	0.1414	0.2599	0.1378	0.2509
∂_{K} [48]	0.3862	0.3867	0.4215	0.3478	0.4238	0.3535
∂_{L}[45]	0.1893	0.1893	0.1774	0.2849	0.1738	0.2963
∂_{Ge} [49]	0.2126	0.2281	0.1288	0.3140	0.1195	0.2635
∂_{*}	0.0791	0.0831	0.0565	0.0969	0.0524	0.1009

Table 11: Comparative analysis for closeness coefficients.

FFDMs	$\Delta(C_1)$	$\Delta(C_2)$	$\Delta(C_3)$	$\Delta(C_4)$	$\Delta(C_5)$	$\Delta(C_6)$
∂_{SY} [11]	0.5534	0.5603	0.6548	0.3322	0.6662	0.3668
∂_{G1}[29]	0.4984	0.4191	0.5797	0.4165	0.5982	0.4715
∂_{G2} [29]	0.8490	0.3546	1.000	0.4701	0.9233	0.5824
∂_{DW1} [46]	0.5808	0.6335	0.5426	0.2465	0.5354	0.2940
∂_{DW2}[46]	0.5534	0.5602	0.6548	0.2951	0.6661	0.3667
∂_{DW3}[46]	0.5808	0.6335	0.5426	0.2465	0.5354	0.2940
∂_{G3} [29]	0.5808	0.4080	0.6338	0.3816	0.6575	0.4468
∂_{G4} [29]	0.5000	0.5000	0.5000	0.5000	0.5000	0.5000
∂_{OE} [47]	0.5535	0.5603	0.6549	0.3322	0.6662	0.3667
∂_{K} [48]	0.4800	0.4771	0.4483	0.5522	0.4439	0.5424
∂_{L}[45]	0.5770	0.6292	0.5884	0.2863	0.5945	0.3111
∂_{Ge} [49]	0.5454	0.4946	0.5884	0.2740	0.7392	0.3738
∂_{*}	0.5441	0.5474	0.6288	0.3708	0.6557	0.3859

The relative closeness of each vehicle alternative to the ideal solution is quantified using the closeness coefficient as shown in Table 11. The new method shows a well-differentiated set of closeness coefficient that is consistent and interpretable showing a good performance.

Table 12 goes on to compare the final rankings of the vehicle models, thereby confirming the reliability and consistency of the presented FFDM. C_4 is ranked as the best vehicle model by the new FFDM, which is the same with 11 out of the 13 methods examined. In contrast, methods like ∂_K [48] are significantly different, placing C_5 as the best and C_4 last. This suggests a potential deficiency in alignment with decision preferences, as it differs from the general opinion. In addition, ∂_{G4} [29] could be problematic for decision-making in real world, as it does not differentiate between any alterna-

Table 12: Comparative vehicle ranking.

FFDMs	Ranking	Best ranked vehicle model
$\eth_{SY}$ [11]	$C_4 \prec C_6 \prec C_1 \prec C_2 \prec C_3 \prec C_5$	C_4
$\eth_{G1}$ [29]	$C_4 \prec C_6 \prec C_2 \prec C_1 \prec C_3 \prec C_5$	C_4
$\eth_{G2}$ [29]	$C_4 \prec C_2 \prec C_6 \prec C_1 \prec C_5 \prec C_3$	C_4
$\eth_{DW1}$ [46]	$C_4 \prec C_6 \prec C_5 \prec C_3 \prec C_1 \prec C_2$	C_4
$\eth_{DW2}$ [46]	$C_4 \prec C_6 \prec C_1 \prec C_2 \prec C_3 \prec C_5$	C_4
$\eth_{DW3}$ [46]	$C_4 \prec C_6 \prec C_5 \prec C_3 \prec C_1 \prec C_2$	C_4
$\eth_{G3}$ [29]	$C_4 \prec C_2 \prec C_6 \prec C_1 \prec C_3 \prec C_5$	C_4
$\eth_{G4}$ [29]	$C_4 = C_6 = C_1 = C_2 = C_3 = C_5$	–
$\eth_{OE}$ [47]	$C_4 \prec C_6 \prec C_1 \prec C_2 \prec C_3 \prec C_5$	C_4
$\eth_{K}$ [48]	$C_5 \prec C_3 \prec C_2 \prec C_1 \prec C_6 \prec C_4$	C_5
$\eth_{L}$ [45]	$C_4 \prec C_6 \prec C_1 \prec C_3 \prec C_5 \prec C_2$	C_4
$\eth_{Ge}$ [49]	$C_4 \prec C_6 \prec C_2 \prec C_1 \prec C_3 \prec C_5$	C_4
$\eth_{*}$	$C_4 \prec C_6 \prec C_1 \prec C_2 \prec C_3 \prec C_5$	C_4

tives, assigning equal closeness to all vehicle types. The comparative analysis thereby shows that the proposed FFDM performs better than other existing methods in its ability to produce stable and interpretable closeness coefficients and evaluate sensitive and discriminative alternatives, as well as yielding consistent rankings that align with the domain expectation. As a result of these qualities, the new FFDM is a practical and robust tool for decision-makers handling complex fuzzy situations.

6 Conclusion

With the integration of FFDM to TOPSIS technique, this study has developed and applied a novel FFDM to a decision-making problem of vehicle selection based on power sources. The shortcomings of existing FF-based distance methods have been addressed effectively and significantly by the proposed method and the method satisfies all the axioms of a distance metric by including all the parameters of FFSs, as well as feature space weighting. The approach exhibited its reliability and robustness in handling subjectivity and uncertainty, which are essential and important in decision-making in real-world situations. The new model supports better informed and eco-conscious decision that are in line with global sustainability goals by optimizing vehicle selection under uncertain conditions. Additionally, comparing the new method with other existing methods goes on to show that the proposed FFDM offers an improved precision and interpretability. The study improves the application in intelligent decision support systems and enriches the theoretical foundation of FF systems. The methodological structure laid out here has the potential for broader application across numerous domains in need of complex decision analysis under uncertainty, even though the study was limited in scope to the vehicle selection problem. In the future, researchers may extend this model to dynamic decision environments and explore its integration with other MCDM techniques for improved flexibility and scalability. This novel FFDM can be expanded to q-rung orthopair FSs, which can be used in medical emergency [50] and energy source selection [51]. Even with the robustness of the proposed FFDM, there are still some lingering limitations. There may be higher computational complexity when the approach is applied to large-scale decision problems, which could have an impact on the efficiency in real-time applications. Furthermore, its performance has been primarily been validated in the vehicle selection context; broader empirical studies would be required to confirm if it can be applicable to numerous other domains.

References

[1] Kempton W, Letendre SE. Electric vehicles as a new power source for electric utilities. Transportation Research Part D: Transport and Environment. 1997;2(3):157–175.

[2] Lui C, Chau KT, Wu D, Gao S. Opportunities and challenges of vehicle-to-home, vehicle-to-vehicle and vehicle-to-grid technologies. Proceedings of the IEEE. 2013;101(11):2409–2427.

[3] Ren Y, Hu Y, Jiang L. Configuration and control of traction motors for electric vehicles; A review. Chinese Journal of Electronic Engineering. 2017;3(3):1–17.

[4] Ntombela M, Musasa K. Load profile and load flow analysis for a grid system with electric vehicles using a hybrid optimization algorithm. Sustainability. 2023;15(12):9390.

[5] Scholz RW. Introduction to decision making under uncertainty: biases, fallacies, and the development of decision making. Advances in Psychology. 1983;16:3–18.

[6] Djulbegovic B, Hozo I, Greenland S. Uncertainty in clinical medicine. Handbook of the Philosophy of Science, Philosophy of Medicine. 2011;16:299–356.

[7] Kochenderfer MJ, Amato C, Chowdhary G, How JP, Reynolds HJD, Thornton JR, Torres-Carrasquillo PA, Üre NK, Vian J. Decision making under uncertainty: theory and application. The MIT Press. 2015.

[8] Zadeh LA. Fuzzy sets. Information and Control. 1965;8(3):338–353.

[9] Atanassov K. Intuitionistic fuzzy sets. Fuzzy Sets and Systems. 1986;20(1):87–96.

[10] Yager RR. Pythagorean membership grades in multi-criteria decision making. New Rochelle, New York, USA: Technical Report MII–3301; Machine Intelligence Institute Iona College, 2013.

[11] Senapati T, Yager RR. Fermatean fuzzy sets. Journal of Ambient Intelligence and Humanized Computing. 2020;11:663–674.

[12] Ejegwa PA, Kausar N, Aydin N, Feng Y, Olanrewaju OA. A new Fermatean fuzzy Spearman-like correlation coefficient and its application in evaluating insecurity problem via multi-criteria decision-making approach. Heliyon. 2024;10(22):e40403.

[13] Dilshad A, Aqsa N, Hanan A, Hamiden AEK, Alhanouf A, Songsong D, Abdul R. A comprehensive study for selecting optimal treatment modalities for blood cancer in a Fermatean fuzzy dynamic environment. Scientific Reports. 20241896;14.

[14] Ejegwa PA, Anum MT, Kausar N, Nwokoro CO, Aydin N, Yu H. New Fermatean fuzzy distance metric and its utilization in the assessment of security crises using the MCDM technique. Mathematics. 2024;12:3214.

[15] Zuakwagh Z, Ejegwa PA, Akwu AD. Fermatean fuzzy sets in multiset framework with application in diagnostic process based on composite relations. In: K. Kumar, R. Arora, G. Kaur (eds). Strategic Fuzzy Extensions and Decision-making Techniques. CRC Press. 2024. pp. 195–210

[16] Adel FA, Pratibha R, Arunodaya RM, Ahmad MA, Fausto C. Fermatean fuzzy distance and Sugeno–Weber operators-based SPC-MARCOS approach for sustainable supplier evaluation in the healthcare supply chain. Scientific Reports. 2024;14:27373.

[17] Chen W, Zhang J, Xu L. Generating fuzzy rules for classification accuracy improvement. Expert Systems with Applications. 2001;20(3):215–228.

[18] Lai H, Liu Y. Medical diagnostics applications of IFS. Journal of Healthcare Engineering. 2012;3(2):85–96.

[19] Akram M, Sitara N. Multi-criteria decision-making in social networks using Pythagorean fuzzy information. Journal of Intelligent and Fuzzy Systems. 2017;32(3):1887–1896.

[20] Vega-Mejía CA, Montoya-Torres JR, Islam SMN. Consideration of triple bottom line objectives for sustainability in the optimization of vehicle routing and loading operations, a systematic literature review. Annals of Operations Research. 2019;273:311–375.

[21] Park H, Jun M. Project prioritization in public sector management using intuitionistic fuzzy AHP-TOPSIS. Journal of Project Management. 2022;40(4):113–125.

[22] Hussain Z, Rahman SU, Nasir S, Hussain R. Novel distance and similarity measure on Pythagorean fuzzy sets and its application to multi-criteria decision making with ELECTRE methods for selection of best mobile phones. Journal of Electric Machines and Control. 2023;16(4):38–64.

[23] Ahemen S, Ujah R, Ejegwa PA. New cosine similarity measures for intuitionistic fuzzy sets with application in decision-making. In: I. Ali, UM Modibbo, A. L. Bolaji, H. Garg (eds). Optimization and Computing using Intelligent Data-Driven Approaches for Decision-Making: Optimization Applications. CRC Press. 2024. pp. 55–69

[24] Anum MT, Zhang H, Ejegwa PA, Feng Y. Tendency coefficient-based weighted distance measure for intuitionistic fuzzy sets with applications. In: Proceedings of the 12th International Conference on Intelligent Control and Information Processing (ICICIP 2024). Nanjing, China. 2024. pp. 54–61.

[25] Zhang Y, Wei Y, Wu Q, Zhao P, Niu S, Huang J. Collaborative unsupervised domain adaptation for medical image diagnosis. IEEE Transactions on Image Processing. 2020;29:7834–7844.

[26] Xu W, Fu YL, Wang Y. ResNet and its application to medical image processing: Research progress and challenges. Neurocomputing. 2023;518:334–351.

[27] Yan D, Wu K, Ejegwa PA, Xie X, Feng Y. Pythagorean fuzzy partial correlation measure and its application. Symmetry. 2023;15:216.

[28] Bhatia, M., Arora, H. D., and Naithani, A. Some new correlation coefficient measures based on Fermatean fuzzy sets using decision making approach in pattern analysis and supplier selection. International Journal of Mathematical, Engineering and Management Sciences. 2023;8(2):245–263.

[29] Ganie AH. Multicriteria decision-making based on distance measures and knowledge measures of Fermatean fuzzy sets. Granula Computing. 2022;7:979–998.

[30] Deng Z, Wang J. Evidential Fermatean fuzzy multicriteria decision-making based on Fermatean fuzzy entropy. International Journal of Intelligent Systems. 2021;36(10):5866–5886.

[31] Ejegwa P, Zuakwagh D. Fermatean fuzzy modified composite relation and its application in pattern recognition. Journal of Fuzzy Extension and Applications. 2022;3:140–151.

[32] Zeb A, Juniad M, Izhar M, Khan A. Fermatean fuzzy soft aggregation operators and their application in symptomatic treatment of COVID-19 (case study of patients' identification). Journal of Ambient Intelligence and Humanized Computing. 2022;25(2):11607–11624.

[33] Ejegwa P, Sarkar A. Fermatean fuzzy approach of diseases diagnosis based on new correlation coefficient operators. Deep Learning in Personalized Healthcare and Decision Support. Academic Press. 2023. pp. 2, 23–38.

[34] Hwang CL, Yoon K. Multiple attribute decision making: Methods and Applications. In: Lecture Notes in Economics and Mathematical Systems. Berlin, Heidelberg: Springer, 1981. pp. 186.

[35] Poongavanam, G., Sivalingam, V., Prabakaran, R., Salman, M., and Kim, S. C. Selection of the best refrigerant for replacing R134a in automobile air conditioning system using different MCDM methods: A comparative study. Case Studies in Thermal Engineering. 2021;27:101344.
[36] Gul, M., Lo, H. W. and Yucesan, M. Fermatean fuzzy TOPSIS-based approach for occupational risk assessment in manufacturing. Complex Intelligence Systems. 2021;7:2635–2653.
[37] Aydoğan H, Çetinsaya ÖV. A Fermatean fuzzy MCDM method for selection and ranking problems: Case Studies. Expert Systems with Applications. 2023;237:121628.
[38] Sindhu MS, Siddique I, Ahsan M, Jarad F, Altunok T, Di Bona G. An approach of decision-making under the framework of Fermatean fuzzy sets. Mathematical Problems in Engineering. 2022;8442123:1–9.
[39] Brito FC, Saba H, Panizio RM, Nobre CP, Guarieiro LLN, Ferreira Cv, Ferreira P, Santos AÁB, Filho ASN. Consumer-driven selection of low-emission vehicles for sustainable urban centers: An AHP-based approach. Science of The Total Environment. 2024;955(176993):0048–9697.
[40] Dwivedi PP, Sharma DK. Evaluation and ranking of battery electric vehicles by Shannon's entropy and TOPSIS methods. Mathematics and Computers in Simulation. 2023;212:457–474.
[41] Ali Y, Khan AU, Huzaifa M, Yasir U, Mehmood B. Development of a new hybrid multi criteria decision-making method for a car selection scenario. Facta Universitatis, Series: Mechanical Engineering. 2020;18(3):357.
[42] Kumar DT, Palaniappan M, Kannan D, Shankar KM. Analyzing the CSR issues behind the supplier selection process using ISM approach. Resources, Conservation and Recycling. 2014;92:268–278.
[43] Luo SJ, Fu YT, Zhou YX. Perceptual matching of shape design style between wheel hub and car type. International Journal of Industrial Ergonomics. 2012;42(1):90–102.
[44] Więckowski J, Wątróbski J, Shkurina A, Sałabun W. Adaptive multi-criteria decision making for electric vehicles: a hybrid approach based on RANCOM and ESP-SPOTIS. Artificial Intelligence Review. 2024;57:270.
[45] Liu Z. A distance measure of Fermatean fuzzy sets based on triangular divergence and its application in medical diagnosis. Journal of Operations Intelligence. 2024;2(1):167–178.
[46] Deng Z, Wang J. New distance measure for Fermatean fuzzy sets and its application. International Journal of Intelligent System. 2022;37:1903–1930.
[47] Onyeke IC, Ejegwa PA. Modified Senapati and Yager's Fermatean fuzzy distance and its application in students' course placement in tertiary institution. In: Sahoo, L., Senapati, T., Yager, R.R. (eds). Real Life Applications of Multiple Criteria Decision-Making Techniques in Fuzzy Domain. Singapore: Springer Nature. 2023. pp. 237–253.
[48] Kirişci M. New cosine similarity and distance measures for Fermatean fuzzy sets and TOPSIS approach. Knowledge and Information Systems. 2023;65:855–868.
[49] Ganie AH, Gheith NEM, Al-Qudah Y, Sharma SK, Aqlan AM, Khalaf MM. An innovative Fermatean fuzzy distance metric with its application in classification and bidirectional approximate reasoning. IEEE Access. 2024;12:4780–4791.
[50] Ejegwa PA, Anum MT, Kausar N, Nwokoro CO, Cagin TA. A novel q-rung orthopair fuzzy distance metric for energy source selection. Journal of Applied Research on Industrial Engineering. 2025;12(3):429–464.
[51] Ejegwa PA, Anum MT, Kausar N, Vrioni B. Some new q-rung orthopair fuzzy distance techniques and their application to medical analysis based on decision-making methods. Journal of Decisions and Operations Research. 2025;e 226485. doi: 10.22105/dmor.2025.530184.1967.

Dristi Rajbonshi, Palash Dutta

A Hybrid MCDM Approach Based on Hyperbolic Fuzzy Sets for Supplier Selection in Sustainable Supply Chain

Abstract: In the modern competitive and sustainability-oriented business environment, selecting the appropriate suppliers plays a major role in an organization's long-term success. This study introduces a hybrid framework for sustainable supplier selection of decision-making that combines hyperbolic fuzzy sets (HyFS) with two multicriteria decision-making (MCDM) methods, namely Technique for Order Preference by Similarity to Ideal Solution (TOPSIS) and Evaluation based on Distance from Average Solution (EDAS). A new score function is also introduced to make the evaluation process more reliable and accurate. In the HyFS-TOPSIS method, a new closeness coefficient is introduced to refine ranking results and reduce bias. Both the methods HYFS-TOPSIS and HyFS-EDAS rank the suppliers based on sustainability factors such as economy, society, and environment. To combine the ranking results obtained from these two methods, the Borda count method is used, which produces an integrated and balanced supplier ranking. The proposed framework provides a reliable and adaptable tool for evaluating suppliers in uncertain environments.

Keywords: sustainable supply chain management, supplier selection, hyperbolic fuzzy set, score function, HyFS-TOPSIS, HyFS-EDAS, Borda count

1 Introduction

In recent years, sustainable supply chain management (SSCM) has gained significant importance in both industry and academia, as it aims to enhance competitiveness while minimizing negative social and environmental impacts. Global enterprises, particularly resource-intensive sectors like automotive and manufacturing, are under growing governmental, customer, and competitive pressure to adopt sustainable practices. Sustainable supplier selection (SSS) has evolved into a strategic process within this framework since suppliers substantially impact the supply chain's overall sustainability performance through environmental efforts, resource usage, innovation, and social behaviours [1].

Dristi Rajbonshi, Department of Mathematics, Dibrugarh University, Dibrugarh, Assam, e-mail: dristirajbonshi4@gmail.com
Palash Dutta, Department of Mathematics, Dibrugarh University, Dibrugarh, Assam, e-mail: palashdutta@dibru.ac.in

 | https://doi.org/10.1515/9783112222249-002

Given the complex and contradictory nature of sustainability criteria, multicriteria decision-making (MCDM) methodologies have been widely employed to aid supplier evaluation and ranking. The popular techniques Technique for Order Preference by Similarity to Ideal Solution (TOPSIS) and Evaluation based on Distance from Average Solution (EDAS) assess alternatives according to how close they are to ideal solutions and how far they deviate from the average outcome.

However, decision results are frequently influenced by the method used, and rankings might differ when various procedures are used. To address this issue, the Borda count aggregation has been proposed as a robust mechanism for combining ranks from multiple MCDM approaches, enhancing decision reliability and mitigating method-specific bias.

In SSS, uncertainty and vagueness in expert judgements remain a common issue. The traditional fuzzy sets and their extensions are effective to some extent, but may not be able to express the complex nature of membership relationships in real-life decision-making situations. Hyperbolic fuzzy sets (HyFS) offer greater flexibility and also provides a more accurate and detailed representation of uncertainty and decision-maker preferences. This makes the HyFS suitable for sustainability studies, where subjective opinions and incomplete information are very common. Considering these factors, this study provides an integrated MCDM framework under the HyFS environment to promote SSS [2].

SSS is an important decision-making area within SSCM. Many researchers have proposed different decision-making methods to support this process, but have been unable to deal with the uncertain or imprecise experts' opinions. To address this issue, fuzzy set and hybrid MCDM methods are increasingly utilized, offering greater accuracy and flexibility in supplier evaluation. Wang et al. [3] considered the "fuzzy analytic hierarchy process (F-AHP)" method for determining the criteria weights with fuzzy TOPSIS to evaluate the green suppliers. The study highlighted that fuzzy methods give better decision quality in uncertain environments than traditional approaches. Following this contribution, more advanced and hybrid MCDM approaches have been developed in later research. Later on, Matić et al. [4] considered 21 criteria across social, economic, and environmental domains and based on them evaluated the supplier ranking using the rough "complex proportional assessment" (COPRAS). To integrate the experts' opinions, the study used a rough Dombi aggregator, which made the model effective for the collective decision-making process. Around the same time, Petrović et al. [5] conducted a study on the evaluation of suppliers in Serbia by utilizing the fuzzy "step-wise weight assessment ratio analysis (SWARA)" method to derive the criteria weights and compared the supplier rankings evaluated by the three methods, namely TOPSIS, "weighted aggregated sum product assessment (WASPAS)", and "additive ratio assessment (ARAS)". The study demonstrated that the choice of MCDM methods may affect the ranking results, thus providing valuable insights for managers in selecting a suitable MCDM method for SSS.

Later on, Kao et al. [6] applied F-AHP to determine the criteria weights and WASPAS to evaluate the rankings of suppliers in the garment industry. Nguyen et al. [7] proposed a two-phase model that combined both qualitative and quantitative factors. To identify the efficient suppliers, "data envelopment analysis (DEA)" was considered, and to determine the subjective weights, spherical F-AHP was applied, and to rank the suppliers, the spherical fuzzy WASPAS method was utilized. A sensitive analysis was also conducted to analyse the consistency and reliability of the proposed framework.

In recent years, sustainability has gained more importance in SSS. Štreimikienė et al. [8] applied the TOPSIS method to evaluate suppliers in the Iranian automotive industry based on the 12 criteria, such as eco-design, waste reduction, and innovation. Similarly, Tronnebati et al. [9] proposed a hybrid framework that combines F-AHP, WASPAS, and TOPSIS to evaluate green suppliers. Their approach proved that multiple MCDM methods can provide more reliable outcomes. Ardra and Barua [10] proposed a three-phase model for SSS in the supermarket industry. They considered the "best-worst method (BWM)" for weighting the criteria and used the fuzzy TOPSIS method for ranking the suppliers.

Kamran et al. [11] applied FAHP and fuzzy TOPSIS methods to evaluate suppliers in an oil and gas industry. Their findings demonstrate that fuzzy methods can handle such complex decision problems effectively. Raveena and Umamaheswari [12] also used TOPSIS to evaluate suppliers in the Iranian automotive industry based on twelve social, economic, and environmental criteria. These studies show an increasing focus on SSS using fuzzy MCDM methods.

Maden [13] introduced a hybrid fuzzy cognitive map-EDAS approach for selecting low-carbon suppliers in the textile sector. Goswami et al. [14] compared "CRiteria Importance Through Intercriteria Correlation (CRITIC)", "Combined Compromise Solution (CoCoSo)", and "Combinative Distance-based Assessment (CODAS)", and the EDAS methods for selecting a renewable energy power plant in India. They found that the EDAS technique offered the most stable and consistent results. Ghanbari et al. [15] proposed a modified EDAS method that could handle both qualitative and quantitative data and found that wind electrolysis is the most efficient option for green hydrogen production. Dey et al. [16] also contributed by developing the "Logarithmic Percentage Change-driven Objective Weighting-Proximity Index Value (LOPCOW-PIV)" framework, combined with the "Borda count method" and "rank index method (RIM)" to select firm performance in mergers and acquisitions.

The following briefly describes the primary contributions of this chapter:

- A novel score function that guarantees monotonicity, consistency, and better discriminatory ability over the existing score functions is proposed for the HyFS framework.
- An integrated MCDM framework combining HyFS-TOPSIS and HyFS-EDAS is developed, capturing different perspectives of supplier performance evaluation.

The outline of the study is structured as follows. Section 1 provides an introduction and a detailed review of existing literature related to SSCM. Section 2 contains the basic definitions of fuzzy sets and their different extensions. Section 3 defines an entirely new score function and compares its properties with the existing score functions. Section 4 describes the methodology used to address MCDM problems using HyFS-TOPSIS and HyFS-EDAS methods, and also explains the Borda count method. Section 5 combines these methods to select the most suitable supplier. Section 6 provides the conclusion for this chapter and its future scope.

2 Preliminaries

Definition 2.1 (Intuitionistic Fuzzy Set) [17]. Let $Y = \{y_1, y_2, \ldots, y_n\}$ be a universe of discourse. An intuitionistic fuzzy set (IFS) A in Y is given by:

$$A = \{(y, P_A(y), Q_A(y)) : y \in Y\} \text{ where } P_A : Y \to [0,1] \text{ and } Q_A : Y \to [0,1]$$

represents the membership function (MF) and non-membership function (NF), respectively, with the condition $0 \le P_A(y) + Q_A(y) \le 1$.

$\pi_A(y) = 1 - (P_A(y) + Q_A(y))$ represents the degree of indeterminacy (DI).

Definition 2.2 (Pythagorean Fuzzy Set) [18]. Let, $Y = \{y_1, y_2, \ldots, y_n\}$ be a universe of discourse. A Pythagorean fuzzy set (PFS) A in Y is given by:

$$A = \{(y, P_A(y), Q_A(y)) : y \in Y\} \text{ where } P_A : Y \to [0,1] \text{ and } Q_A : Y \to [0,1]$$

represent the MF and NF, respectively, with the condition $0 \le P_A^2(y) + Q_A^2(y) \le 1$.

$\pi_{(A)}(y) = \sqrt[2]{1 - (P_A^2(y) + Q_A^2(y))}$ represents the DI.

Definition 2.3 (*Q*-Rung Orthopair Fuzzy Set) [19]. Let $Y = \{y_1, y_2, \ldots, y_n\}$ be a universe of discourse. A *Q*-Rung Orthopair Fuzzy Set (*Q*-ROFS) A in Y is given by:

$$A = \{(y, P_A(y), Q_A(y)) : y \in Y\} \text{ where } P_A : Y \to [0,1] \text{ and } Q_A : Y \to [0,1]$$

represents the MF and NF, respectively, with the condition $0 \le P_A^q(y) + Q_A^q(y) \le 1, q \ge 1$.

$\pi_A(y) = \sqrt[q]{1 - (P_A^q(y) + Q_A^q(y))}$ represents the DI.

Definition 2.4 (Fermatean Fuzzy Set) [20]. Let, $Y = \{y_1, y_2, \ldots, y_n\}$ be a universe of discourse. A Fermatean fuzzy set (FFS) A in Y is given by:

$$A = \{(y, P_A(y), Q_A(y)) : y \in Y\} \text{ where } P_A : Y \to [0,1] \text{ and } Q_A : Y \to [0,1]$$

represent the MF and NF, respectively, with the condition $0 \le P_A^q(y) + Q_A^q(y) \le 1, q \ge 1$.

$$\pi_A(y) = \sqrt[3]{1 - (P_A{}^3(y) + Q_A{}^3(y))} \text{ represents the DI.}$$

Definition 2.5 (Hyperbolic Fuzzy Set) [21]. Let, $Y = \{y_1, y_2, \ldots, y_n\}$ be a universe of discourse. A hyperbolic fuzzy set (HyFS) A in Y is given by

$$A = \{(y, P_A(y), Q_A(y)) : y \in Y\} \text{ where } P_A : Y \rightarrow [0,1] \text{ and } Q_A : Y \rightarrow [0,1]$$

represent the optimistic degree and pessimistic degree of $y \in Y$, respectively, with the condition $0 \leq P_A(y).Q_A(y) \leq 1$.

2.1 Some Existing Score Functions

A few of the scoring functions that are now available in IFS, PFS, FFS, and HyFS are given as follows:

1. Luo & Mo [22]: $S_{lm}(A) = P - Q$
2. Feng et al. [23]: $S_f(A) = (PQ + 1)/2$
3. Peng [24]: $S_p(A) = \frac{e^{(P-Q)}}{(1+\pi)}, \pi = 1 - (P + Q)$
4. Peng and Garg [25]: $S_{pg}(A) = \frac{P - Q + 1 - \ln(1+\pi)}{2}$
5. Wu and Wei [26]: $S_{ww}(A) = \frac{1}{2}(P^2 - Q^2 + 1)$
6. Peng and Dai [27]: $S_{pd}(A) = \frac{e^{P^2 - Q^2}}{\pi^2 + 1}$
7. Senapati and Yager [20]: $S_{sp}(A) = P^3 - Q^3$
8. Peng et al. [28]: $S_{pxd}(A) = (P^2 - Q^2) - \ln(1 + \pi^2)$
9. Dutta and Borah [21]: $S_{db}(A) = 2P - P \times Q$
10. Dutta and Banik [29]: $S_{da}(A) = \frac{2 \times P - 2 \times Q + 2 + P \times Q}{4}$

2.1.1 Drawbacks of Existing Score Functions

The existing score functions for IFS, PFS, FFS, and HyFS mentioned above have several drawbacks. For example, consider $A_1 = (0.5, 0.5)$ and $A_2 = (0, 0)$, the score functions proposed by Luo and Mo [22], Wu and Wei [26], and Senapati and Yager [20] are unable to distinguish between A_1 and A_2. The score functions proposed by Feng et al. [23], Dutta and Borah [21] fail to distinguish between $A_1 = (0, 0.2)$, $A_2 = (0, 0.3)$. Furthermore, when considering $A_1 = (0, 0.2)$ and $A_2 = (0, 0.3)$, where $A_1 > A_2$, but score the functions proposed by Peng and Dai [27] show the relation $S_{pd}(A_1) < S_{pd}(A_2)$, and Peng et al. [28] show the relation as $S_{pxd}(A_1) < S_{pxd}(A_2)$, which contradicts the condition of a valid score measure. A comparison table highlighting the drawbacks of some of these score functions is presented in Section 3.1 (Table 1).

3 Novel Score Function for HyFS

In the framework of HyFS, consider P and Q are the MF and NF, respectively, satisfying the condition $0 \le PQ \le 1$, with $P, Q \in [0, 1]$. We propose the subsequent score function to ensure effective ranking and discrimination among HyFS elements:

Definition 3.1. Let $A = \{ < Y_j,\ P_A(Y_j),\ Q_A(Y_j) >; Y_j \in Y\}$ be a HyFS, then the score is denoted as $\mathrm{S(A)} \in [0, 3]$.

$$S\ (A) = 1 + 2\ {P_A}^3(Y_j) - {Q_A}^5(Y_j) + \frac{{P_A}^7(Y_j) \times {Q_A}^{11}(Y_j)}{3} \tag{1}$$

The design of $S(A)$ follows several rational considerations. The cubic term $2{P_A}^3(Y_j)$ ensures a monotonic increase with respect to the membership degree, while the negative fifth-power term $-{Q_A}^5(Y_j)$ enforces a strong monotonic decrease with respect to the non-membership degree. To enhance separability, the asymmetric high-order interaction term $\frac{{P_A}^7(Y_j) \times {Q_A}^{11}(Y_j)}{3}$ is introduced. The use of unequal exponents prevents symmetric cancellations that often occur in lower-order interactions such as $P_A(Y_j) \times Q_A(Y_j)$, thereby yielding more distinctive score values. Within the feasible domain defined by $P_A(Y_j) \times Q_A(Y_j) \le 1$. The interaction term remains bounded while amplifying sensitivity in the interior of the domain. As a result, the proposed score function reduces the likelihood of different (P, Q) pairs producing identical or near-identical scores, which improves the discriminative capability of HyFS in ranking and decision-making applications.

Definition 3.2. Let $A = \{ < Y_j, P_A(Y_j), Q_A(Y_j) >; Y_j \in Y\}$be a HyFS, then the accuracy is denoted as

$$K(A) = 1 + 2{Q_A}^3(Y_j) - {P_A}^5(Y_j) + \frac{{Q_A}^7(Y_j) \times {P_A}^{11}(Y_j)}{3}, \tag{2}$$

where $K(A) \in [0, 3]$.

Theorem 3.1. Let $A \in$ HyFS(Y). Then

(i) $S\ (A) = 0$ iff $P = 0$ and $Q = 1$.

(ii) $S\ (A) = 3$ iff $P = 1$ and $Q = 0$.

Proof:

(i) Let $A \in$ HyFS (Y) and $S(A) = 1 + 2P^3 - Q^5 + 1/3\, P^7Q^{11}$ with $P, Q \in [0, 1]$
Let $S(A) = 0$

$$\Rightarrow 1 + 2P^3 - Q^5 + 1/3\, P^7Q^{11} = 0$$

$$\Rightarrow 3 + 6P^3 - 3Q^5 + P^7Q^{11} = 0$$

$$\Rightarrow 3 + 6P^3 + P^7Q^{11} = 3Q^5$$

As $P, Q \in [0,1]$, the left side satisfies

$$3 + 6P^3 + P^7Q^{11} \geq 3$$

Because, $6P^3 \geq 0$ and $P^7Q^{11} \geq 0$. The right side $3Q^5 \geq 3$ (since $1 \geq Q^5$). For equality to hold, we must have $3 + 6P^3 + P^7Q^{11} = 3$ and $3Q^5 = 3$.

The latter forces $Q^5 = 1 \Rightarrow Q = 1$.

Then, putting $Q = 1$ in the equation, we will get:

$$6P^3 + P^7 = 0,$$

And since both terms are non-negative on [0, 1], this implies $P = 0$.

Therefore, the only solution is $P = 0$ and $Q = 1$.

Conversely, suppose $P = 0$ and $Q = 1$,

Then $S(A) = 1 + 0 - 1^5 + 0/3 = 1 - 1 = 0$. Hence proved.

(ii) Let, $A \in$ HyFS (Y) and $S(A) = 1 + 2P^3 - Q^5 + 1/3\, P^7Q^{11}$ with $P, Q \in [0,1]$
Let, $S(A) = 3$

$$\Rightarrow 1 + 2P^3 - Q^5 + 1/3\, P^7Q^{11} = 3$$

$$\Rightarrow 3 + 6P^3 - 3Q^5 + P^7Q^{11} = 9$$

$$\Rightarrow 3 + 6P^3 - 3Q^5 + P^7Q^{11} = 9 \tag{3}$$

$$\Rightarrow 6P^3 + P^7Q^{11} = 3Q^5 + 6$$

As $6P^3 \leq 6$ and $P^7Q^{11} \leq 1$, so $6P^3 + P^7Q^{11} \leq 7 \Rightarrow 3Q^5 + 6 \leq 7 \Rightarrow 3Q^5 \leq 1$.

Again, from eq. (3) we have:

$$\Rightarrow \mathbf{6\,P^3 - 6 = 3Q^5 - P^7Q^{11}} \tag{4}$$

$$\Rightarrow 6\left(P^3 - 1\right) = 3Q^5 - P^7Q^{11}$$

$$\text{LHS} = 6\left(P^3 - 1\right) \leq 0, \text{ as } P \in [0, 1].$$

$$\text{RHS} = 3Q^5 - P^7Q^{11}$$

Since $Q^5 \geq 0$ and $P^7Q^{11} \geq 0$, the RHS can be either negative or positive, but is bounded by $3Q^5 \leq 1$.

For equality, we need both sides ≤ 0.
So, $6(P^3 - 1) \leq 0$, $3Q^5 - P^7Q^{11} \leq 0 \Rightarrow 3Q^5 \leq P^7Q^{11}$.
$\Rightarrow$ If $Q = 0$, then eq. (4) becomes $6P^3 = 6 \Rightarrow P = 1$
If $Q > 0$, then from $3Q^5 \leq P^7Q^{11}$, we can divide Q^5: $P^7Q^6 \geq 3$.
But $P^7 \leq 1$, $Q^6 \leq 1$, so $P^7Q^6 \leq 1$. Impossible to reach 3.
So, $Q = 0$ and $P = 1$ is the only solution.
Conversely, If $Q = 0$ and $P = 1$, then
$S(A) = 1 + 2 - 0 - 0 = 3$. Hence proved.

Theorem 3.2. If $A \in$ HyFS(Y), then the score function $S(A)$ exhibits a monotonic increase with respect to P and a monotonic decrease with respect to Q.

Proof: Let $A \in$ HyFS and $S(A) = 1 + 2P_A{}^3(Y_j) - Q_A{}^5(Y_j) + \dfrac{P_A{}^7(Y_j) \times Q_A{}^{11}(Y_j)}{3}$,

$$\frac{\partial S}{\partial P_A} = 6 \times P_A{}^2(Y_j) + \frac{7 \times P_A{}^6(Y_j) \times Q_A{}^{11}(Y_j)}{3} \geq 0, \text{ monotonically increasing.}$$

$$\frac{\partial S}{\partial Q_A} = -5Q_A{}^4(Y_j) + \frac{11 \times P_A{}^7(Y_j) \times Q_A{}^{10}(Y_j)}{3} \leq 0, \text{ monotonically decreasing.}$$

Hence the result.

Theorem 3.3. Let A_1 and $A_2 \in$ HyFS (Y). Then the following properties hold:
(i) $S(A_1) < S(A_2)$ iff $A_1 < A_2$.
(ii) $S(A_1) > S(A_2)$ iff $A_1 > A_2$.
(iii) $S(A_1) = S(A_2)$ iff $A_1 = A_2$.

3.1 Comparative Evaluation of the Proposed Score Function with Existing Ones

To show the reliability and flexibility of the suggested scoring function on HyFS, a comparison of a few existing score functions with the proposed score function is provided in this section.

Table 1: Comparative evaluation of the proposed score function with existing ones.

Score functions	$A_1 = (0, 1)$, $A_2 = (1, 0)$	$A_1 = (0, 0.2)$, $A_2 = (0, 0.3)$	$A_1 = (0.5, 0.5)$, $A_2 = (0, 0)$	$A_1 = (0.5, 0.5)$, $A_2 = (0.4, 0.4)$	$A_1 = (0.3, 0.7)$, $A_2 = (0.4, 0.5)$
$S_{lm}(A)$	$A_1 < A_2$	$A_1 > A_2$	$A_1 = A_2$	$A_1 = A_2$	$A_1 < A_2$
$S_f(A)$	$A_1 = A_2$	$A_1 = A_2$	$A_1 > A_2$	$A_1 > A_2$	$A_1 > A_2$
$S_p(A)$	$A_1 < A_2$	$A_1 > A_2$	$A_1 > A_2$	$A_1 > A_2$	$A_1 < A_2$
$S_{pg}(A)$	$A_1 < A_2$	$A_1 > A_2$	$A_1 > A_2$	$A_1 > A_2$	$A_1 < A_2$
$S_{ww}(A)$	$A_1 < A_2$	$H_1 > A_2$	$A_1 = A_2$	$A_1 > A_2$	$A_1 < A_2$

Table 1 (continued)

Score functions	$A_1 = (0, 1)$, $A_2 = (1, 0)$	$A_1 = (0,0.2)$, $A_2 = (0, 0.3)$	$A_1 = (0.5, 0.5)$, $A_2 = (0, 0)$	$A_1 = (0.5, 0.5)$, $A_2 = (0.4, 0.4)$	$A_1 = (0.3, 0.7)$, $A_2 = (0.4, 0.5)$
$S_{pd}(A)$	$A_1 < A_2$	$A_1 < A_2$	$A_1 > A_2$	$A_1 > A_2$	$A_1 < A_2$
$S_{pxd}(A)$	$A_1 < A_2$	$A_1 < A_2$	$A_1 > A_2$	$A_1 > A_2$	$A_1 < A_2$
$S_{sp}(A)$	$A_1 < A_2$	$A_1 > A_2$	$A_1 = A_2$	$A_1 = A_2$	$A_1 > A_2$
$S_{db}(A)$	$A_1 < A_2$	$A_1 = A_2$	$A_1 > A_2$	$A_1 > A_2$	$A_1 < A_2$
$S_{da}(A)$	$A_1 < A_2$	$A_1 > A_2$	$A_1 > A_2$	$A_1 > A_2$	$A_1 < A_2$
Proposed score, $S(A)$	$A_1 < A_2$	$A_1 > A_2$	$A_1 > A_2$	$A_1 > A_2$	$A_1 < A_2$

4 Methodology

This section describes the methodological framework, which includes HyFS-TOPSIS and HyFS-EDAS, for ranking alternatives based on different criteria in the presence of uncertainty and imprecision. Finally, the Borda count method is explained, which aids in aggregating the rankings provided by the HyFS-TOPSIS and HyFs-EDAS methods into a balanced final ranking.

4.1 Algorithm for the HyFS-TOPSIS Method

The TOPSIS was originally introduced by Hwang and Yoon [30] in 1981. The core principle of this technique is to determine the most appropriate alternative by identifying the option that exhibits the minimum distance from the positive ideal solution (PIS) and, maximum distance from the negative ideal solution (NIS) [31, 32]. In the present work, the TOPSIS approach is adapted within the framework of HyFS by incorporating a novel closeness coefficient measure. The detailed procedure of the proposed method is described as follows:

Step 1: Formulate the decision matrix
Given a set of alternatives $Y = \{Y_1, Y_2, Y_3, Y_4, \ldots, Y_m\}$ and criteria $C = \{C_1, C_2, C_3, C_4, \ldots, C_n\}$ with weight vector $w = (w_1, w_2, w_3, w_4, \ldots, w_n)^T$, satisfying $0 \le w_j \le 1$ and $\sum_{j=1}^{n} w_j = 1$.

The decision matrix $R = (A_{kl})_{m\times n}$, where $A_{kl} = (\tilde{P}_{kl}, \tilde{Q}_{kl})$ is a hyperbolic fuzzy number with optimistic degree $\tilde{P}_{kl}$ and pessimistic degree $\tilde{Q}_{kl}$:

$$R = \begin{bmatrix} A(P_{11}, Q_{11}) & \cdots & A(P_{1n}, Q_{1n}) \\ \vdots & \ddots & \vdots \\ A(P_{m1}, Q_{m1}) & \cdots & A(P_{mn}, Q_{mn}) \end{bmatrix}$$

Step 2: Normalize the decision matrix
Normalization depends on the type of criterion:

$$A_{kl} = \begin{cases} (P_{kl}, Q_{kl}), & \text{for benefit criteria} \\ (1 - P_{kl}, 1 - Q_{kl}), & \text{for cost criteria.} \end{cases}$$

Step 3: Determine PIS and NIS
Positive ideal alternative (PIA):

$$\mathrm{Y}^{+} = \{\mathrm{C}_l, \max \mathrm{S}(\mathrm{A}_{kl}) \,|\, l = 1, 2, \ldots, n\} = \left\{\langle \mathrm{C}_1, \mathrm{A}\left(\tilde{P}_1^{\,+}, \tilde{Q}_1^{\,+}\right)\rangle, \ldots, \langle \mathrm{C}_n, \mathrm{A}\left(\tilde{P}_n^{\,+}, \tilde{Q}_n^{\,+}\right)\rangle\right\}.$$

Negative ideal alternative (NIA):

$$\mathrm{Y}^{-} = \{\mathrm{C}_l, \min \mathrm{S}(\mathrm{A}_{kl}) \,|\, l = 1, 2, \ldots, n\} = \{\langle \mathrm{C}_1, \mathrm{A}\left(\tilde{P}_1^{\,-}, \tilde{Q}_1^{\,-}\right)\rangle, \ldots, \langle \mathrm{C}_n, \mathrm{A}\left(\tilde{P}_n^{\,-}, \tilde{Q}_n^{\,-}\right)\rangle\}.$$

where $S\,(A_{kl})$ is the score function defined in eq. (1).

Step 4: Calculate distances
Distance from Y_k to PIA:

$$\mathrm{D}(Y_k, Y^{+}) = \frac{1}{2}\left[\sum_{l=1}^{n} w_l\left(\left|\widetilde{P_{kl}} - \tilde{P}^{+}{}_{l}\right| + \left|\widetilde{Q_{kl}} - \tilde{Q}^{+}{}_{l}\right|\right)\right]$$

Distance from Y_k to NIA:

$$\mathrm{D}(Y_k, Y^{-}) = \frac{1}{2}\left[\sum_{l=1}^{n} w_l\left(\left|\widetilde{P_{kl}} - \tilde{P}^{-}{}_{l}\right| + \left|\widetilde{Q_{kl}} - \tilde{Q}^{-}{}_{l}\right|\right)\right]$$

Step 5: Compute the closeness coefficient
The relative closeness coefficient of an alternative Y_k to the ideal solution is given by:

$$\zeta(Y_k) = \frac{\mathrm{D}(Y_k, Y^{-})}{\mathrm{D}(Y_k, Y^{-}) + \mathrm{D}(Y_k, Y^{+})}$$

where $\zeta(Y_k) \in [0,1]$.

Later, in 2014, Hadi-Vencheh and Mirjaberi [33] proved that there are some situations where the closeness coefficient cannot provide a proper output. To handle such types of problems, they introduced a revised closeness coefficient, given by:

$$\zeta(Y_k) = \left[\frac{\mathrm{D}(Y_k, Y^{-})}{D_{\max}(Y_k, Y^{-})} - \frac{\mathrm{D}(Y_k, Y^{+})}{D_{\min}(Y_k, Y^{+})}\right]$$

Despite the revised closeness coefficient being commonly used in fuzzy MCDM research due to its effectiveness in differentiating between alternatives, it often generates values that are either negative or zero. To resolve the interpretability and negative values associated with normalized-based closeness coefficients, we introduced a modified revised closeness coefficient:

$$\zeta(Y_k) = \left| \left[\frac{D(Y_k, Y^-)}{D_{max}(Y_k, Y^-)} - \frac{D(Y_k, Y^+)}{D_{min}(Y_k, Y^+)} \right] \right| + \left(\frac{D(Y_k, Y^-)}{D_{max}(Y_k, Y^-)} - \frac{D(Y_k, Y^+)}{D_{min}(Y_k, Y^+)} \right)$$

where $\zeta(Y_k)$ provides non-negative values and [.] refers as greatest integer function.

Step 6: Rank the alternatives
Rank the alternatives based on the descending order of $\zeta(Y_k)$.
The best alternative is determined as: $Y^* = \{Y_k: \zeta(Y_k) = \max \zeta(Y_k)\}$.

4.2 Algorithm for the HyFS-EDAS Method

The EDAS method was first introduced by Ghorabaee et al. [34] to address decision problems. This method considers the ideal choice, which is determined by its distance from the average solution (AS). The method calculates two key measures: positive distance from the average (PDA) and negative distance from the average (NDA). Higher PDA scores and lower NDA values give better results [35, 36]. We proposed an advanced EDAS method with a new score function to improve decision-making accuracy.

Step 1: Construct the decision matrix
Given a set of alternatives $Y = \{Y_1, Y_2, Y_3, Y_4, \ldots, Y_m\}$ and criteria $C = \{C_1, C_2, C_3, C_4, \ldots, C_n\}$ with weight vector $w = (w_1, w_2, w_3, w_4, \ldots, w_n)^T$, satisfying $0 \le w_j \le 1$ and $\sum_{j=1}^{n} w_j = 1$.

The decision matrix $R = (A_{kl})_{m\times n}$, where $A_{kl} = (\tilde{P}_{kl}, \tilde{Q}_{kl})$ is a hyperbolic fuzzy number with optimistic degree $\tilde{P}_{kl}$ and pessimistic degree $\tilde{Q}_{kl}$:

$$R = \begin{bmatrix} A(P_{11}, Q_{11}) & \cdots & A(P_{1n}, Q_{1n}) \\ \vdots & \ddots & \vdots \\ A(P_{m1}, Q_{m1}) & \cdots & A(P_{mn}, Q_{mn}) \end{bmatrix}$$

Step 2: Then, to normalize the decision matrices, we need to convert the cost type into a benefit type with the help of the proposed score function as follows:

$$S_{kl} = \begin{cases} 1 + 2{P_{kl}}^3 - {Q_{kl}}^5 + \frac{{P_{kl}}^7 \times {Q_{kl}}^{11}}{3}, \text{for benefit criteria} \\ 1 + 2(1 - {P_{kl}}^3) - (1 - {Q_{kl}}^5) + \frac{(1-{P_{kl}}^7)\times(1-{Q_{kl}}^{11})}{3}, \text{for cost criteria} \end{cases}$$

Then, we have the evaluation matrix as:

$$[S_{kl}]_{m\times n} = \left[\frac{S_{kl}}{S_{max}}\right]_{m\times n}, \quad \text{where } S_{max} = 3.$$

Step 3: Average solution

The AS is calculated using the following equation:

$$\mathrm{AS}_l = \frac{1}{m}\sum_{k=1}^{m} S_{kl}, l = 1, 2, \ldots, n$$

Step 4: Calculate the measure of PDA and NDA:

$$\mathrm{PDA}_{kl} = \frac{\max\left(0,\ \left(S_{kl} - \mathrm{AS}_l\right)\right)}{\mathrm{AS}_l},$$

$$\mathrm{NDA}_{kl} = \frac{\max\left(0,\ \left(\mathrm{AS}_l - S_{kl}\right)\right)}{\mathrm{AS}_l}$$

Step 5: Determination of the weighted positive and negative distances:

$$\mathrm{WSP}_k = \sum_{l=1}^{n} w_l.\mathrm{PDA}_{kl}$$

$$\mathrm{WSN}_k = \sum_{l=1}^{n} w_l.\,\mathrm{NDA}_{kl}$$

Step 6: Determination of the normalized values of WSP_k and WSN_k:

$$\mathrm{NSP}_k = \frac{\mathrm{WSP}_k}{\max_k\ (\mathrm{WSP}_k)}$$

$$\mathrm{NSN}_k = 1 - \frac{\mathrm{WSN}_k}{\max_k\ (\mathrm{WSN}_k)}$$

Step 7: Calculate the performance scores of alternatives:

$$\mathrm{PS}_k = \tfrac{1}{2}\left(\mathrm{NSP}_k + \mathrm{NSN}_k\right)$$

Step 8: Rank the alternatives

By lowering the values of the options' performance scores, the final ranking is determined.

4.3 Borda Count Method

Borda [37] introduced the Borda count method to combine the rankings of various models. Let us look at a decision problem with m alternatives. The best option is given (m–1) points using this method. In the same way, the next best option (m–3), the second-best option (m–2), etc. Alternatives are arranged from the highest to lowest scores [38].

5 Numerical Study

For the strategic development of industry *XY*, it has been recognized that strengthening SSCM requires identifying long-term, stable, and sustainable suppliers. Following preliminary screening, four possible suppliers – A_1, A_2, A_3, and A_4 – were chosen for further consideration. Ten key criteria (Table 2) were considered to determine the most appropriate supplier. Consider that the weights of each criterion are equal, to preserve impartiality and eliminate bias in this investigation.

Consider, weight $w = (0.1, 0.1, 0.1, 0.1, 0.1, 0.1, 0.1, 0.1, 0.1, 0.1)$.

Table 2: Sustainable supplier evaluation criteria.

Main criteria		Sub-criteria	Type
Economic	C_1	Cost	Cost
	C_2	Innovation capability	Benefit
	C_3	Quality	Benefit
	C_4	Service capability	Benefit
	C_5	Long-term cooperation	Benefit
Environmental	C_6	Environmental management system	Benefit
	C_7	Reduce pollution	Benefit
	C_8	Green image	Benefit
Social	C_9	Social responsibility	Benefit
	C_{10}	Employment practices	Benefit

This work used linguistic variables and then translated them into their corresponding hyperbolic fuzzy numerical values to aid in quantitative analysis. Table 3 shows the mapping of linguistic variables to their associated fuzzy values. Rather than gathering direct expert assessments, the decision matrix was developed using hypothetical data from previous MCDM research in the SSS sector.

Table 3: Linguistic factors and their associated fuzzy quantities.

Linguistic factors	Fuzzy quantities
Extremely good (EG)	1.00
Very good (VG)	0.90
Good (G)	0.80
Nearly good (NG)	0.70
Above average (AA)	0.60
Average (A)	0.50
Below average (BA)	0.40

Table 3 (continued)

Linguistic factors	Fuzzy quantities
Nearly bad (NB)	0.30
Bad (B)	0.20
Very bad (VB)	0.10
Extremely bad (EB)	0.00

To present expert-like comments on each supplier based on the given criteria, optimistic and pessimistic assessments were expressed using linguistic phrases, as shown in Table 3. These linguistic assessments (Table 4) were converted into hyperbolic fuzzy numerical values for further processing. Table 5 shows the numerical decision matrix generated by this procedure.

Table 4: Decision matrix in linguistic form.

Criteria \ Supplier	A_1	A_2	A_3	A_4
C_1	(VG, B)	(G, B)	(G, AA)	(G, B)
C_2	(VG, A)	(VG, NG)	(G, B)	(VG, G)
C_3	(EG, NG)	(G, NG)	(NG, NB)	(AA, BA)
C_4	(VG, VG)	(NG, NB)	(G, NB)	(NG, BA)
C_5	(A, BA)	(AA, A)	(G, BA)	(VG, BA)
C_6	(NG, VB)	(G, B)	(G, NB)	(G, B)
C_7	(G, B)	(G, VB)	(VG, NB)	(VG, NB)
C_8	(A, BA)	(AA, BA)	(A, NB)	(A, A)
C_9	(BA, NB)	(A, BA)	(AA, BA)	(A, A)
C_{10}	(AA, BA)	(NG, BA)	(A, NB)	(AA, BA)

Table 5: Decision matrix.

Criteria \ Supplier	A_1	A_2	A_3	A_4
C_1	(0.90, 0.20)	(0.80, 0.20)	(0.80, 0.60)	(0.80, 0.20)
C_2	(0.90, 0.50)	(0.90, 0.70)	(0.80, 0.20)	(0.90, 0.80)
C_3	(1.00, 0.70)	(0.80, 0.30)	(0.70, 0.30)	(0.60, 0.40)
C_4	(0.90, 0.90)	(0.70, 0.30)	(0.80, 0.30)	(0.70, 0.40)
C_5	(0.50, 0.40)	(0.60, 0.50)	(0.80, 0.40)	(0.90, 0.40)
C_6	(0.70, 0.10)	(0.80, 0.20)	(0.80, 0.30)	(0.80, 0.20)
C_7	(0.80, 0.2)	(0.8, 0.1)	(0.9, 0.3)	(0.9, 0.3)
C_8	(0.50, 0.4)	(0.6, 0.4)	(0.5, 0.3)	(0.5, 0.5)
C_9	(0.40, 0.3)	(0.5, 0.4)	(0.6, 0.4)	(0.5, 0.5)
C_{10}	(0.60, 0.4)	(0.7, 0.4)	(0.5, 0.3)	(0.6, 0.4)

The performance of each supplier was first assessed using the HyFS-TOPSIS method. In this approach, the distance of each alternative from the PIS and the NIS was calculated. Each supplier's closeness coefficient (ζ) was derived based on these distances. The computed closeness values and the corresponding rankings are summarized in Table 6.

Table 6: Results from the HyFS-TOPSIS Method.

Supplier	$D(Y_k, Y^+)$	$D(Y_k, Y^-)$	$\zeta(Y_k)$	Rank
A_1	0.1870	0.1760	0.7192	4
A_2	0.1570	0.1750	0.9190	2
A_3	0.1460	0.1500	0.8523	3
A_4	0.1505	0.1705	0.9379	1

The alternatives were then evaluated using the HyFS-EDAS approach. This method determines the weighted positive distance (WSP_k), the weighted negative distance (WSN_k), and the resulting performance score (PS_k). These measures collectively capture the overall effectiveness of an alternative under less favourable conditions. The computed values WSP_k, WSN_k, and PS_k, along with the final rankings of the four suppliers, are provided in Table 7.

Table 7: Results from the HyFS-EDAS method.

Supplier	WSP_k	WSN_k	Performance score	Rank
A_1	0.0579	0.1087	0.2970	4
A_2	0.0359	0.0409	0.5013	3
A_3	0.0957	0.0408	0.8135	1
A_4	0.0519	0.0510	0.5380	2

To combine the rankings obtained from the individual methods (HyFS-TOPSIS and HyFS-EDAS), the Borda count method was applied. This approach assigns scores to each alternative according to its respective positions in the individual rankings and aggregates them to generate a final overall ranking. The combined results are presented in Table 8.

Table 8: Final ranking using the Borda count method.

Supplier	HyFS-TOPSIS	HyFS-EDAS	Borda score	Final ranking
A_1	4	4	0	**4**
A_2	2	3	3	**3**
A_3	3	1	4	**2**
A_4	1	2	5	**1**

As shown in Table 8, supplier A_4 has the highest Borda score, so it is the most favourable option for selection. Supplier A_3 ranked second, while supplier A_1 received the lowest overall score.

This final ranking provides a more reliable decision-making outcome for SSS in industry *XY*.

5.1 Bar Diagram Between the Closeness Coefficient of HyFS-TOPSIS (CC-HyFS-TOPSIS) and the Performance Score HyFS-EDAS (PS-HyFS-EDAS)

Figure 1 shows the comparison between the CC-HyFS-TOPSIS and the PS-HyFS-EDAS values of the four suppliers (A_1–A_4).

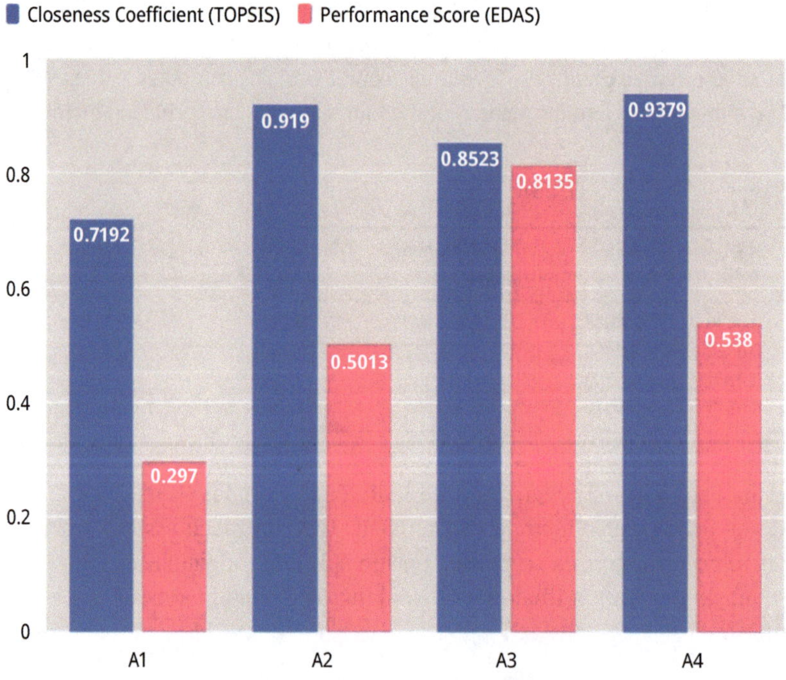

Figure 1: Comparison between closeness coefficient (CC-HyFS-TOPSIS) and performance score (PS-HyFS-EDAS).

Supplier A_1 has a good HyFS-TOPSIS score of 0.7192, but has a poor score in HyFS-EDAS (0.2970), making it the worst choice. Supplier A_2 has a high HyFS-TOPSIS score (0.9190) but a medium HyFS-EDAS score of 0.5013. Supplier A_3 performs consistently and well in both techniques, has a HyFS-TOPSIS score of 0.8523 and a HyFS-EDAS score of 0.8135, making it a stable and balanced option. Supplier A_4 has the greatest HyFS-TOPSIS score of 0.9379 and has its HyFS-EDAS score as 0.5380. Overall, while A_4 is the best choice according to HyFS-TOPSIS, A_3 leads in HyFS-EDAS.

6 Conclusion

In this study, we developed a completely new score function in HyFS frameworks. We use three methods such as TOPSIS, EDAS, and the Borda count methods to choose the best sustainable supplier. A numerical examination of four suppliers for the industry *XY* was carried out, and we found that supplier A_4 was the best choice among the four suppliers, followed by A_3, and A_1 was the least suitable. This shows that the proposed methodology can work well even when the information is uncertain.

The development of a new HyFS-based score function and inclusion of ranking aggregation not only promote the improvement of the fuzzy MCDM technique but also add some practical value to industries with an organized strategy for selecting sustainable suppliers.

In the future, this methodology can be tried with real industry data, bigger supply chains, or other fuzzy extensions like FFS or Q-ROFS. Other than choosing a supplier, the approach can be used in many fields where sustainability and uncertainty are important considerations: renewable energy, healthcare, or finance. Therefore, this work offers a new idea to fuzzy decision-making.

References

[1] Fan D, Xiao C, Zhang X, Guo Y. Gaining Customer Satisfaction through Sustainable Supplier Development: The Role of Firm Reputation and Marketing Communication. Transp Res Part E Logist Transp Rev. 2021;154. 102453.

[2] Shang Z, Yang X, Barnes D, Wu C. Supplier Selection in Sustainable Supply Chains: Using the Integrated BWM, Fuzzy Shannon Entropy, and Fuzzy MULTIMOORA Methods. Expert Syst Appl. 2022;195. 116567.

[3] Wang Chen HM, Chou SY, Luu QD, Yu THK. A Fuzzy MCDM Approach for Green Supplier Selection from the Economic and Environmental Aspects. Math Probl Eng. 2016;2016(1):8097386.

[4] Matić B, Jovanović S, Das DK, Zavadskas EK, Stević Ž, Sremac S. et al A New Hybrid MCDM Model: Sustainable Supplier Selection in a Construction Company. Symmetry. 2019 Mar;11(3):353.

[5] Petrović G, Mihajlović J, Ćojbašić Ž, Madić M, Marinković D. Comparison of Three Fuzzy MCDM Methods for Solving the Supplier Selection Problem. Facta Univ Ser Mech Eng. 2019 Nov 29;17(3):455–69.

[6] Kao JC, Wang CN, Nguyen T, Husain VST. A Fuzzy MCDM Model of Supplier Selection in Supply Chain Management. Intell Autom Soft Comput. 2022;31(3):1451–66.

[7] Nguyen TL, Nguyen PH, Pham HA, Nguyen TG, Nguyen DT, Tran TH. et al A Novel Integrating Data Envelopment Analysis and Spherical Fuzzy MCDM Approach for Sustainable Supplier Selection in Steel Industry. Mathematics. 2022 Jan;10(11):1897.
[8] Štreimikienė D, Bathaei A, Streimikis J. MCDM Approaches for Supplier Selection in Sustainable Supply Chain Management. Sustainability. 2024 Jan;16(23):10446.
[9] Tronnebati I, Jawab F, Frichi Y, Arif J. Green Supplier Selection Using Fuzzy AHP, Fuzzy TOPSIS, and Fuzzy WASPAS. . . A Case Study of the Moroccan Automotive Industry. Sustainability. 2024 Jan;16 (11):4580.
[10] Ardra S, Barua MK. Sustainable supplier selection among supermarket's fresh fruits and vegetable supply chains based on circular practices in India. Environ Dev Sustain. 2024 Oct;22. 1–37.
[11] Kamran MA, Afsharfar S, Al Mawali F, Babazadeh R, Al Balushi M. Application of Fuzzy MCDM Methods to Optimize Supplier Selection in Oil and Gas Industry of Oman. Int J Ind Eng. 2025;36(2):52–67.
[12] Raveena R, Umamaheswari S. Integrated MCDM framework for sustainable pharmacy supplier selection using pioneering criteria with fuzzy TOPSIS SVR and GRA. Sci Rep. 2025 May 31;15(1):1.
[13] Maden A. Low-Carbon Supplier Selection Using a Scenario-Based Integrated FCM-EDAS Approach: A Textile Company Case Study. Int J Fuzzy Syst. 2025 May 7; 1–20.
[14] Goswami SS, Tapankumar T, Naik NC, Gowrishankar J, Bhosle N, Singh A. et al Multi-Model MCDM Framework for Sustainable Renewable Energy Selection in India: Integrating CRITIC-EDAS-CODAS-CoCoSo. Discov Sustain. 2025 Jun 4;6(1):500.
[15] Ghanbari H, Seiti H, Mohammadi E, Elkamel A. Selecting a Sustainable Hydrogen Production Method Using a Novel Dual Evaluation EDAS Approach. Ann Oper Res. 2025 May 23; 1–42.
[16] Dey S, Mitra G, Biswas S, Pamucar D. Evaluation of Firm Performance under Merger and Acquisition Effect: An Integrated LOPCOW-PIV Approach. Decis Mak Appl Manag Eng. 2025 May 25;8(1):588–614.
[17] Atanassov KT. Intuitionistic Fuzzy Sets. Fuzzy Sets Syst. 1986;20. 87–96.
[18] Yager RR. Pythagorean Membership Grades in Multicriteria Decision Making. IEEE Trans Fuzzy Syst. 2013;22(4):958–65.
[19] Yager RR. Generalized Orthopair Fuzzy Sets. IEEE Trans Fuzzy Syst. 2016;25(5):1222–30.
[20] Senapati T, Yager RR. Fermatean Fuzzy Sets. J Ambient Intell Humaniz Comput. 2020;11. 663–74.
[21] Dutta P, Borah G. Construction of Hyperbolic Fuzzy Set and Its Applications in Diverse COVID-19 Associated Problems. New Math Nat Comput. 2023;19(1):217–88.
[22] Luo S, Mo S. A Method of Decision-Making Based on Intuitionistic Fuzzy Soft Sets and Its Applications for Enterprises' Decisions under Asymmetric Information. Syst Eng. 2015;33(6):137–41.
[23] Feng FE, Yan M, Liu X. Intuitionistic fuzzy soft decision-making method based on expectation score functions. J Jilin Univ. 2018;56. 1084–90.
[24] Peng X. Some Novel Decision-Making Algorithms for Intuitionistic Fuzzy Soft Set. J Intell Fuzzy Syst. 2019;37(1):1327–41.
[25] Peng X, Garg H. Intuitionistic Fuzzy Soft Decision-Making Method Based on CoCoSo and CRITIC and CCN Cache Placement Strategy Selection. Artif Intell Rev. 2021;54. 1–38.
[26] Wu SJ, Wei GW. Pythagorean Fuzzy Hamacher Aggregation Operators and Their Application to Multiple Attribute Decision-Making. Int J Knowl-Based Intell Eng Syst. 2017;21(3):189–201.
[27] Peng X, Dai J. Approaches to Pythagorean Fuzzy Stochastic Multi-Criteria Decision-Making Based on Prospect Theory and Regret Theory with New Distance Measure and Score Function. Int J Intell Syst. 2017;32(11):1187–214.
[28] Peng X, Zhang X, Luo Z. Pythagorean Fuzzy MCDM Method Based on CoCoSo and CRITIC with Score Function for 5G Industry Evaluation. Artif Intell Rev. 2020;53(5):3813–47.

[29] Dutta P, Banik AK. Hyperbolic Fuzzy TOPSIS Method for Multi-Criteria Decision-Making Problems. In: Fuzzy Optimization, Decision-Making and Operations Research: Theory and Applications. Singapore, Springer; 2023. 319–41.

[30] Hwang C, Yoon K. TOPSIS (Technique for Order Preference by Similarity to Ideal Solution) – A Multiple Attribute Decision Making. In: Multiple Attribute Decision Making – Methods and Applications: A State-of-the-Art Survey. Berlin, Springer-Verlag; 1981. 128–91.

[31] Erdebilli B, Gecer E, Yılmaz İ, Aksoy T, Hacıoglu U, Dinçer H. et al Q-ROF Fuzzy TOPSIS and VIKOR Methods for the Selection of Sustainable Private Health Insurance Policies. Sustainability. 2023;15(12):9229.

[32] Siahkoohi MEN. Predicting the Factors Affecting the Bankruptcy of Companies Using AHP-TOPSIS Technique and an Approach to Artificial Neural Networks. Trans Quant Financ Beyond. 2024;1(2):226–32.

[33] Hadi-Vencheh A, Mirjaberi M. Fuzzy Inferior Ratio Method for Multiple Attribute Decision-Making Problems. Inf Sci. 2014;277. 263–72.

[34] Ghorabaee MK, Zavadskas EK, Olfat L, Turskis Z. Multi-Criteria Inventory Classification Using a New Method of Evaluation Based on Distance from Average Solution (EDAS). Informatica. 2015;26(3):435–51.

[35] Ghorabaee MK, Zavadskas EK, Amiri M, Turskis Z. Extended EDAS Method for Fuzzy Multi-Criteria Decision-Making: An Application to Supplier Selection. Int J Comput Commun Control. 2016;11(3):358–71.

[36] Garg H, Ünver M, Olgun M, Türkarslan E. An Extended EDAS Method with Circular Intuitionistic Fuzzy Value Features and Its Application to Multi-Criteria Decision-Making Process. Artif Intell Rev. 2023 Dec 1;56(3):3173–204.

[37] Borda J. Mémoire sur les Élections au Scrutin. Histoire de l'Académie Royale des Sciences pour 1781. Paris, Imprimerie Royale; 1784. English Transl. by Grazia A. Isis. 1953; 44:42–51.

[38] Dutta P, Rajbonshi D, Bahrami A. Cryptocurrency Investment Decision-Making: A Hyperbolic Fuzzy MCDM Approach. Int J Res Ind Eng. 2025 Sep 1;14(3):445–65.

Raina Ahuja

Inappropriateness of Geometric Approach for Ranking Generalized Trapezoidal Fuzzy Numbers

Abstract: It has been asserted that employing the existing ranking methods (RgMs) to compare generalized trapezoidal fuzzy numbers (GrTFzNs) is unsuitable, as these RgMs are unable to differentiate between two distinct GrTFzNs. To address this limitation, a new ranking method (RgM) was introduced for comparing GrTFzNs. Similarly, it has also been stated elsewhere that using existing RgMs for comparing GrTFzNs is inappropriate because they fail to distinguish two distinct GrTFzNs. To overcome this shortcoming, another RgM was proposed for comparing GrTFzNs. It is pertinent to mention that neither any drawback of the first RgM is pointed out in the published work introducing the second RgM, nor any drawback of the second RgM is pointed out in the published work introducing the first RgM. So, it is a challenging problem for researchers to decide which of the existing RgMs should be used for solving real-life problems (RLPs). The objective of this chapter is to inform researchers that the first RgM is unable to differentiate between a GrTFzN and its additive inverse. Therefore, using the first RgM to solve RLPs is unsuitable. In contrast, the second RgM consistently distinguishes between any two distinct GrTFzNs, making it suitable for solving RLPs.

Keywords: GrTFzN, GrTFzNs, generalized triangular fuzzy numbers, RgM

1 Introduction

Bihari et al. [1] claimed that:

(i) The existing ranking methods (RgMs) [2–4] are unable to differentiate between the distinct generalized trapezoidal fuzzy numbers (GrTFzNs) $\tilde{P} = (-1, 0, 1, 2; 1)$ and $\tilde{Q} = (0, 0, 1, 1; 1)$. Therefore, these existing RgMs are not appropriate for solving real-life problems (RLPs).

(ii) The existing RgMs [5–7] are unable to differentiate between the distinct GrTFzNs $\tilde{P} = (1.5, 3, 5, 7, 5; 0.6)$ and $\tilde{Q} = (1.5, 3, 5, 7.5; 1)$. Therefore, these existing RgMs are not appropriate for solving RLPs.

Raina Ahuja, Department of Mathematics, Thapar Institute of Engineering & Technology (Deemed to be University), Patiala, Punjab, India, e-mail: rahuja_phd21@thapar.edu

 | https://doi.org/10.1515/9783112222249-003

Additionally, to address the inadequacy of existing RgMs, Bihari et al. [1] proposed an RgM for comparing GrTFzNs.

Bihari et al. [9] also claimed that the existing RgMs [2, 4, 10, 11] are unable to differentiate between the distinct GrTFzNs $\tilde{P} = (2, 4, 6; 1)$ and $\tilde{Q} = \left(1, 5, 6; \frac{17}{18}\right); \tilde{P} = (0, 4a, 5a, 10a; 1)$ and $\tilde{Q} = (2a, 2a, 6a, 9a; 1); \tilde{P} = (2, 4, 6; 1)$ and $\tilde{Q} = (2, 4, 6; 0.6)$; and $\tilde{P} = (a, b, c, d; w_1)$ and $\tilde{Q} = (a, b, c, d; w_2)$, respectively. Therefore, these existing RgMs [2, 4, 10, 11] are not appropriate for solving RLPs.

Also, to resolve inappropriateness of existing RgMs, Bihari et al. [9] proposed an RgM for comparing GrTFzNs.

In this paper, it is pointed out that Bihari et al.'s [1] RgM is unable to differentiate between the GrTFzN $\tilde{A} = (a_1, b_1, c_1, d_1; w_1)$ and its additive inverse $-\tilde{A} = (-d_1, -c_1, -b_1, -a_1; w_1)$. Hence, Bihari et al.'s [1] RgM is not suitable for solving RLPs. Conversely, Bihari et al.'s [9] RgM successfully distinguishes between any two distinct GrTFzNs $\tilde{A} = (a_1, b_1, c_1, d_1; w_1)$ and $\tilde{B} = (a_2, b_2, c_2, d_2; w_2)$. Therefore, it is appropriate to use Bihari et al.'s [9] RgM to solve RLPs.

This chapter is organized as follows: Section 2 presents the preliminaries. Section 3 discusses Bihari et al.'s [1] first RgM, while Section 4 focuses on their second RgM. Section 5 examines the inappropriateness of the first RgM, and Section 6 highlights the appropriateness of the second RgM. Finally, Section 7 concludes the chapter.

2 Preliminaries

Definition 2.1 ([12]). A generalized fuzzy number $\tilde{A} = (a_1, b_1, c_1, d_1; w)$ is said to be a GrTFzN if its MFn $\mu_{\tilde{A}}(x)$ is defined as:

$$\mu_{\tilde{A}}(x) = \begin{cases} w\left(\frac{x - a_1}{b_1 - a_1}\right), & a_1 < x \le b_1 \\ w, & b_1 \le x \le c_1 \\ w\left(\frac{u_4 - x}{u_4 - d_1}\right), & c_1 \le x < d_1 \\ 0, & otherwise \end{cases}$$

where w represents the height and satisfies the condition $0 < w \le 1$.

When $b_1 = c_1$, the trapezoidal fuzzy number converts into a triangular fuzzy number and can be denoted by $\tilde{A} = (a_1, b_1, d_1; w)$; w ∈ (0,1].

3 Bihari et al.'s First Ranking Method

To compare two GrTFzNs $\tilde{A}_1 = (a_1, b_1, c_1, d_1; w_1)$ and $\tilde{A}_2 = (a_2, b_2, c_2, d_2; w_2)$, Bihari et al. [1] proposed the following RgM:

Step 1: Find the value of $D_{\tilde{A}_i}$, where

$$D_{\tilde{A}_i} = \sqrt{\left(\frac{2a_i + b_i - 3c_i}{6}\right)^2 + \left(\frac{w_i}{6}\right)^2} + \sqrt{\left(\frac{2d_i + c_i - 3b_i}{6}\right)^2 + \left(\frac{w_i}{6}\right)^2}$$
$$+ \sqrt{\left(\frac{2a_i + 2b_i}{3}\right)^2 + \left(\frac{w_i}{6}\right)^2} + \sqrt{\left(\frac{2c_i + 2d_i}{3}\right)^2 + \left(\frac{w_i}{6}\right)^2}, i = 1, 2. \quad (1)$$

Step 2: Check that $D_{\tilde{A}_i} > D_{\tilde{A}_j}$ or $D_{\tilde{A}_i} < D_{\tilde{A}_j}$ or $D_{\tilde{A}_i} = D_{\tilde{A}_j}$.
Case (i): If $D_{\tilde{A}_i} > D_{\tilde{A}_j}$, then $\tilde{A}_i \succ \tilde{A}_j$.
Case (ii): If $D_{\tilde{A}_i} < D_{\tilde{A}_j}$, then $\tilde{A}_i \prec \tilde{A}_j$.
Case (iii): If $D_{\tilde{A}_i} = D_{\tilde{A}_j}$, then $\tilde{A}_i \sim \tilde{A}_j$.

4 Bihari et al.'s Second Ranking Method

To compare two GrTFzNs $\tilde{A}_1 = (a_1, b_1, c_1, d_1; w_1)$ and $\tilde{A}_2 = (a_2, b_2, c_2, d_2; w_2)$, Bihari et al. [9] proposed the following RgM:

Step 1: Find the value of $M(\tilde{A}_i)$, where

$$M(\tilde{A}_i) = \frac{w_i(a_i + b_i + c_i + d_i + a_i c_i + b_i d_i)}{4}, i = 1, 2.$$

Step 2: Check that $M(\tilde{A}_i) > M(\tilde{A}_j)$ or $M(\tilde{A}_i) < M(\tilde{A}_j)$ or $M(\tilde{A}_i) = M(\tilde{A}_j)$.
Case (i): If $M(\tilde{A}_i) > M(\tilde{A}_j)$ then $\tilde{A}_i \succ \tilde{A}_j$.
Case (ii): If $M(\tilde{A}_i) < M(\tilde{A}_j)$, then $\tilde{A}_i \prec \tilde{A}_j$.
Case (iii): If $M(\tilde{A}_i) = M(\tilde{A}_j)$, then go to Step 3.

Step 3: Find the value of $R_{\tilde{A}_i}$, where

$$R_{\tilde{A}i} = \frac{\left(a_i + b_i + c_i + d_i \sqrt{(d_i - b_i)^2 + w_i^2} + \sqrt{(c_i - a_i)^2 + w_i^2}\right)}{6}, i = 1, 2.$$

Step 4: Check that $R_{\tilde{A}_i} > R_{\tilde{A}_j}$, or $R_{\tilde{A}_i} < R_{\tilde{A}_j}$ or $R_{\tilde{A}_i} = R_{\tilde{A}_j}$.
Case (i): If $R_{\tilde{A}_i} > R_{\tilde{A}_j}$, then $\tilde{A}_i \succ \tilde{A}_j$.
Case (ii): If $R_{\tilde{A}_i} < R_{\tilde{A}_j}$, then $\tilde{A}_i \prec \tilde{A}_j$.
Case (iii): If $R_{\tilde{A}_i} = R_{\tilde{A}_j}$, then $\tilde{A}_i \sim \tilde{A}_j$.

5 Inappropriateness of Bihari et al.'s First Ranking Method

In this section, it is shown that an existing RgM [1] is unable to differentiate between the GrTFzN $\tilde{A} = (a_1, b_1, c_1, d_1; w_1)$ and its additive inverse $-\tilde{A} = (-d_1, -c_1, -b_1, -a_1; w_1)$. Therefore, it is not appropriate to use that existing RgM [1] for solving RLPs.

Based on the existing RgM [1], the comparison between $\tilde{A}$ and $-\tilde{A}$ can be carried out as follows:

Step 1: According to Step 1 discussed in Section 3, there is a need to evaluate $D_{\tilde{A}}$ and $D_{-\tilde{A}}$.

Using expression (1):

$$D_{\tilde{A}} = \sqrt{\left(\frac{2a_1 + b_1 - 3c_1}{6}\right)^2 + \left(\frac{w_1}{6}\right)^2} + \sqrt{\left(\frac{2d_1 + c_1 - 3b_1}{6}\right)^2 + \left(\frac{w_1}{6}\right)^2} + \sqrt{\left(\frac{2a_1 + 2b_1}{3}\right)^2 + \left(\frac{w_1}{3}\right)^2} + \sqrt{\left(\frac{2c_1 + 2d_1}{3}\right)^2 + \left(\frac{w_1}{3}\right)^2} \quad (2)$$

$$\begin{aligned} D_{-\tilde{A}} &= \sqrt{\left(\frac{-2d_1 - c_1 + 3b_1}{6}\right)^2 + \left(\frac{w_1}{6}\right)^2} + \sqrt{\left(\frac{-2a_1 - b_1 + 3c_1}{6}\right)^2 + \left(\frac{w_1}{6}\right)^2} \\ &\quad + \sqrt{\left(\frac{-2d_1 - 2c_1}{3}\right)^2 + \left(\frac{w_1}{3}\right)^2} + \sqrt{\left(\frac{-2b_1 - 2a_1}{3}\right)^2 + \left(\frac{w_1}{3}\right)^2} \\ &= \sqrt{\left(\frac{(-1)\ (2d_1 + c_1 - 3b_1)}{6}\right)^2 + \left(\frac{w_1}{6}\right)^2} + \sqrt{\left(\frac{(-1)\ (2a_1 + b_1 - 3c_1)}{6}\right)^2 + \left(\frac{w_1}{6}\right)^2} \\ &\quad + \sqrt{\left(\frac{(-1)\ (2d_1 + 2c_1)}{3}\right)^2 + \left(\frac{w_1}{3}\right)^2} + \sqrt{\left(\frac{(-1)\ (2b_1 + 2a_1)}{3}\right)^2 + \left(\frac{w_1}{3}\right)^2} \\ &= \sqrt{\left(\frac{2d_1 + c_1 - 3b_1}{6}\right)^2 + \left(\frac{w_1}{6}\right)^2} + \sqrt{\left(\frac{2a_1 + b_1 - 3c_1}{6}\right)^2 + \left(\frac{w_1}{6}\right)^2} \\ &\quad + \sqrt{\left(\frac{2d_1 + 2c_1}{3}\right)^2 + \left(\frac{w_1}{3}\right)^2} + \sqrt{\left(\frac{2b_1 + 2a_1}{3}\right)^2 + \left(\frac{w_1}{3}\right)^2} \\ &= \sqrt{\left(\frac{2a_1 + b_1 - 3c_1}{6}\right)^2 + \left(\frac{w_1}{6}\right)^2} + \sqrt{\left(\frac{2d_1 + c_1 - 3b_1}{6}\right)^2 + \left(\frac{w_1}{6}\right)^2} \\ &\quad + \sqrt{\left(\frac{2b_1 + 2a_1}{3}\right)^2 + \left(\frac{w_1}{3}\right)^2} + \sqrt{\left(\frac{2d_1 + 2c_1}{3}\right)^2 + \left(\frac{w_1}{3}\right)^2}. \end{aligned} \quad (3)$$

Step 2: According to Step 2 discussed in Section 3, it is necessary to verify whether, $D_{\tilde{A}} > D_{-\tilde{A}}$ or $D_{\tilde{A}} < D_{-\tilde{A}}$ or $D_{\tilde{A}} = D_{-\tilde{A}}$.

Clearly, from eq. (2) and eq. (3), we have, $D_{\tilde{A}} = D_{-\tilde{A}}$. Therefore, according to Case (iii) of Step 2 of Bihari et al.'s [1] RgM, $\tilde{A} \sim -\tilde{A}$, i.e., Bihari et al.'s [1] RgM is unable to differentiate between $\tilde{A}$ and $-\tilde{A}$.

6 Appropriateness of Bihari et al.'s Second Ranking Method

It is appropriate to use an existing RgM [9] due to the following reasons:

(i) It can be easily verified that the existing RgM [9] does not fail to distinguish the GrTFzN $\tilde{A} = (a_1, b_1, c_1, d_1; w_1)$ and its additive inverse $-\tilde{A} = (-d_1, -c_1, -b_1, -a_1; w_1)$.

(ii) The drawbacks, pointed out by the existing RgM [1] and the existing RgM [9] in other existing RgMs, does not occur in the existing RgM [9].

7 Conclusion

Based on the present study, it can be concluded that the existing RgM [1] is unable to distinguish between two distinct GrTFzNs. Hence, it is unsuitable for solving RLPs. In contrast, the existing RgM [9] successfully distinguishes between two distinct GrTFzNs; therefore, it is suitable for solving RLPs.

References

[1] Bihari R, Jeevaraj S, Kumar A. Geometric approach for ranking generalized trapezoidal fuzzy numbers and its application in selecting security guard service company. Expert Syst Appl. 2023;234:121052.

[2] Cheng CH. A new approach for ranking fuzzy numbers by distance method. Fuzzy Sets Syst. 1998;95(3):307–317.

[3] Chu TC, Tsao CT. Ranking fuzzy numbers with an area between the centroid point and original point. Comput Math Appl. 2002;43(1–2):111–117.

[4] Chi HTX, Vincent FY. Ranking generalized fuzzy numbers based on centroid and rank index. Appl Soft Comput. 2018;68:283–292.

[5] Abbasbandy S, Hajjari T. A new approach for ranking of trapezoidal fuzzy numbers. Comput Math Appl. 2009;57(3):413–419.

[6] Asady B. The revised method of ranking LR fuzzy number based on deviation degree. Expert Syst Appl. 2010;37(7):5056–5060.

[7] Dinagar DS, Kamalanathan S. A method for ranking of fuzzy numbers using new area method. Int J Fuzzy Math Arch. 2015;9(1):61–71.

[8] Yager RR. Ranking fuzzy subsets over the unit interval. In: 1978 IEEE Conference on Decision and Control including the 17th symposium on adaptive processes. pp. 1435–1437. IEEE.

[9] Bihari R, Jeevaraj S, Kumar A. A new ranking principle for ordering generalized trapezoidal fuzzy numbers based on diagonal distance, mean and its applications to supplier selection. Soft Comput. 2023b. https://doi.org/10.1007/s00500-023-08749-x.

[10] Liou TS, Wang MJJ. Ranking fuzzy numbers with integral value. Fuzzy Sets Syst. 1992;50(3):247–255.

[11] Kumar A, Singh P, Kaur P, Kaur A. Rm approach for ranking of lr type generalized fuzzy numbers. Soft Comput. 2011;15(7):1373–1381.

[12] Chen SH. Operations on fuzzy numbers with function principal. Tamkang J Manag Sci. 1985;6(1):13–26.

Ajendra Kumar, Vishal Kumar

Fuzzy Intelligence for Circular Economy: A Systematic Review of Soft and Metaheuristic Models

Abstract: The shift to a circular economy requires what we may term smart and adaptive decision-making frameworks which in turn have to deal with issues of uncertainty, complexity, and multiple sometimes conflicting goals. This review explores the forefront of fuzzy intelligence and soft computing, which includes traditional fuzzy logic and also metaheuristic optimization. We examine recent hybrid models that put together fuzzy clustering, causal fuzzy evaluation, and algorithms such as particle swarm and genetic, which are increasingly utilized in fields like task scheduling, resource optimization, and sustainable system design. We identify key methodological themes and application fields alongside emerging trends. We also highlight key research issues related to real-time adaptability, interpretability, and the combination of fuzzy logic with evolutionary computation.

This chapter illustrates currently undocumented findings to guide researchers and practitioners in building intelligent, adaptive, and circular economy-integrated sustainable systems, which help to facilitate the design of decision-support systems essential for achieving long-term sustainability goals.

Keywords: circular economy, fuzzy logic, metaheuristic algorithms, intelligent decision-making, task scheduling, hybrid optimization models

1 Introduction

In times when supplies are shrinking and nature's under pressure, the circular economy (CE) shows up as a shifting idea that lets economies grow without using more stuff. Instead of draining resources, focus moves to renewing them – waste fades as reuse takes center stage, flipping how we make and use things with green thinking built in. Shifting from old straight-line methods to circular loops brings messy challenges, especially in areas where facts are shaky, aims clash, and systems act in unpredictable ways.

In everyday situations, it is clear that things like sorting waste better, picking eco-friendly suppliers, judging contractors by their environmental impact, or planning

Ajendra Kumar, Department of Mathematics and Statistics, Gurukula Kangri (Deemed to be University), Haridwar 249404, Uttarakhand, India, e-mail: drajendra@gkv.ac.in
Vishal Kumar, Department of Mathematics and Statistics, Gurukula Kangri (Deemed to be University), Haridwar 249404, Uttarakhand, India, e-mail: vishal27121996@gmail.com

 | https://doi.org/10.1515/9783112222249-004

jobs to save energy – often called CE uses – are rarely straightforward. It turns out these cases need systems able to handle fuzzy judgments, patchy information, and balancing multiple goals at once. Meanwhile, old-school models built for neat numbers and rigid logic tend to fall short when faced with such messy realities.

To tackle this issue, experts are adopting smarter decision tools combining two useful computing styles: fuzzy logic that handles unclear written info and mimics human thinking when things are not clear and fuzzy AHP merged with DEMATEL instead of using them separately. Approaches such as particle swarm optimization (PSO) or genetic algorithms (GAs) stand out due to their trial-based tactics for solving complex problems with many goals.

In every single method, there is something useful. Yet when combined into hybrid systems, their real strength shows up – thanks to fuzzy reasoning paired with sharp search abilities. These blends shift fluidly across vast choices, respond instantly to live data, while balancing ecological, financial, and performance demands with precision.

This chapter explores the integration of fuzzy logic with soft computing within the decision-making frameworks of a CE. Scanning nine solid studies, we explore how hybrid artificial intelligence (AI) setups work in sharing tasks across networks, planning resource-smart timelines, or boosting efficiency with green goals in mind. Even though none of them outright label themselves as CE, each carries the core logic and approach of loop-based thinking – so they fit right into this picture.

The goals here break into four parts:

1. To sketch out the terrain of blended methods using fuzzy and soft computing that relate to CE issues
2. To break down their makeup along with how they function
3. To spot main uses plus see how well they work
4. To point out unresolved research issues while sketching ahead paths for smart, responsive decision tools within circular environments

This section aims to give a starting point for those looking to build flexible, efficient, less rigid decision methods that fit better with how CEs actually work – whether they are studying or applying them in real situations.

2 Methodology of the Review

This review examines the integration of fuzzy logic and soft computing alongside metaheuristic techniques, which is used for tackling decision challenges that reflect real-world CE setups. Since this area keeps changing and pulls from many fields, a storytelling-style review was chosen, so it could cover different models, applications, and ways to optimize them without rigid rules.

2.1 Research Objective and Review Focus

What's at the heart of this review is:

> How do smart decision systems use fuzzy logic along with soft computing methods like metaheuristics? How are they set up in circular economy models? What's their structure when put into real-world use? How does deployment work within sustainable loops?

This covers studies that directly tackle CE uses as well as work in related areas – distributed task planning, grouping resources, or designing systems with low costs – in which reasoning methods can shift to fit CE situations such as green supply networks, handling waste flows better, or picking contractors wisely.

2.2 Search and Selection Strategy

To keep things focused and meaningful, we selected sources carefully through a specific search approach over online libraries like SpringerLink, ScienceDirect, IGI Global, or Google Scholar.

Search terms included:
- "Fuzzy logic + metaheuristic optimization"
- "PSO-GA hybrid with fuzzy system"
- "Decision-making in distributed systems"
- "Task scheduling using fuzzy clustering"
- "Fuzzy optimization in circular economy"

Inclusion criteria:
- Came out from 2019 up until 2024
- Systems that use fuzzy logic together with a metaheuristic method
- Areas where CE aims apply – like cutting waste, recycling materials, or smarter distribution
- Includes step-by-step logic design along with tests showing how fast it runs

Exclusion criteria:
- Just abstract ideas or number stuff with no real-world use
- Research applying strict multi-criteria decision-making (MCDM) approaches alone, like analytic hierarchy process (AHP) or technique for order preference by similarity to ideal solution (TOPSIS) – without fuzzy logic or learning-based adjustments

This approach led to picking nine solid papers – these make up the main analysis here.

2.3 Classification Framework

Each selected paper was examined and classified using the following lenses.

Table 1: Classification framework for review.

Dimension	Description
Fuzzy techniques	Fuzzy inferences systems (FIS), fuzzy AHP, DEMATEL, fuzzy *C*- means
Metaheuristics	PSO, GA, PSO-GA hybrid, GA-based clustering
Application area	Task scheduling, system clustering, cost optimization, and contractor evaluation
Decision metrics	System cost (SC), response time (RT), reliability (RS), communication overhead, and sustainability

This classification (Table 1) enables structured comparison and highlights common patterns and innovations across methodologies.

2.4 Analytical Procedure

The review process had three main steps:

Step 1: Break down the model using code-driven methods – examine its setup and outputs, check how mixed techniques move through it, also look at where fuzzy logic plugs in.

Step 2: Performance mapping – check how scores line up, how fast things settle, also where results get fine-tuned.

Step 3: CE mapping – look at how each model matches up with circular economy ideas like smart use of resources, recycling, reusing stuff, also cutting waste through smarter planning.

This step led to side-by-side comparisons, which then set the stage for spotting missing pieces in the research.

2.5 Rationale for Narrative Review Approach

A structured overview, such as systematic literature review, usually requires extensive volume of research papers and number crunching – that can weaken focus on narrow, new topics. Given this analysis looks at how methods come together (fuzzy logic mixed with soft computing) in one particular setting (systems tied to CE goals), a storytelling-style review felt like a better fit.

This setup makes it possible to:
- Richer understanding from in-depth analysis
- Grouping methods by topic themes
- Focus on how the model's built, rather than just what it does when used
- Fitting nearby systems into CE targets can be done in different ways, depending on needs or conditions instead of fixed rules.

This review takes a close look – sometimes detailed, sometimes more flexible – at how blending soft computing methods can boost smart decision tools for eco-friendly solutions. Instead of just listing them, it lines up fuzzy logic and mixed intelligent systems from ten years' worth of studies, showing what's solid and where gaps remain. Through this side-by-side breakdown, this chapter builds a clear starting point – not flashy, but practical – for future work on smarter CEs.

3 Theoretical Foundations of Fuzzy and Soft Computing in CE Context

Classic decision tools often do not work well when dealing with CE setups – they are too rigid for situations with many goals and constant surprises. Instead of handling unclear targets or shifting benchmarks smoothly, old-school methods like fixed scoring and step-by-step rules fall short. On the flip side, fuzzy logic and adaptive computing are stepping in, offering smarter ways to weigh options, adjust on the fly, and manage messy real-world inputs without breaking a sweat.

Zadeh kicked off fuzzy logic, giving a math setup to handle vague or incomplete info. When data's missing, choices rely on expert hunches, or when balancing social, economic, and ecological targets, it plays a key role in CE systems. For sustainability-driven uses, tools like fuzzy AHP, fuzzy DEMATEL, and fuzzy inference systems (FIS) often map cause-effect ties and assign importance scores.

Fuzzy logic fits under soft computing – a broader approach that brings in smart tools such as neural networks, along with GA and evolutionary programming. This chapter highlights how fuzzy reasoning works with optimization methods, including metaheuristics such as PSO, GA (genetic algorithms), or hybrid versions of them. These approaches suit CE environments well since systems need ongoing self-adjustment; they help navigate complex choices, handle unpredictability, while updating responses in real time. Together, fuzzy systems paired with soft computing offer two perks – clear logic that is easy to explain to stakeholders, while staying adaptable in fast-moving decision scenarios.

4 Architectural Strategies in Fuzzy-Soft Computing Decision Models

The integration of fuzzy logic within soft computing frameworks has led to different designs aimed at smart choices in CE setups. Beyond varying in structure and parts, these setups also change based on how they rank factors, deal with uncertainty, or deliver clear outcomes.

On a simple level, you will find single-layer fuzzy setups where inputs link straight to choices through rule-driven thinking. Since they are clear and easy to follow, these work well when decisions must be explained – like picking suppliers, rating contractor performance, or checking organizational maturity. Often, they rely on methods like fuzzy AHP, fuzzy TOPSIS, or fuzzy best-worst method (BWM). Even though they are built from expert views and team input, these models do not need heavy computing resources.

Much smarter setups use step-by-step decision layouts, where unclear or messy inputs move through stages – first grouped, then sorted, later fine-tuned. Karishma and Kumar [3] proposed a hybrid PSO-GA framework in which fuzzy-ranked task grids go into *K*-means grouping, followed by evolutionary tweaks. Similarly, Kumar and Tyagi [8] developed a two-part setup – the first part clusters tasks using fuzzy logic tuned by a chaos-modified GA (HCMGA (hybrid clustering multi-objective genetic algorithm)), while the second part assigns processors using a hybrid of a GA and branch-and-bound algorithm. Such layered designs reveal how CE-focused schedulers balance clarity with performance tuning. As shown in Kreutzer's research [4], layered fuzzy rule setups form another design pattern. Here, CE maturity gets assessed through stacked rule checks over several aspects.

The links between factors that push sustainability are mapped by Sarker et al. [2] through fuzzy DEMATEL, building a cause-effect setup that reflects real circular processes. In contrast, frameworks relying on weighted decisions appear in the works by Alavi et al. [6] and Demirbağ et al. [1], where fuzzy methods assign importance to standards – via BWM or AHP – and then sort options with tools such as TOPSIS. Such approaches fit mid-sized decision settings with modest computing capacity, mainly due to their clarity and plug-in design.

5 Hybrid Fuzzy-Metaheuristic Decision Models

The growing mess and unpredictability in CE setups have pushed the shift toward smart decision systems that are both clear and flexible. On this front, fuzzy logic – which handles unclear data and word-based ambiguity – is often paired with soft computing methods such as grouping techniques, cause-effect analysis, or optimization tricks. Although tools like PSO and GA bring strong search power across wide areas, not many CE-focused models actually combine them into mixed designs. Lately, though, a clearer pattern has shown up: lighter fuzzy decision styles are gaining ground – one type relies on rules, another on scoring outputs, another on grouping data – all giving usable results without needing heavy processing.

This part analyzes nine new studies in depth, grouped by their design along with fuzzy method, optimization approach, and where they're used.

5.1 Structural Patterns Across Models

A typical setup seen across many models uses layers – fuzzy logic parts handle unclear, messy inputs first, then optimization or sorting choices kicks in. Sometimes, before tasks get assigned, fuzzy grouping happens; other times, rule-based results come from using fuzzy scores instead.

Figure 1 shows the broad-layer design, which is based on the mixed setup from Karishma and Kumar [3]; however, in this version, fuzzy matrices are ranked using robust techniques, while clustering improves via a PSO-GA combo.

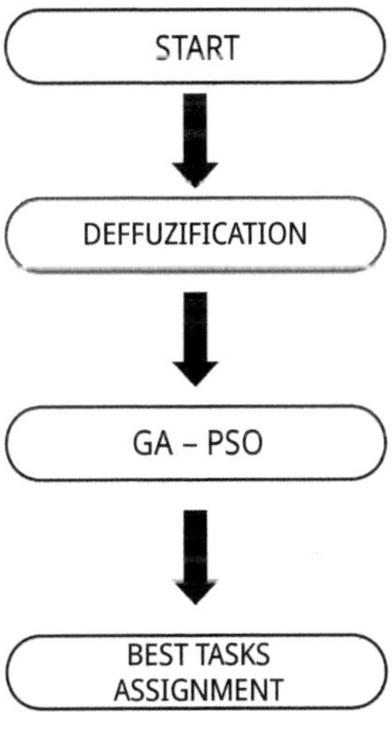

Figure 1: Generalized fuzzy-metaheuristic framework [3].

5.2 Detailed Review of Models

These observations pull together how the nine models differ in their approaches:

Demirbağ et al. [1] introduced a mixed approach that uses fuzzy AHP to figure out how important each factor is, then applies fuzzy TOPSIS to line up contractors when decisions are unclear. Instead of exact numbers, it turns expert opinions into fuzzy values, linking these with a step-by-step method to handle uncertainty without losing fairness in scoring. Even though it does not use advanced search algorithms for optimization, the system fits well within CE needs – especially when picking green suppliers or builders, since it balances price, trustworthiness, and eco-impact in a clear way.

Sarker et al. [2] used a fuzzy DEMATEL setup to study how CE tactics in fashion connect with one another. Instead of clear-cut logic, their method turns expert opinions into fuzzy diagrams that show which strategies drive others versus get driven. Because of this, leaders can spot high-impact moves – like choosing eco-friendly fabrics or improving return systems – that shape broader CE efforts. What stands out is how it handles unclear inputs while still revealing cause-effect patterns, offering useful clarity without complex math. Even though it does not use advanced search algorithms, the technique proves fuzzy modeling can support real-world planning in sustainability-stressed sectors, fitting well within CE strategy tools.

Karishma and Kumar [3] built a mix of PSO with GA to improve task scheduling in varied computing setups. Their method uses fuzzy rankings and robust ranked matrices, then groups inputs through K-means before kicking off optimization. Rather than sticking to standard FIS, it applies math-driven fuzzification along with centroid clustering – helping manage uncertainty during scheduling. What makes it stand out is the step-by-step combo: rank fuzzy values, cluster them, then use hybrid evolutionary tricks – boosting efficiency and room to scale. Even though it doesn't mention CE outright, its way of cutting resource waste and adjusting allocations fits well within CE ideas, like sharing processing loads or managing recycled materials across networks.

Kreutzer et al. [4] came up with a step-by-step fuzzy logic setup to score how well manufacturing companies apply CE methods. Instead of just listing metrics like material reuse, trash minimization, or product lifespan planning, it combines them through organized fuzzy reasoning – one way this turns subjective judgments into clear numerical scores. What stands out is its flexible layout and transparency, so leaders can review individual workflows or gauge total CE preparedness even when data's shaky. Although it skips heavy number crunching in favor of simplicity, the approach still delivers a hands-on tool that handles ambiguity well, useful for factories shifting toward circular models.

Karishma and Kumar [5] compared K-means with fuzzy C-means clustering when assigning tasks in distributed real-time setups (DRTS). Although it skips metaheuristic optimization, the focus on fuzzy logic allows looser groupings when task boundaries are not clear, supporting more responsive allocation. Because fuzzy membership values show partial belonging, decisions become easier to follow versus rigid clusters – this helps systems focused on CE principles where resources shift often and need smart sharing. Even if less complex than combined methods, the paper shows fuzzy clustering alone can act like a lightweight smart tool that boosts performance.

Alavi et al. [6] built a fuzzy BWM paired with rule-based logic for picking sustainable suppliers more effectively. Instead of strict comparisons, stakeholder views are turned into flexible weights, cutting down mismatches seen in older techniques. Though it skips metaheuristic optimization, its real power comes from softening preferences and using interpretable rules, so evaluations stay clear even when data is shaky. Because of this, decisions become more open and adjustable under unclear conditions. It fits well with CE goals by pushing eco-friendly, resource-smart sourcing choices – showing fuzzy decision tools can shape greener supply networks directly.

Paul et al. [7] pulled together a summary of fuzzy MCDM techniques used in sustainable supply chains – focusing less on code or math, more on how fuzzy thinking shapes choices in picking suppliers, cutting waste, and managing green transport. Instead of offering step-by-step models, it sketches out how these methods differ in design yet connect to broader CE goals like resource efficiency, flexibility, and long-term sustainability. The value: It shows a range – from fuzzy AHP to fuzzy TOPSIS – not as isolated tools but as parts of a bigger picture where uncertainty is handled thoughtfully. Even without running simulations, the paper acts like a link, showing how fuzzy logic fits into smarter, greener decisions.

Kumar and Tyagi [8] came up with a split-stage setup for spreading tasks across systems. The first stage tweaks fuzzy *C*-means grouping by blending it with a mixed genetic method (HCMGA), sharpening how jobs are bundled when conditions are not clear. The next stage uses a combo of genetic algo and branch-and-bound (HBBGA) to hand these grouped jobs to processors, cutting imbalance. Instead of treating clustering as rigid, it wraps fuzzy logic into a two-layer search routine, boosting flexibility while lowering run costs and squeezing more out of available hardware. Out of all similar works checked, this one pushes further by merging uncertain-data reasoning with evolving solution strategies, aiming straight at leaner resource use and longer-lasting system output.

Kumar and Tyagi [9] compared *K*-means with fuzzy *C*-means (FCM) for spreading tasks across systems. Though they did not use advanced optimization methods, their focus was on how fuzzy clustering handles unclear or mixed task areas better. Showing that FCM creates smoother, more adaptable clusters compared to rigid groupings, the study points out real benefits from using fuzzy logic. Because it helps assign resources smarter when conditions are unpredictable, this approach fits well within CE contexts – proving fuzzy clustering can be simple but effective for smarter decisions.

5.3 Patterns and Trends

The reviewed studies show a shifting pattern in how fuzzy and soft computing methods are used for decisions in the CE setup. A clear trend is the heavier use of fuzzy

Table 2: Comparative summary of reviewed fuzzy and soft computing models.

S. no.	Authors	Fuzzy technique	Metaheuristic/logic	Application domain
1	Demirbağ et al.	Fuzzy AHP + TOPSIS	Fuzzy AHP + hierarchical ranking via TOPSIS	Contractor evaluation in CE
2	Sarker et al.	Fuzzy DEMATEL	Fuzzy causal prioritization using DEMATEL	Fast fashion strategy
3	Karishma and Kumar	Robust ranking (no FIS)	Hybrid PSO-GA with *K*-means clustering	DRTS task scheduling
4	Kreutzer et al.	Hierarchical fuzzy logic	Expert rule-based fuzzy system	CE maturity model in manufacturing
5	Karishma and Kumar	Fuzzy *K*-means	Comparative clustering only	DRTS Task Allocation
6	Alavi et al.	Fuzzy BWM + FIS	Fuzzy weighted scoring using BWM + rule base	Supplier selection in circular SCM
7	Paul et al.	Fuzzy MCDM	Fuzzy MCDM review synthesis	Sustainable supply chain management
8	Kumar and Tyagi	Fuzzy *C*-means + linguistic ranking	Two-phase optimization (HCMGA + HBBGA)	Real-time task scheduling
9	Kumar and Tyagi	FCM versus *K*-means	Comparative evaluation of FCM and *K*-means	Load balancing in DRTS

logic models instead of complex hybrid metaheuristic systems. Out of nine papers analyzed (Table 2), just two – including the work by **Karishma and Kumar [3]** along with **Kumar and Tyagi [8]** – apply two-phase metaheuristic optimization approaches; others depend mostly on soft computing tools like fuzzy reasoning, cause effect mapping, or grouping algorithms.

A clear pattern in several studies: More folks are turning to clustering methods – especially fuzzy types – to group tasks or resources before handing them out. In the studies of **Karishma and Kumar [5] and Kumar and Tyagi [9]**, their setups use clustering not just upfront, but also to bake flexibility and modular design into choices – a fit for tangled, circular systems.

Another common pattern is choosing rule-based or scoring methods that value clarity instead of chasing high precision. The studies of **Demirbağ et al. [1], Kreutzer et al. [4], Alavi et al. [6]**, and **Sarker et al. [2]** show fuzzy logic setups guided by experts – like rule sets or causal maps built with DEMATEL – can deliver solid results without needing complex trial-and-error algorithms. That shift points to a growing lean toward open, simple decision models, especially where sustainability or policy goals matter most. On top of this, fuzzy tools such as BWM and DEMATEL are getting popular as practical substitutes for strict number-crunching, particularly when weighing multiple criteria at once. Such methods let users build personal judgments and ambiguity right into assessments, keeping choices reliable while skipping heavy computing demands.

It is worth noting that while the papers do not always say they are tied to CE uses, most still tackle issues like eco-friendliness, recycling materials, better workflow performance, or assessing freelancers – all things that fit within CE ideas. That shows the way a study is built and what it aims to do usually counts more than just labeling it as CE.

5.4 Visual Blueprint: Model Architectures Across Studies

Even though just some of the studies looked at use full hybrid intelligence, the variety in fuzzy-based and soft computing setups stands out. These systems differ when it comes to layers, how processes flow, or where fuzzy logic fits in. **Karishma and Kumar [3]** introduced a split two-phase setup; here, fuzzy input matrices get ranked through Robust methods, then move into clustering plus PSO-GA tuning. Their approach combines learning – using mixed algorithms – with fuzzified output meaning. **Kumar and Tyagi [8]** pushed this idea further with a two-part setup (Figure 2), where phase 1 zeroes in on fuzzy clustering (FCM) fine-tuned via GA, whereas phase 2 applies a mix of GA along with branch-and-bound to handle processor allocation. What stands out about this approach is its flexible structure, ability to run tasks at once, plus how detailed the choices get.

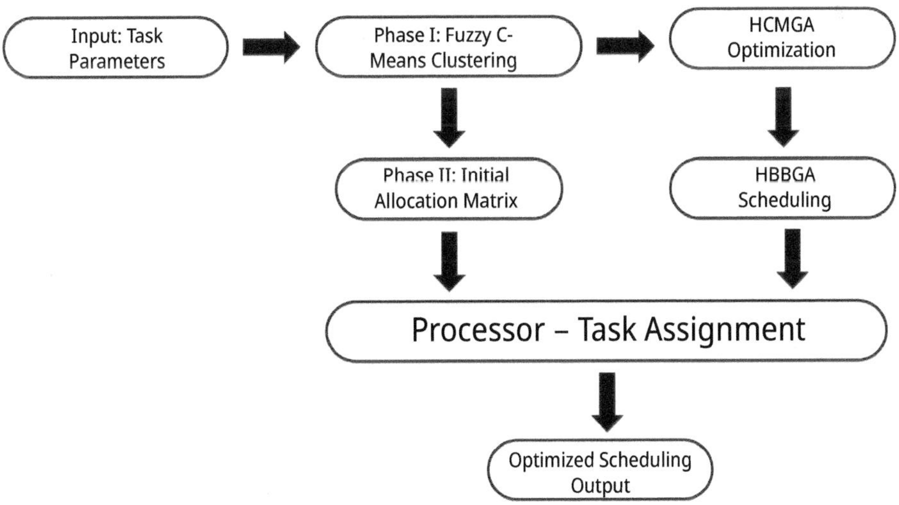

Figure 2: Dual-phase architecture [8].

On the flip side, some models – for example, **Demirbağ et al. [1], Sanker et al. [2], Kreutzer et al. [4], or Alavi et al. [6]** – use basic fuzzy setups that value clear structure more than heavy computation. Instead of complex math, they work with fuzzy logic, cause-effect maps from DEMATEL, or rule weights via BWM – usually skipping any tuning cycles.

Figure 3 shows a basic layout of these non-heuristic fuzzy models, pointing out how they move from word-based inputs toward decisions shaped by weights.

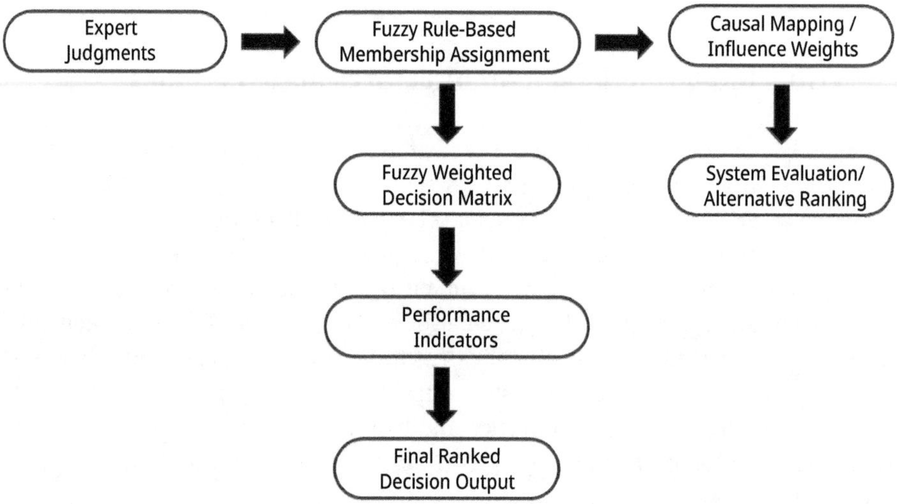

Figure 3: Generalized soft fuzzy logic-based decision system.

These images show CE models do not need heavy computing power all the time; rather, clear soft systems built on fuzzy logic can offer useful results while staying easy to follow and flexible.

5.5 Summary Insight

The look at newer fuzzy and soft computing decision methods shows a field shifting toward smarter structures instead of just complex calculations. Although mixed optimization approaches – like the ones used by Karishma and Kumar [3], along with Kumar and Tyagi [8] – highlight benefits when merging clustering and broad search strategies, most meaningful advances actually come from flexible logic and fuzzy setups that do not need heavy computing power.

Models using fuzzy DEMATEL [2], FIS setups [4], or fuzzy AHP [1] give clear structures – focusing on involving stakeholders while adapting to context, which matters a lot in CE systems where data is shaky and goals pile up. Oddly enough, approaches from [2, 5, 6], and [9] tackle reuse, cut waste, stay green, and shift plans on the fly without even naming the CE. Turns out, circular thinking pops up as a side effect more than by design, and smart setups – be it mapping causes, grouping loosely, or ranking gently – naturally fit circular ideals.

6 Comparative Insights and Research Gaps

In spite of different approaches, comparing the nine models shows uneven focus when handling optimization, adaptability, or unpredictability. Instead, these gaps highlight underlying beliefs about what counts as smart within the CE – beyond mere tech choices. Notably, fuzzy logic appears more often than metaheuristic methods. While some studies prioritize clarity over complex code, blended systems from Karishma and Kumar [3] and Kumar and Tyagi [8] offer expandable designs using dynamic grouping and step-by-step tuning. Still, this fits a broader trend in CE models: when goals center on strategic fit instead of peak numbers, clear reasoning often matters more than fast results.

A second idea involves how fuzzy setups can swap out parts like weights, rule mappings, or clustering methods when used in areas such as picking contractors, choosing suppliers, or organizing tasks; yet most setups miss metaheuristic levels, which means they rarely adapt or fix themselves once running. It turns out CE-alignment works better in practice than on paper. Although models aren't labeled CE-focused, plenty still target sustainability, efficiency, reuse, along with balanced decision criteria. That raises a doubt – could the limited use of CE terms in current studies be hiding its broader relevance across fields?

Lastly, not adjusting in real time is a noticeable weakness in the method. Because models struggle to capture random shifts, reactions from individual agents, or live data updates. On top of that, hardly any models include social effects or carbon measurements – even though these are key for genuine CE assessment – and interest in green solutions keeps growing.

These comparisons show a field full of potential – though some parts are still rough – weak when it comes to live adaptation, yet strong in flexible frameworks. What follows digs into how this gap opens doors for targeted progress and tweaks.

7 Conclusions

A close look at fuzzy and soft computing methods tied to CE setups was covered here. Most real-world uses focus on clarity, flexible parts, together with lean smart systems – traits that fit well with messy, uncertain CE conditions – even though mixed optimization approaches exist, shown through nine key studies. Fuzzy logic still comes out on top, comparison revealed, serving both as a base for building blocks together with a tool for handling doubt. While only a few models apply search techniques such as PSO or GA, there is a growing move toward blending soft computing with knowledge-based rules.

This study showed a key finding – CE alignment usually works in practice, not just in name. Even when CE terms are not used outright, plenty of models quietly mirror its goals: long-term use, resource care, recycling, or flexible design. What this hints at is a gap ready to be filled – smarter methods matched with clearer focus in the field ahead. Moving forward, the section points toward building fuzzy models that evolve on their own, weaving CE metrics straight into their core purpose. Folks could build smarter next-gen models – ones that behave better, care about nature, maybe even learn like living things – by weaving in earth-focused data while linking clear decision-making to self-adjusting features. Truth is, progress means crafting tech that evolves naturally, explains itself, fits into lasting cycles instead of just cranking out stronger code.

References

[1] Demirbağ AT. Circular Economy-Based Decision-Making Model for Contractor Selection. Buildings. 2025;15:1665. https://doi.org/10.3390/buildings15101665.

[2] Sarker AR et. al. Circular economy strategies: a fuzzy DEMATEL decision framework for the fast fashion footwear manufacture. Discover Sustainability. 2024;5:260. doi: https://doi.org/10.1007/s43621-024-00484-z.

[3] Karishma, Kumar H. A novel hybrid model for task scheduling based on particle swarm optimization and genetic algorithms. Math Eng. 2024;6(4):559–606. https://doi.org/10.3934/mine.2024023.

[4] Kreutzer D et. al. Fuzzy Logic Approach to Circular Economy Maturity Assessment of Manufacturing Companies. Eng Proc. 2024;76:4. doi: https://doi.org/10.3390/engproc2024076004.

[5] Karishma, Kumar H. A new hybrid particle swarm optimisation algorithm for optimal tasks scheduling in distributed computing system. Intell Syst Appl. 2023;18:200219. https://doi.org/10.1016/j.iswa.2023.200219.

[6] Alavi B et. al. A Dynamic Decision Support System for Sustainable Supplier Selection in Circular Economy. Sustain Prod Consum. 2021;27:905–920. doi: https://doi.org/10.1016/j.spc.2021.02.015.

[7] Paul A et. al. Sustainable Supply Chain Management and Multi-Criteria Decision-Making Methods: A Systematic Review. Sustainability. 2021;13:7104. doi: https://doi.org/10.3390/su13137104.

[8] Kumar H, Tyagi I. Hybrid model for tasks scheduling in distributed real time system. J Ambient Intell Hum Comput. 2020;12:2881–2903. https://doi.org/10.1007/s12652-020-02445-6.

[9] Kumar H, Tyagi I. Implementation and Comparative Analysis of k-means and Fuzzy c-means Clustering Algorithms for Tasks Allocation in Distributed Real Time System. Int J Embedded Real-time Commun Syst. 2020;10(2):66–86. https://doi.org/10.4018/IJERTCS.2019040105.

[10] Yadav PK, Pradhan P, Singh PP. A fuzzy clustering method to minimize the inter task communication effect for optimal utilization of processor's capacity in distributed real time systems. In: Deep K, Nagar, A, Pant, M, Bansal, J. Proceedings of the International Conference on Soft Computing for Problem Solving (SocProS 2011) December 20–22, 2011, Advances in Intelligent and Soft Computing, Springer; 2012. Vol. 130 pp. 159–168. https://doi.org/10.1007/978-81322-0487-9_16.

[11] Yadav PK, Singh MP, Kumar A, Agarwal B. An efficient tasks scheduling model in distributed processing systems using ANN. Int J Circuits Syst. 2011;1(1–2):53–66.

[12] Kumar A, Yadav PK. Task management algorithm for distributed system. In: 15th International conference of international academy of physical science. 2012.

[13] Kaushal U, Kumar A. Improving the performance of DRTS by optimal allocation of multiple tasks under dynamic load sharing scheme. Int J Sci Eng Res. 2013;4(7):1316–1321.

Ashu Redhu, Reeta Bhardwaj, Kamal Kumar, Rishu Arora*

Identification of Optimal Hybrid Renewable Energy System for Rural Electrification Based on *p*, *q*-Quasirung Orthopair Fuzzy Archimedean Weighted Geometric Aggregation Operator

Abstract: Access to reliable electricity remains a fundamental challenge in rural areas, particularly in developing countries where grid infrastructure is underdeveloped or poor. Hybrid renewable energy systems (HRESs), which combine several renewable energy sources, such as solar, wind power, and biomass, have gained recognition as an effective solution to address the problem of rural electrification while promoting sustainable development. In this chapter, we present two novel operation rules for p, q-quasirung orthopair fuzzy numbers (p, q-QOFNs) build upon Archimedean norms. Next, we propose the p, q-QOF Archimedean weighted geometric (p, q-QOFAWG) aggregation operator (AO) for the aggregation of the p, q-QOFNs. Additionally, we examine some key desirable characteristics of the suggested p, q-QOFAWG AO. Moreover, by employing the suggested p, q-QOFAWG AO and the "CRiteria Importance Through Intercriteria Correlation" (CRITIC) method, we introduce a multi-criteria decision-making (MCDM) methodology for p, q-QOFNs to handle decision-making issues with unknown criteria weights. To highlight the practicality and usefulness of the proposed MCDM methodology, we apply it to an MCDM problem in selecting the optimal HRES for rural electrification. Finally, the outcomes confirm that the proposed MCDM methodology is very efficient and provides reliable and consistent outcomes.

Keywords: decision-making, MCDM, p, q-quasirung orthopair fuzzy set, rural electrification

***Corresponding author: Rishu Arora**, Department of Mathematics, Amity Institute of Applied Sciences, Amity University, Noida 201313, Uttar Pradesh, India, e-mail: rarora4@gmail.com
Ashu Redhu, Department of Mathematics, Amity School of Applied Sciences, Amity University, Gurugram 122413, Haryana, India, e-mail: ashuredhu17@gmail.com
Reeta Bhardwaj, Department of Mathematics, Amity School of Applied Sciences, Amity University, Gurugram 122413, Haryana, India, e-mail: bhardwajreeta84@gmail.com
Kamal Kumar, Department of Mathematics, Amity School of Applied Sciences, Amity University, Gurugram 122413, Haryana, India, e-mail: kamalkumarrajput92@gmail.com

 | https://doi.org/10.1515/9783112222249-005

1 Introduction

Rural electrification plays a significant part in the socio-economic development of a country. Reliable access to electricity is essential for education, transport, agriculture, healthcare, and overall improvement of quality of life. However, electrifying rural areas through conventional grid expansion technologies is not a simple task and is often considered as a non-viable option [1]. Here, hybrid renewable energy systems (HRESs) act as an attractive alternative. HRESs integrate several renewable energy sources, such as solar, wind power, and biomass, to offer sustainable and reliable energy solution. However, the process of identifying an optimal HRES for rural electrification is very complex and involves uncertainty and fuzziness in data. To deal with such situations, fuzzy set (FS) emerges as an effective mathematical framework. Zadeh [2] introduced the concept of FS where membership degree (MD) is allocated to an element in the range of [0, 1]. Over time, several extensions of FS have been proposed, such as intuitionistic fuzzy set (IFS) [3], Pythagorean fuzzy set (PFS) [4], and q-rung orthopair fuzzy set (q-ROFS) [5]. These frameworks provide greater flexibility and capture more complex forms of uncertainty as compared to FS. Several researchers have applied these frameworks in many areas of decision-making [6–11]. Bajaj and Kumar [6] introduced correlation coefficient for IFSs and developed a multi-criteria decision-making (MCDM) methodology. Jabeen et al. [8] presented weighted aggregation operator (AO) using power Bonferonni mean operator and Aczel-Alsina norms and developed an MCDM model in the PFS framework. Mishra et al. [11] proposed novel similarity measure for q-ROFS and an MCDM methodology using the proposed measure and COPRAS method to select site for electric vehicle charging station. Peng et al. [12] introduced a decision-making model, known as weighted distance-based approximation (WDBA) within the q-ROFN environment. Peng et al. [13] proposed an MCDM technique utilizing the proposed generalized mean AOs and Weights by ENvelope and SLOpe (WENSLO) methodology.

Recently, Seikh and Mandal [14] introduced a novel framework known as p, q-quasirung orthopair fuzzy set (p, q-QOFS), which is a generalization of existing fuzzy frameworks. This framework provides greater flexibility in modeling uncertainty and hesitancy in decision-making. In p, q-QOFS, the addition of pth power of MD and qth power of non-membership degree (NMD) is at most 1, with $p, q \geq 1$. Numerous researches have utilized the concept of p, q-QOFSs in several decision-making problems [15–20]. Ali and Naeem [17] proposed AOs on the basis of Aczel-Alsina norms and an MCDM methodology in the p, q-QOFS framework. Redhu et al. [16] introduced the Schweizer-Sklar AOs and possibility degree measure for p, q-QOFNs and by using these developed MCDM methodology. Boz [18] developed a hybrid MCDM methodology based on SIWEC (simple weight calculation) and AROMAN (alternative ranking order method accounting for two-step normalization) approaches to select the drone for spraying agricultural lands. Rahim et al. [20] presented sine-trigonometric AOs under the p, q-QOFN framework.

1.1 Novelty and Contribution

An AO is essential for ranking all available alternatives by integrating the preferences of decision-maker into a single value and thereby identifying the best alternative. To address these decision-making issues, in this chapter, we first present novel operation rules for p, q-QOFNs build upon Archimedean's norms [21]. Next by employing these proposed operation rules, we present the p, q-quasirung orthopair fuzzy Archimedean weighted geometric (p, q-QOFAWG) AO with the purpose of accumulating the p, q-QOFNs and examine its desirable properties. Furthermore, we develop a novel MCDM methodology by utilizing the suggested AO and CRiteria Importance Through Intercriteria Correlation (CRITIC) [22] method. The CRITIC method is used to obtain the weights of the attributes by considering both the contrast intensity of each criterion and the correlation among them. Consequently, to demonstrate the practical efficacy of the suggested MCDM methodology, we solve a case study of the selection of optimal HRES for rural electrification. Additionally, to validate the proposed MCDM methodology, we conduct a comparative analysis. The results demonstrate the reliability and validity of the proposed MCDM methodology and highlight its effectiveness in solving complex decision-making problems.

The framework of this chapter is organized in the following order: Section 2 reviews some definitions and findings utilized in this chapter. Section 3 presents two new operational rules for p, q-QOFNs. Section 4 introduces the p, q-QOFAWG AO. In Section 5, we propose a novel MCDM methodology by integrating the proposed operator and the CRITIC technique. In Section 6, we solve a case study of identifying the best HRES for rural electrification. Additionally, we conduct a comparative study to illustrate the effectiveness of the proposed MCDM methodology. Lastly, Section 7 presents the conclusion of this chapter and outlines its future scope.

2 Preliminaries

This section reviews some definitions and findings utilized in the subsequent sections of this study.

Definition 2.1 ([14]). Let $\forall$ be a p, q-QOFS defined over a finite universal set E as follows:

$$\forall = \{\langle e, \gamma_{\forall}(e), \delta_{\forall}(e)\rangle | e \in \mathrm{E}\}, \tag{1}$$

where $\gamma_{\forall}(e): \mathrm{E} \to [0,1]$ signifies the MD and $\delta_{\forall}(e): \mathrm{E} \to [0,1]$ signifies the NMD of $e \in E$ with $0 \leq (\gamma_{\forall}(e))^p + (\delta_{\forall}(e))^q \leq 1$, and p and q being natural numbers. The indeterminacy degree of $e \in E$ is represented as $(\pi_{\forall}(e))^l = 1 - (\gamma_{\forall}(e))^p - (\delta_{\forall}(e))^q$ where l is the least common multiple (lcm) of p and q. The pair $\langle \gamma_{\forall}, \delta_{\forall} \rangle$ is signified as a p, q-QOFN.

Definition 2.2 ([14]). Consider $\forall_1 = \langle \gamma_1, \delta_1 \rangle$, $\forall_2 = \langle \gamma_2, \delta_2 \rangle$, and $\forall = \langle \gamma, \delta \rangle$ as three p, q-QOFNs. For $\rho > 0$, their operations are defined as follows:

(i) $\forall_1 \oplus \forall_2 = \langle \sqrt[p]{\gamma_1^p + \gamma_2^p - \gamma_1^p \gamma_2^p}, \delta_1 \delta_2 \rangle$.

(ii) $\forall_1 \otimes \forall_2 = \langle \gamma_1 \gamma_2, \sqrt[q]{\delta_1^q + \delta_2^q - \delta_1^q \delta_2^q} \rangle$.

(iii) $\rho \forall = \langle \sqrt[p]{1 - (1 - \gamma^p)^\rho}, \delta^\rho \rangle$.

(iv) $\forall^\rho = \langle \gamma^\rho, \sqrt[q]{1 - (1 - \delta^q)^\rho}$.

Definition 2.3 ([14]). Let $\forall = \langle \gamma, \delta \rangle$ be a p, q-QOFN. Then, score function § and accuracy function H of the p, q-QOFN $\forall$, are stated as follows:

$$\S(\forall) = \frac{1 + \gamma^p - \delta^q}{2}, \tag{2}$$

$$H(\forall) = \gamma^p + \delta^q, \tag{3}$$

where $p, q \geq 1$ and $0 \leq \S(\forall), H(\forall) \leq 1$.

Definition 2.4 ([14]). Let $\forall_1 = \langle \gamma_1, \delta_1 \rangle$ and $\forall_2 = \langle \gamma_2, \delta_2 \rangle$ be two p, q-QOFNs, then we have

(i) If $\S(\forall_1) < \S(\forall_2)$ then $\forall_1 < \forall_2$.

(ii) If $\S(\forall_1) = \S(\forall_2)$ then $\forall_1 = \forall_2$ and

   (a) If $H(\forall_1) < H(\forall_2)$ then $\forall_1 < \forall_2$.

   (b) If $H(\forall_1) = H(\forall_2)$ then $\forall_1 \sim \forall_2$.

Definition 2.5 ([21]). Let a be decreasing generator and b be an increasing generator. Then strict Archimedean t norm A_N and strict Archimedean t-conorm A_C are stated as:

$$A_N(t_1, t_2) = a^{-1}(a(t_1) + a(t_2)),$$

$$A_C(t_1, t_2) = b^{-1}(b(t_1) + b(t_2)),$$

where $a\colon [0,1] \to R$ is strictly decreasing continuous function and $b\colon [0,1] \to R$ is strictly increasing continuous function, with $a(1) = 0$ and $b(0) = 0$.

3 Proposed Operational Rules Based on Archimedean's Norms for *p*, *q*-QOFNs

By employing Archimedean's norms as stated in Definition 2.5, this section introduces novel multiplication and scalar power operations for *p*, *q*-QOFNs.

Definition 3.1. Let $\forall_1 = \langle \gamma_1, \delta_1 \rangle$ and $\forall_2 = \langle \gamma_2, \delta_2 \rangle$ be two *p*, *q*-QOFNs. The operational rules for *p*, *q*-QOFNs build upon Archimedean's norms stated in Definition 2.5 are formulated below:

(i) Multiplication operation:

$$\forall_1 \otimes \forall_2 = \left\langle \sqrt[p]{a^{-1}\left(a\left(\gamma_1^p\right) + a\left(\gamma_2^p\right)\right)}, \sqrt[q]{b^{-1}\left(b\left(\delta_1^q\right) + b\left(\delta_2^q\right)\right)} \right\rangle, \tag{4}$$

where p and $q \geq 1$.

(ii) Scalar power operation:

$$\forall_1^{\vartheta} = \left\langle \sqrt[p]{a^{-1}\left(\vartheta a\left(\gamma_1^p\right)\right)}, \sqrt[q]{b^{-1}\left(\vartheta b\left(\delta_1^q\right)\right)} \right\rangle, \tag{5}$$

where $p, q \geq 1$ and $\vartheta > 0$

Theorem 3.1. Let $\forall_1 = \langle \gamma_1, \delta_1 \rangle$ and $\forall_2 = \langle \gamma_2, \delta_2 \rangle$ be two *p*, *q*-QOFNs. The operational rules presented in Definition 3.1 satisfy the following characteristics:

(i) $\forall_1 \otimes \forall_2 = \forall_2 \otimes \forall_1$,

(ii) $\forall_1^{\vartheta_1} \otimes \forall_1^{\vartheta_2} = \forall_1^{(\vartheta_1 + \vartheta_2)}$,

(iii) $\forall_1^{\vartheta} \otimes \forall_2^{\vartheta} = (\forall_1 \otimes \forall_2)^{\vartheta}$,

where $\vartheta_1, \vartheta_2,$and $\vartheta > 0$.

Proof. Suppose, $\forall_1 = \langle \gamma_1, \delta_1 \rangle$ and $\forall_2 = \langle \gamma_1, \delta_1 \rangle$ be two *p*, *q*-QOFNs

(i) By employing eq. (4), we have:

$$\begin{aligned} \forall_1 \otimes \forall_2 &= \left\langle \sqrt[p]{a^{-1}\left(a\left(\gamma_1^p\right) + a\left(\gamma_2^p\right)\right)}, \sqrt[q]{b^{-1}\left(b\left(\delta_1^q\right) + b\left(\delta_2^q\right)\right)} \right\rangle \\ &= \left\langle \sqrt[p]{a^{-1}\left(a\left(\gamma_2^p\right) + a\left(\gamma_1^p\right)\right)}, \sqrt[q]{b^{-1}\left(b\left(\delta_2^q\right) + b\left(\delta_1^q\right)\right)} \right\rangle = \forall_2 \otimes \forall_1. \end{aligned}$$

(ii) By utilizing eqs. (4) and (5), we have:

$$\forall_1^{\vartheta_1} \otimes \forall_1^{\vartheta_2} = \left\langle \sqrt[p]{a^{-1}\left(\vartheta_1 a\left(\gamma_1^p\right)\right)}, \sqrt[q]{b^{-1}\left(\vartheta_1 b\left(\delta_1^q\right)\right)} \right\rangle \otimes \left\langle \sqrt[p]{a^{-1}\left(\vartheta_2 a\left(\gamma_1^p\right)\right)}, \sqrt[q]{b^{-1}\left(\vartheta_2 b\left(\delta_1^q\right)\right)} \right\rangle$$

$$= \left\langle \sqrt[p]{a^{-1}((\vartheta_1 + \vartheta_2)a(\gamma_1^p))}, \sqrt[q]{b^{-1}((\vartheta_1 + \vartheta_2)b(\delta_1^q))} \right\rangle = \forall_1^{(\vartheta_1 + \vartheta_2)}.$$

(iii) By utilizing eqs. (4) and (5), we have:

$$\forall_1^{\vartheta} \otimes \forall_2^{\vartheta} = \left\langle \sqrt[p]{a^{-1}(\vartheta a(\gamma_1^p))}, \sqrt[q]{b^{-1}(\vartheta b(\delta_1^q))} \right\rangle \otimes \left\langle \sqrt[p]{a^{-1}(\vartheta a(\gamma_2^p))}, \sqrt[q]{b^{-1}(\vartheta b(\delta_2^q))} \right\rangle$$

$$= \left\langle \sqrt[p]{a^{-1}(\vartheta(a(\gamma_1^p) + a(\gamma_2^p)))}, \sqrt[q]{b^{-1}(\vartheta(b(\delta_1^q) + b(\delta_2^q)))} \right\rangle$$

$$= \left\langle \sqrt[p]{a^{-1}(a(\gamma_1^p) + a(\gamma_2^p))}, \sqrt[q]{b^{-1}(b(\delta_1^q) + b(\delta_2^q))} \right\rangle^{\vartheta} = (\forall_1 \otimes \forall_2)^{\vartheta}.$$

4 Proposed *p*, *q*-Quasirung Orthopair Fuzzy Archimedean Weighted Geometric Aggregation Operator of *p*, *q*-QOFNs

Utilizing the operation rules proposed in Definition 3.1, this section introduces the *p*, *q*-QOFAWG AO to aggregate the *p*, *q*-QOFNs.

Definition 4.1. Suppose $\forall_1 = \langle \gamma_1, \delta_1 \rangle, \forall_2 = \langle \gamma_2, \delta_2 \rangle, \ldots, \forall_n = \langle \gamma_n, \delta_n \rangle$ be a collection of *p*, *q*-QOFNs. The proposed *p*, *q*-QOFAWG AO to aggregate the *p*, *q*-QOFNs $\forall_1 = \langle \gamma_1, \delta_1 \rangle, \forall_2 = \langle \gamma_2, \delta_2 \rangle, \ldots, \forall_n = \langle \gamma_n, \delta_n \rangle$ is stated as:

$$p, q - QOFAWG(\forall_1, \forall_2, \ldots, \forall_n) = \otimes_{i=1}^{n} \forall_i^{\omega_i}$$

$$= \left\langle \sqrt[p]{a^{-1}\left(\sum_{i=1}^{n} \omega_i a(\gamma_i^p)\right)}, \sqrt[q]{b^{-1}\left(\sum_{i=1}^{n} \omega_i b(\delta_i^q)\right)} \right\rangle, \tag{6}$$

where $p, q \geq 1$, and ω_i indicates the weight assigned to the *p*, *q*-QOFN $\forall_i$, with $\omega_i \in [0, 1], i = 1, 2, \ldots, n$ and $\sum_{i=1}^{n} \omega_i = 1$.

Next, to validate the suggested *p*, *q*-QOFAWG AO defined in Definition 4.1, we present some of its desirable properties.

Property 1 (Idempotency). Suppose $\forall_i = \langle \gamma_i, \delta_i \rangle$ be a collection of *p*, *q*-QOFNs associated with weight ω_i, where $\omega_i > 0, \ i = 1, 2, \ \ldots, n$ and $\sum_{i=1}^{n} \omega_i = 1$. If $\forall_i = \forall = \langle \gamma, \delta \rangle$, then:

$$p, q - QOFAWG(\forall_1, \forall_2, \ldots, \forall_n) = \forall.$$

Proof. Because $\forall_i = \forall = \langle \gamma, \delta \rangle, \forall \ i = 1, 2, \ldots, n$ then by using eq. (6), we have

$$p,q-QOFAWG(\forall_1,\forall_2,\ldots,\forall_n)=\left\langle\sqrt[p]{a^{-1}\left(\sum_{i=1}^{n}\omega_i a(\gamma_i^p)\right)},\sqrt[q]{b^{-1}\left(\sum_{i=1}^{n}\omega_i b(\delta_i^q)\right)}\right\rangle$$
$$=\left\langle\sqrt[p]{a^{-1}\left(\sum_{i=1}^{n}\omega_i a(\gamma^p)\right)},\sqrt[q]{b^{-1}\left(\sum_{i=1}^{n}\omega_i b(\delta^q)\right)}\right\rangle$$
$$=\left\langle\sqrt[p]{a^{-1}(a(\gamma^p))},\sqrt[q]{b^{-1}(b(\delta^q))}\right\rangle=\langle\gamma,\delta\rangle=\forall.$$

Property 2 (Monotonicity). Suppose $\forall_i=\langle\gamma_i,\delta_i\rangle$ and $\forall_i'=\langle\gamma_i',\delta_i'\rangle$ be two collections of p, q-QOFNs, where $i=1,2,\ldots,n$. If $\forall_i\le\forall_i'$, then:

$$p,q-QOFAWG(\forall_1,\forall_2,\ldots,\forall_n)\le p,q-QOFAWG(\forall_1',\forall_2',\ldots,\forall_n').$$

Proof. Because $\forall_i=\langle\gamma_i,\delta_i\rangle$ and $\forall_i'=\langle\gamma_i',\delta_i'\rangle$ be two collections of p, q-QOFNs, where $i=1,2,\ldots,n$. If $\forall_i\le\forall_i'$, this indicates that $\gamma_i\le\gamma_i'$ and $\delta_i\ge\delta_i'$, we have:

$$\gamma_i\le\gamma_i'\Rightarrow\gamma_i^p\le\gamma_i'^p\Rightarrow a(\gamma_i^p)\ge a(\gamma_i'^p)\Rightarrow\omega_i a(\gamma_i^p)\ge\omega_i a(\gamma_i'^p)$$
$$\Rightarrow\sum_{i=1}^{n}\omega_i a(\gamma_i^p)\ge\sum_{i=1}^{n}\omega_i a(\gamma_i'^p)\Rightarrow a^{-1}\left(\sum_{i=1}^{n}\omega_i a(\gamma_i^p)\right)\le a^{-1}\left(\sum_{i=1}^{n}\omega_i a(\gamma_i'^p)\right)$$
$$\Rightarrow\sqrt[p]{a^{-1}\left(\sum_{i=1}^{n}\omega_i a(\gamma_i^p)\right)}\le\sqrt[p]{a^{-1}\left(\sum_{i=1}^{n}\omega_i a(\gamma_i'^p)\right)}.$$

Similarly, since $\delta_i\ge\delta_i'$, we have $\Rightarrow\sqrt[q]{b^{-1}\left(\sum_{i=1}^{n}\omega_i b(\delta_i^q)\right)}\ge\sqrt[q]{b^{-1}\left(\sum_{i-1}^{n}\omega_i b(\delta_i'^p)\right)}$.

Thus, $p,q-QOFAWG(\forall_1,\forall_2,\ldots,\forall_n)\le p,q-QOFAWG(\forall_1',\forall_2',\ldots,\forall_n')$.

Property 3 (Boundedness). Suppose $\forall_i=\langle\gamma_i,\delta_i\rangle$ be a collection of p, q-QOFNs, where $i=1,2,\ldots,n$. Let $\forall^+=max\{\forall_1,\forall_2,\ldots,\forall_n\}$ and $\forall^-=min\{\forall_1,\forall_2,\ldots,\forall_n\}$, then:

$$\forall^-\le p,q-QOFAWG(\forall_1,\forall_2,\ldots,\forall_n)\le\forall^+.$$

Proof. Since, $\forall^+=max\{\forall_1,\forall_2,\ldots,\forall_n\}$ and $\forall^-=min\{\forall_1,\forall_2,\ldots,\forall_n\}$, then by using Properties 1 and 2, we obtain:

$$p,q-QOFAWG(\forall_1,\forall_2,\ldots,\forall_n)\le p,q-QOFAWG\left(\forall_1^+,\forall_2^+,\ldots,\forall_n^+\right)=\forall^+,$$
$$p,q-QOFAWG(\forall_1,\forall_2,\ldots,\forall_n)\ge p,q-QOFAWG\left(\forall_1^-,\forall_2^-,\ldots,\forall_n^-\right)=\forall^-.$$

Thus, $\forall^-\le p,q-QOFAWG(\forall_1,\forall_2,\ldots,\forall_n)\le\forall^+$.

5 Novel MCDM Methodology Based on Proposed p, q-QOFAWG AO of p, q-QOFNs

This section presents an MCDM methodology that utilizes the CRITIC technique to evaluate the unknown criteria weights and the suggested p, q-QOFAWG AO to aggregate the information expressed in form of p, q-QOFN.

Let $ß_1, ß_2, \ldots, ß_m$ be m alternatives and $C_1, C_2, \ldots, C_n$ be n criteria with corresponding weights $\omega_1, \omega_2, \ldots, \omega_n$, such that $\omega_i \in [0,1]$, $\sum_{i=1}^{n} \omega_i = 1$ and $i = 1, 2, \ldots, n$. The decision expert examines the alternative $ß_h$ against each criterion C_i by utilizing p, q-QOFN $\tilde{\forall}_{hi} = \langle \widetilde{\gamma_{hi}}, \widetilde{\delta_{hi}} \rangle$ to assemble the decision evaluation matrix $\tilde{\Delta} = \left(\widetilde{\forall_{hi}}\right)_{m\times n}$, shown as follows:

$$
\tilde{\Delta} = \begin{matrix} & C_1 & C_2 & \cdots & C_n \\ ß_1 \\ ß_2 \\ \vdots \\ ß_m \end{matrix}
\begin{pmatrix} \tilde{\forall}_{11} & \tilde{\forall}_{12} & \cdots & \tilde{\forall}_{1n} \\ \tilde{\forall}_{21} & \tilde{\forall}_{22} & \cdots & \tilde{\forall}_{2n} \\ \vdots & \vdots & \ddots & \vdots \\ \tilde{\forall}_{m1} & \tilde{\forall}_{m2} & \cdots & \tilde{\forall}_{mn} \end{pmatrix}
$$

The proposed MCDM methodology comprises of the subsequent working steps:

Step 1: Convert the decision matrix $\tilde{\Delta} = \left(\widetilde{\forall_{hi}}\right)_{m\times n} = \left(\left\langle \widetilde{\gamma_{hi}}, \widetilde{\delta_{hi}} \right\rangle\right)_{m\times n}$ into the normalized decision matrix $\Delta = (\forall_{hi})_{m\times n} = (\langle \gamma_{hi}, \delta_{hi} \rangle)_{m\times n}$ where $h = 1, 2, \ldots, m$ and $i = 1, 2, \ldots, n$, as follows:

$$
\forall_{hi} = \begin{cases} \langle \tilde{\gamma}_{hi}, \tilde{\delta}_{hi} \rangle, & \textit{if } C_i \textit{ is a benefit type criterion} \\ \langle \tilde{\delta}_{hi}, \tilde{\gamma}_{hi} \rangle, & \textit{if } C_i \textit{ is a cost type criterion} \end{cases} \tag{7}
$$

Step 2: By using eq. (2), compute the score value $\S(\forall_{\mathrm{hi}})$ of p, q-QOFN $\forall_{hi}$, to form the score matrix $S' = (\S(\forall_{\mathrm{hi}}))_{m\times n}$, as follows:

$$
\S(\forall_{\mathrm{hi}}) = \frac{1 + \gamma_{hi}^{p} - \delta_{hi}^{q}}{2}. \tag{8}
$$

Step 3: Compute the standard deviation σ_i of the criterion C_i, as follows:

$$
\sigma_i = \sqrt{\frac{\sum_{h=1}^{m} \left(\S(\forall_{hi}) - \bar{S}_i\right)^2}{m}}. \tag{9}
$$

where $\overline{S_i} = \frac{\sum_{h=1}^{m} \S(\forall_{hi})}{m}$, $h = 1, 2, \ldots, m$ and $i = 1, 2, \ldots, n$.

Step 4: Compute the correlation coefficient among criteria pairs:

$$c_{ik} = \frac{\sum_{h=1}^{m}\left[\left(\S(\forall_{hi}) - \overline{S_i}\right)\left(\S(\forall_{hk}) - \overline{S_k}\right)\right]}{\sqrt{\sum_{h=1}^{m}\left(\S(\forall_{hi}) - \overline{S_i}\right)^2 \sum_{h=1}^{m}\left(\S(\forall_{hk}) - \overline{S_k}\right)^2}}. \tag{10}$$

Step 5: Compute the weight ω_i of the criterion C_i, as follows:

$$\omega_i = \frac{r_i}{\sum_{i=1}^{n} r_i}, \tag{11}$$

where $r_i = \sigma_i \sum_{k=1}^{n}(1 - c_{ik})$ is the quantity of information.

Step 6: By using the suggested p, q-QOFAWG AO stated in eq. (6), aggregate the p, q-QOFNs $\forall_{h1}, \forall_{h2}, \ldots, \forall_{hn}$ present in the hth row of the normalized decision matrix $\Delta = (\forall_{hi})_{m\times n}$ to evaluate the comprehensive p, q-QOFN $\forall_h = \langle \gamma_h, \delta_h \rangle$ of the alternatives ß_h, as given below:

$$\forall_h = p, q - QOFAWG(\forall_{h1}, \forall_{h2}, \ldots, \forall_{hn})$$

$$= \left\langle \sqrt[p]{a^{-1}\left(\sum_{i-1}^{n} \omega_i a(\gamma_i^p)\right)}, \sqrt[q]{b^{-1}\left(\sum_{i-1}^{n} \omega_i b(\delta_i^q)\right)} \right\rangle. \tag{12}$$

Step 7: By using eq. (2), compute the score value $\S(\forall_h)$ of the p, q-QOFN $\forall_h$ of the alternative ß_h, as given below:

$$\S(\forall_h) = \frac{1 + \gamma_h^p - \delta_h^q}{2}, \tag{13}$$

Step 8: By using Definition 2.4, compute the preference order (PO) of the alternatives $\text{ß}_h (h = 1, 2, \ldots, m)$ and choose the optimal option.

6 Case Study of the Selection of Optimal Hybrid Renewable Energy Systems for Rural Electrification

Sustainable electrical energy is one of the most critical factors that drives socio-economic and industrial development in every country that wants to advance in development. Its shortage in rural and remote areas contributes significantly to slow their socio-economic development, as electricity is essential for education, transportation, agriculture, and healthcare. This highlights the importance of rural electrification, which refers to providing reliable and affordable power to underserved areas where grid extension is often unfeasible. HRESs that integrate multiple renewable energy sources, such as solar, wind power, and biomass, have become a viable solution for rural electrification. Unlike conventional electrification methodologies that depend on fossil fuels, harming the climate, environment and human health, HRESs provide cleaner, more reliable, and cost-effective alternatives that can drive rural development. However, selecting the most suitable HRES is complex as it involves balancing multiple conflicting criteria, including system reliability, environmental sustainability, system efficiency, and operation and maintenance cost. Thus, it is a suitable case of MCDM problem.

In the following, we have considered five HRESs: Photovoltaic- ($ß_1$), Photovoltaic-Biomass ($ß_2$), Hydro-Wind ($ß_3$), Wind-Solar ($ß_4$), and Biomass-Wind-Fuel cell ($ß_5$), as alternatives and they are assessed by the decision-making expert under four criteria: System reliability (C_1), Environmental sustainability (C_2), System efficiency (C_3), and Operation and maintenance cost (C_4). The expert expressed the assessment information in terms of a p, q-QOFN $\tilde{\forall}_{hi} = \langle \widetilde{\gamma_{hi}}, \widetilde{\delta_{hi}} \rangle$ for each alternative $ß_h$ corresponding to criterion C_i, to construct the decision matrix $\tilde{\Delta} = \left(\widetilde{\forall_{hi}} \right)_{m \times n}$, where

$$\tilde{\Delta} = \begin{matrix} \\ ß_1 \\ ß_2 \\ ß_3 \\ ß_4 \\ ß_5 \end{matrix} \begin{matrix} C_1 & C_2 & C_3 & C_4 \\ \left(\begin{matrix} \langle 0.1, 0.6 \rangle & \langle 0.4, 0.3 \rangle & \langle 0.3, 0.4 \rangle & \langle 0.5, 0.1 \rangle \\ \langle 0.4, 0.2 \rangle & \langle 0.5, 0.3 \rangle & \langle 0.4, 0.5 \rangle & \langle 0.1, 0.3 \rangle \\ \langle 0.5, 0.2 \rangle & \langle 0.4, 0.1 \rangle & \langle 0.4, 0.2 \rangle & \langle 0.3, 0.4 \rangle \\ \langle 0.3, 0.5 \rangle & \langle 0.4, 0.3 \rangle & \langle 0.5, 0.1 \rangle & \langle 0.1, 0.4 \rangle \\ \langle 0.5, 0.2 \rangle & \langle 0.4, 0.1 \rangle & \langle 0.5, 0.3 \rangle & \langle 0.1, 0.5 \rangle \end{matrix} \right) \end{matrix}$$

To resolve this MCDM problem, we employ the proposed MCDM methodology as given below:

Step 1: As criterion C_4 is cost type and criteria $C_1, C_2,$ and C_3 are benefit type, thus by employing eq. (7), we derive the normalized decision matrix, where:

$$\Delta = \begin{array}{c} \\ ß_1 \\ ß_2 \\ ß_3 \\ ß_4 \\ ß_5 \end{array} \begin{array}{c} \begin{array}{cccc} C_1 & C_2 & C_3 & C_4 \end{array} \\ \begin{pmatrix} \langle 0.1, 0.6\rangle & \langle 0.4, 0.3\rangle & \langle 0.3, 0.4\rangle & \langle 0.1, 0.5\rangle \\ \langle 0.4, 0.2\rangle & \langle 0.5, 0.3\rangle & \langle 0.4, 0.5\rangle & \langle 0.3, 0.1\rangle \\ \langle 0.5, 0.2\rangle & \langle 0.4, 0.1\rangle & \langle 0.4, 0.2\rangle & \langle 0.4, 0.3\rangle \\ \langle 0.3, 0.5\rangle & \langle 0.4, 0.3\rangle & \langle 0.5, 0.1\rangle & \langle 0.4, 0.1\rangle \\ \langle 0.5, 0.2\rangle & \langle 0.4, 0.1\rangle & \langle 0.5, 0.3\rangle & \langle 0.5, 0.1\rangle \end{pmatrix} \end{array}$$

Step 2: By employing eq. (8), we obtain the score matrix S', where $p=2, q=1$:

$$S' = \begin{pmatrix} 0.2050 & 0.4300 & 0.3450 & 0.2550 \\ 0.4800 & 0.4750 & 0.3300 & 0.4950 \\ 0.5250 & 0.5300 & 0.4800 & 0.4300 \\ 0.2950 & 0.4300 & 0.5750 & 0.5300 \\ 0.5250 & 0.5300 & 0.4750 & 0.5750 \end{pmatrix}.$$

Step 3: By using eq. (9), we obtain the standard deviation σ_i of the criterion C_i, where $i=1, 2, \ldots, n$; $\sigma_1 = 0.1315, \sigma_2 = 0.0448, \sigma_3 = 0.0918$, and $\sigma_4 = 0.1116$.

Step 4: By employing eq. (10), we obtain the correlation coefficient matrix c_{ik}, where

$$c_{ik} - \begin{pmatrix} 1 & 0.9214 & 0.0810 & 0.6288 \\ 0.9214 & 1 & 0.1121 & 0.4329 \\ 0.0810 & 0.1121 & 1 & 0.5449 \\ 0.6288 & 0.4329 & 0.5449 & 1 \end{pmatrix}$$

Step 5: By using eq. (11), we obtain the weight ω_i of the criterion C_i, where $i=1, 2, \ldots, n$; $\omega_1 = 0.2943, \omega_2 = 0.1122, \omega_3 = 0.3395$, and $\omega_4 = 0.2540$.

Step 6: By employing eq. (12), we evaluate the overall p, q-QOFN ∇_h of the alternatives for $a(t) = \log(t)$, where $h=1,2,\ldots,m$; $\nabla_1 = \langle 0.1696, 0.4827\rangle, \nabla_2 = \langle 0.3812, 0.3077\rangle, \nabla_3 = \langle 0.4271, 0.2164\rangle, \nabla_4 = \langle 0.3965, 0.2640\rangle$ and $\nabla_5 = \langle 0.4876, 0.2018\rangle$.

Step 7: By using eq. (13), we obtain the score values $\S(\nabla_1) = 0.2730, \S(\nabla_2) = 0.4188, \S(\nabla_3) = 0.4830, \S(\nabla_4) = 0.4466$ and $\S(\nabla_5) = 0.5180$.

Step 8: Since, $\S(\nabla_5) > \S(\nabla_3) > \S(\nabla_4) > \S(\nabla_2) > \S(\nabla_1)$, it follows that the PO of the alternatives is: "$ß_5 \succ ß_3 \succ ß_4 \succ ß_2 \succ ß_1$".

Table 1: POs obtained by the different MCDM methodologies.

MCDM methodologies	POs
Peng et al.'s MCDM methodology [12]	$ß_5 \succ ß_3 \succ ß_4 \succ ß_2 \succ ß_1$
Peng et al. 's MCDM methodology [13]	$ß_5 \succ ß_4 \succ ß_2 \succ ß_3 \succ ß_1$
Proposed MCDM methodology	$ß_5 \succ ß_3 \succ ß_4 \succ ß_2 \succ ß_1$

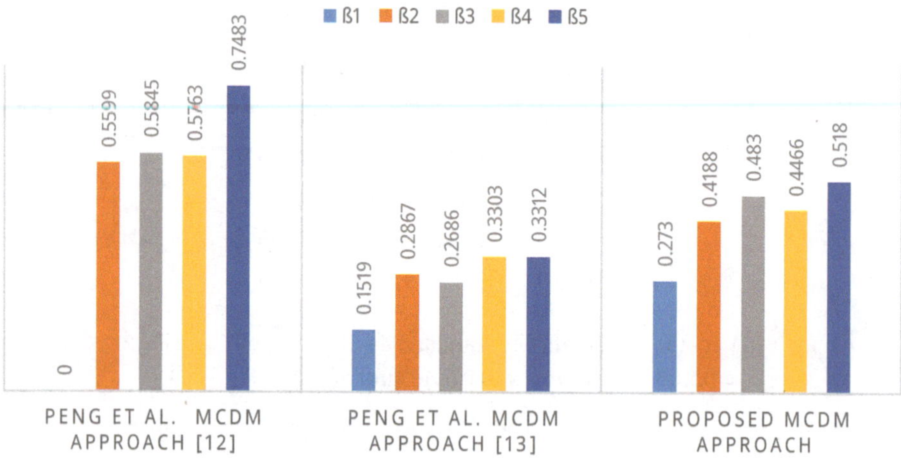

Figure 1: Graphical comparison of POs derived by various MCDM methodologies.

6.1 Comparison Study

Next, we compare the RO of the HRESs $ß_1, ß_2, ß_3, ß_4$, and $ß_5$ obtained using proposed MCDM methodology with Peng et al. [12, 13] to highlight the efficacy and practical validity of the suggested methodology.

Table 1 and Figure 1 illustrate a comparative analysis of the PO of the alternatives $ß_1, ß_2, ß_3, ß_4$, and $ß_5$ as derived using the proposed MCDM methodology and the existing methodologies defined in [12, 13]. The outcomes in Table 1 and Figure 1 clearly show that Peng et al.'s [13] MCDM methodology produces the PO "$ß_5 \succ ß_4 \succ ß_2 \succ ß_3 \succ ß_1$" of the alternatives $ß_1, ß_2, ß_3, ß_4$, and $ß_5$. While, Peng et al.'s [12] MCDM methodology and the suggested MCDM methodology yield the same PO "$ß_5 \succ ß_3 \succ ß_4 \succ ß_2 \succ ß_1$". Although the PO obtained from Peng et al. [13] MCDM methodology and the proposed MCDM methodology differ, but the optimal alternative remains the same, i.e., Biomass-Wind-Fuel cell ($ß_5$) is the most suitable HRES. This highlights the robustness and efficacy of the suggested MCDM methodology. The difference in PO may be due to the variation in criteria weight determination techniques and AOs. Peng et al. [13] MCDM methodology utilizes WENSLO technique to derive the criteria weights, whereas the proposed MCDM methodology utilizes CRITIC technique. This comparison confirms that the proposed methodology not only identifies the same optimal alternative but it is also very reliable and offers greater consistency. Therefore, the proposed MCDM methodology is suitable for handling complex decision-making problems.

7 Conclusion

In this chapter, new operation rules for p, q-QOFNs have been proposed that are built upon Archimedean norms. Next, by employing these proposed operation rules, the p, q-QOFAWG AO has been developed. We have also discussed the key characteristics of the proposed p, q-QOFAWG AO, including idempotency, boundedness, and monotonicity, to confirm its reliability. Furthermore, by integrating the proposed operator and CRITIC technique, we have developed a novel MCDM methodology. Here, the CRITIC technique is used to evaluate the unknown criteria weights and the proposed p, q-QOFAWG AO is used to aggregate the information represented in p, q-QOFNs. To illustrate the practical utility of the proposed MCDM methodology, a case study on identifying the most suitable HRES for rural electrification have been solved. The findings indicate that Biomass-Wind -Fuel cell ($ß_5$) is the optimal HRES among “Photovoltaic-Wind” ($ß_1$), “Photovoltaic-Biomass” ($ß_2$), “Hydro-Wind” ($ß_3$), “Wind-Solar” ($ß_4$), and “Biomass-Wind-Fuel cell” ($ß_5$). Additionally, we have conducted a comparison study, in which proposed MCDM methodology is compared with Peng et al. [12] and Peng et al. [13] MCDM methodologies. The findings illustrate the reliability and validity of the proposed MCDM methodology and highlight its effectiveness in handling complex decision-making problems. Nevertheless, the proposed MCDM methodology has certain restrictions. It is confined to individual decision-making, whereas real-world problems often require group based evaluations. Moreover, the proposed methodology utilizes the p, q-QOFAWG AO and does not integrate any classical MCDM methodologies, such as TOPSIS, VIKOR, and EDAS. In future, we plan to implement the proposed MCDM technique in practical group decision-making problems, such as supplier selection, water management, and sustainable transportation.

References

[1] Mainali B, Silveira S. Alternative pathways for providing access to electricity in developing countries. Renew Energy. 2013;57:299–310.

[2] Zadeh LA. Fuzzy sets. Inf Control. 1965;8(3):338–353.

[3] Atanassov KT. Intuitionistic fuzzy sets. Fuzzy Sets Syst. 1986;20(1):87–96.

[4] Yager RR. Pythagorean membership grades in multicriteria decision making. IEEE Trans Fuzzy Syst. 2013;22(4):958–965.

[5] Yager RR. Generalized orthopair fuzzy sets. IEEE Trans Fuzzy Syst. 2017;25(5):1222–1230.

[6] Bajaj J, Kumar S. A new intuitionistic fuzzy correlation coefficient methodology with applications in multi-criteria decision-making. Decis Anal J. 2023;9:100340.

[7] Aggarwal E, Mohanty BK. An algorithmic-based multiattribute decision making model under intuitionistic fuzzy environment. J Intell Fuzzy Syst. 2022;42(6):5537–5551.

[8] Jabeen K, Ullah K, Pedrycz W, Khan Q, Ali Z, Yin S. Pythagorean fuzzy Aczel-Alsina power Bonferroni mean operators for multi-attribute decision-making. Granul Comput. 2024;9(1):24.

[9] Alkan N, Kahraman C. CODAS extension using novel decomposed Pythagorean fuzzy sets: Strategy selection for IOT based sustainable supply chain system. Expert Syst Appl. 2024;237:121534.

[10] Rao Y, Kosari S, Hameed S, Yousaf Z. Multiattribute decision-making using q-rung orthopair fuzzy Zagreb index. Artif Intell Rev. 2025;58(5):1–31.

[11] Mishra AR, Alrasheedi M, Lakshmi J, Rani P. Multi-criteria decision analysis model using the q-rung orthopair fuzzy similarity measures and the COPRAS method for electric vehicle charging station site selection. Granul Comput. 2024;9(1):1–20.

[12] Peng X, Krishankumar R, Ravichandran KS. Generalized orthopair fuzzy weighted distance-based approximation (WDBA) algorithm in emergency decision-making. Int J Intell Syst. 2019;34 (10):2364–2402.

[13] Peng X, Yu L, Wu X. q-rung orthopair fuzzy multi-criteria decision-making method for Internet of Things platforms selection. Eng Appl Artif Intell. 2025;148:110336.

[14] Seikh MR, Mandal U. Multiple attribute group decision making based on quasirung orthopair fuzzy sets: Application to electric vehicle charging station site selection problem. Eng Appl Artif Intell. 2022;115:105299.

[15] Rahim M, Ahmad S, Younis B, Egami RH, Ahmed MM. Multiple attribute group decision making based on p, q-quasirung orthopair Bonferroni mean operators and their applications. Evol Syst. 2025;16(1):9.

[16] Redhu A, Bhardwaj R, Kumar K, Kaur G. p, q-quasirung orthopair fuzzy Schweizer-Sklar aggregation operators and their application in multi-attribute decision-making. J Appl Math Comput. 2025;URL https://doi.org/10.1007/s12190-025-02449-5.

[17] Ali J, Naeem M. Analysis and application of p, q-quasirung orthopair fuzzy Aczel–Alsina aggregation operators in multiple criteria decisionmaking. IEEE Access. 2023;11:49081–49101.

[18] Boz E. Evaluation for the selection of agricultural spraying drones using p, q-quasirung orthopair fuzzy logic and multi-criteria methods. Sci Rep. 2025;15(1):27606.

[19] Redhu A, Arora R, Kumar K. Selection of the optimal health care waste treatment technology using Yager prioritized arithmetic operator-based p, q-quasirung orthopair fuzzy group decision-making method. Process Integr Optim Sustain. 2025;URL http://dx.doi.org/10.1007/s41660-025-00498-8.

[20] Rahim M, Garg H, Khan S, Alqahtani H, Khalifa HAEW. Group decision-making algorithm with sine trigonometric p, q-quasirung orthopair aggregation operators and their applications. Alex Eng J. 2023;78:530–542.

[21] Klement EP, Mesiar R. Logical, algebraic, analytic and probabilistic aspects of triangular norms. 2005, Elsevier: New York, NY, USA.

[22] Diakoulaki D, Mavrotas G, Papayannakis L. Determining objective weights in multiple criteria problems: The CRITIC method. Comput Oper Res. 1995;22(7):763–770.

Dr. Divya Mol K.T.

Application of Complex Fuzzy Matrices in the Diagnosis and Management of Lung Diseases via Symptom-Based Analysis

Abstract: In this chapter, we delve into the analysis of lung disease symptoms, a crucial aspect of accurate diagnosis. Diagnosing various lung conditions can be quite challenging due to the way symptoms often overlap and present similarly. Since conventional fuzzy membership cannot adequately represent the overlapping of symptoms, we use a complex grade of membership for better representation. We introduce a new algorithm aimed at identifying the most common symptoms among various patients, allowing us to effectively determine the related symptoms. Furthermore, we provide an illustration to exemplify our findings.

Keywords: complex fuzzy matrices, symptom-based analysis, lung diseases, disease management, relation matrices, comparison matrices

1 Introduction

The use of normal set theory or probability distributions is not feasible for scenarios involving partial occurrence, or partial truth. Zadeh [1] introduced fuzzy set theory in 1965 to overcome this crisis. Fuzzy sets are not adequate for modeling real-world problems that have both uncertainty and approximately periodic phenomena. The concept of complex fuzzy sets was introduced by Ramot et al. [2] in 2002 to address this issue. The thought of extending the characteristic function from {0, 1} to [0, 1] leads to the introduction of fuzzy set theory, and the introduced notion equipped with classical matrix theory opens up the world of fuzzy matrix theory by Kim and Rosh [3] in 1980. By broadening the codomain of the membership function from [0, 1] to a unit disc in the complex plane, complex fuzzy sets are produced. These sets are connected to fuzzy matrix theory and were the foundation of the complex fuzzy matrix theory proposed by Zhao [4] in 2015.

The complex fuzzy membership [1] can effectively model numerous real-world problems related to decision-making across various fields, including medicine, engineering, and business, utilizing complex fuzzy matrices [2] and complex fuzzy sets [1]. There are numerous studies in the field of medical decision-making that utilize classical fuzzy set theory, probability statistics, and other methods [3–6]. However, in this chapter, we have chosen to employ complex fuzzy membership [1] as our decision-

Dr. Divya Mol K.T., Department of Basic Science, Amal Jyothi College (Autonomous) of Engineering, Kanjirappally, Kerala, India, e-mail: ktdivyamol@amaljyothi.ac.in

 | https://doi.org/10.1515/9783112222249-006

making tool. This innovative approach provides a nuanced understanding of uncertainty and variability in medical data. By incorporating complex fuzzy membership, we can improve the precision of our decisions, resulting in better patient outcomes and more effective healthcare strategies.

Various lung diseases, such as asthma, bronchitis, chronic obstructive pulmonary disease (COPD), tuberculosis (TB), and lung cancer, share overlapping symptoms, making diagnosis complicated and challenging. There are many confounding and unclear mismatches between the data reported by the patient and the clinical data in the case of these lung diseases. Due to this confusing scenario, we cannot implement usual probability distributions or classical statistical methodologies as tools for modeling. The resolution of these crises can be achieved through the use of classical fuzzy sets [7–9], which enhance the representation of ambiguous data. The modification of fuzzy set theory by intuitionist fuzzy set theory [10, 11] improved the modeling quality. Intuitive fuzzy models cannot model multidimensional uncertainty in various cases, such as the relationship between symptoms and their variation correlations, etc. [6, 5].

To clarify the study, most existing works examine the overlapping symptoms separately. However, it does not provide sufficient accurate and correct decision-making. To address all the aforementioned crises, we have modeled various lung diseases by analyzing their overlapping symptoms with the support of complex fuzzy matrices and fuzzy sets in this chapter.

2 Preliminaries

Definition 2.1 ([1]). A complex fuzzy set defined on the universe of discourse U, characterized by a membership function $\chi_A(x)$, that assigns any element $x \in A$, a complex grade of membership in A.

The value $\chi_A(x)$ may receive all lie within the unit circle in the complex plane, and are thus of the form $r_A(x)e^{i\omega_A(x)}$, $i = \sqrt{-1}$, where $r_A(x)$ and $\omega_A(x)$ both real valued, $r_A(x) \in [0,1]$ and $\omega_A(x) \in [0, 2\pi)$. The complex fuzzy set is also represented as the set of ordered pairs $A = \{(x, \chi_A(x)) : x \in U\}$.

Definition 2.2 ([2]). A complex fuzzy matrix is a matrix, which has its elements from the unit ball in the complex plane. A complex fuzzy matrix A of order $(m \times n)$ is defined as $A = \left[(a_{ij}, a_{ij\chi})\right]_{m \times n}$, where $a_{ij\chi} = r_A(x)e^{i\omega_A(x)}$, where $r_A(x)$ and $\omega_A(x)$ both real valued, $r_A(x) \in [0,1]$ and $\omega_A(x) \in [0, 2\pi)$.

Definition 2.3 ([3]). Relative values are the values that depend on other values. Here, we use the following equation to find the relative values from the known values: $r_{ij} = \left(f\left(\frac{p_i}{d_j}\right)\right); i = 1, 2, 3, 4, \ldots\ j = 1, 2, 3, 4, \ldots$

Definition 2.4 ([3]). A comparison matrix helps to compare attributes and characteristics of items and helps us to conclude the comparative and relative study. Here, we use the following method to create the comparison matrix using the relative values:

$$\text{Comparison matrix, } R = [r_{ij}];\ r_{ij} = \left(f\left(\frac{P_i}{d_j}\right)\right);\ i = 1, 2, 3, 4, \ldots\ j = 1, 2, 3, 4, \ldots$$

2 Use of Complex Fuzzy Matrices in the Diagnosis of Common Lung Diseases

Lung conditions pose significant health risks globally, increasing morbidity and mortality. These include obstructive and restrictive diseases like COPD and pulmonary fibrosis, as well as infections such as lung cancer, pneumonia, and TB. Understanding their biology, risk factors, and symptoms is essential for improving diagnosis, treatment, and prevention. Advances in imaging and biomarkers aid early detection and management. However, the rising prevalence of respiratory diseases exacerbated by smoking, air pollution, and occupational hazards presents ongoing challenges for healthcare systems worldwide.

The most common respiratory diseases and their main symptoms are listed below:

1. Asthma: Symptoms are shortness of breath, wheezing, chest tightness, and coughing (especially at night).
2. COPD: Symptoms are persistent cough with mucus, shortness of breath (especially during exertion), wheezing, and frequent respiratory infection.
3. Pneumonia: Symptoms are fever, chills, cough producing yellow, green, or bloody mucus, chest pain that worsens with breathing or coughing, and shortness of breath.
4. TB: Symptoms are persistent cough (lasting >3 weeks), coughing up blood or sputum, fever, night sweats, and weight loss.
5. Bronchitis: Symptoms are cough (often producing mucus), wheezing, shortness of breath, and chest discomfort.
6. Pulmonary fibrosis: Symptoms are persistent dry cough, shortness of breath (especially during physical activity), fatigue, and chest pain.
7. Lung cancer: Symptoms are persistent cough (may worsen over time), coughing up blood, unexplained weight loss, and chest pain.
8. Acute respiratory distress syndrome (ARDS): Symptoms are severe shortness of breath, rapid breathing, low oxygen levels, and fatigue.
9. Interstitial lung disease: Symptoms are persistent dry cough, progressive shortness of breath, and fatigue.

Given the intricacies involved in diagnosing lung diseases, planning treatment procedures requires careful analysis of various combinations of factors that can lead to complicated outcomes. Initially, sets of symptoms are considered as input variables, each defined by its own complex fuzzy membership function. To effectively analyze the data, we employ the concepts of complex fuzzy sets and complex fuzzy matrices.

The input variables are:

i. Shortness of breath – s_1
ii. Fatigue – s_2
iii. Wheezing – s_3
iv. Coughing (especially at night) – s_4
v. Chest pain – s_5
vi. Persistent cough with mucus – s_6
vii. Frequent respiratory infections – s_7
viii. Fever, chills – s_8
ix. Weight loss – s_9
x. Rapid breathing and low oxygen levels – s_{10}

Considering these input variables, we assign specific ranges to define the membership values for each input. For example, shortness is assigned a range between 0 and 4. Similarly, for other variables where precise measurement methods or instruments are not available, we also define membership values within the range of 0 – 4.

For example, consider the variable s_{10}, which represents rapid breathing and low oxygen levels of a particular patient. If the patient's oxygen level is normal (between 95% and 100%, the value s_{10} is assigned within the range of 0 – 1, depending on their physical condition. If the patient experiences an oxygen level between 90% and 95%, a grade value between 1 and 3 is assigned, placing the variable s_{10} in the medium range. In contrast, if the patient has a low oxygen level, below 90%, a grade value is given within the range of 3 – 4, indicating a severe condition.

The levels and rates of respiratory symptoms can be assessed using standard clinical methods. Shortness of breath, s_1, can be evaluated through functional capacity and oxygen saturation using the 6-min walk test, SPO_2 monitoring, or the Dyspnea Scale. Fatigue, s_3, is commonly measured through fatigue severity assessment using the Chalder Fatigue Questionnaire. Wheezing, s_4, is quantified by evaluating airflow obstruction percentage through spirometry. Coughing, especially at night, s_5, is assessed by cough frequency and severity using the Cough Symptom Score. Tools like the McGill Pain Questionnaire and the Numeric Pain Rating Scale are used to measure the quality and intensity of chest pain. A chest X-ray or blood tests (CRP and WBC count) are used to look for signs of infection and inflammation in a persistent cough with mucus. A digital thermometer is used to measure body temperature in order to assess fever and chills. Changes in weight and body mass index or the Unintentional Weight Loss Questionnaire are used to measure weight loss, s_9. Lastly, oxygen saturation is monitored to determine low oxygen levels, s_{10}, and rapid breathing.

The similar approach can be applied to the remaining input variables, ensuring consistency in defining membership values.

4 Description of the Algorithm

Step 1: Consider all the necessary confounding alternatives in the stated problem.
Step 2: List all the important alternatives as input variables with their membership functions.
Step 3: Using the concept of complex fuzzy sets and complex fuzzy matrices, assign complex grade of membership for each of the input variables. The complex fuzzy function can be formed by writing their degrees of uncertainty as the modulus value $r_A(x)$ and the grade of the uncertainty as the amplitude term $\omega_A(x)$, in the polar form of the complex fuzzy membership.
Step 4: Considering the set of items and finding all the input variable membership values corresponding to their grades.
Step 5: Using these values, construct the input variable-cause of input variable complex fuzzy matrix.
Step 6: Calculate the relative values, $f\left(\frac{P_i}{d_j}\right)$, using the equation:

$$f\left(\frac{P_i}{d_j}\right) = \frac{\left|\chi_{d_i}(P_j)\right| - \left|\chi_{P_i}(d_j)\right|}{\max\left\{\left|\chi_{d_i}(P_j)\right|, \left|\chi_{P_i}(d_j)\right|\right\}}$$

Step 7: Using the relative values, form the comparison matrix:

$$R = [r_{ij}];\ r_{ij} = \left(f\left(\frac{P_i}{d_j}\right)\right);\ i = 1, 2, 3, 4, \ \ldots \ j = 1, 2, 3, 4, \ \ldots$$

Step 8: Using the ranking of the problem, by taking maximum in each row of the comparison matrix, we can identify the major cause of input variable of that problem.

5 Illustrative Example

i. Shortness of breath – s_1

The variable s_1 stands for shortness of breath, which ranges from 0 to 4.

When s_1 range is $0 \leq x \leq 2$, the patient has no breathing difficulties. We use the phase value 0 to represent the range in the fuzzy set.

When s_1 ranges between 2 and 3, the patient has mild breathing issues. We use the phase value $\frac{2\pi}{3}$ to represent the particular range in the fuzzy set.

When s_1 range is $3 \leq x \leq 4$, the patient has severe breathing issues. We use the phase value $\frac{4\pi}{3}$ to represent the range in the fuzzy set.

The defined membership functions are:

$$\chi_{s_1}(x) = \begin{cases} 0 & 0 \le x \le 2 \\ \dfrac{3-x}{2} & 2 < x < 3 \\ 1 & 3 \le x \le 4 \end{cases}$$

$$= \begin{cases} 0e^{i0} \\ \left(\dfrac{3-x}{2}\right)e^{i\frac{2\pi}{3}} \\ 1e^{i\frac{4\pi}{3}} \end{cases}$$

Similarly, for all the remaining input variables, we use the range 0 – 4 and the same membership function:

$$\chi_{s_k}(x) = \begin{cases} 0 & 0 \le x \le 2 \\ \dfrac{3-x}{2} & 2 < x < 3 \\ 1 & 3 \le x \le 4 \end{cases}$$

$$= \begin{cases} 0e^{i0} \\ \left(\dfrac{3-x}{2}\right)e^{i\frac{2\pi}{3}} \\ 1e^{i\frac{4\pi}{3}} \end{cases}$$

Table 1: Symptom severity for 10 patients (0 = none, 4 = severe).

Patient/symptoms	s_1	s_2	s_3	s_4	s_5	s_6	s_7	s_8	s_9	s_{10}
1	3	2	1.5	2.5	2.9	1.5	3.5	2.5	0.5	3
2	3.7	1.5	2	2.5	1.5	2.9	1.2	3.1	1	0
3	1.7	0	0	1.4	0.5	0	2.6	3.9	2	1.9
4	2.5	1.3	2.7	2.8	1	0	3	2.2	1.5	3.5
5	3	2.4	1.3	0.4	0.2	3.3	2.8	3	2.7	3.4
6	0.75	1.8	3.1	0.9	0.5	1	2.6	0.8	0	0
7	2	2	2	2.8	3.1	2.9	1.2	3.1	0	2.5
8	1.4	3.3	2	3	4	2.3	3.5	0.3	3.2	4
9	1.5	1.6	0	0.11	0.5	3.4	1.6	0.5	0	1.5
10	3.5	3.2	2.5	3.5	1.2	2	3.7	2.9	1	3

Here, in Table 1, for patient P_1 has breathing difficulty of scale 3, then the membership value become $\chi_{P_1}(s_1) = 1e^{i\frac{4\pi}{3}}$, obtained by substituting the value 1 in the corresponding membership function:

$$\text{Fatigue } = 2 \ \Rightarrow \chi_{P_1}(s_2) = 0.5e^{i\frac{2\pi}{3}}$$

$$\text{Wheezing} = 1.5 \ \Rightarrow \chi_{P_1}(s_3) = 0.75e^{i\frac{2\pi}{3}}$$

$$\text{Coughing (especially at night)} = 2.5 \ \Rightarrow \chi_{P_1}(s_4) = 0.25e^{i\frac{2\pi}{3}}$$

$$\text{Chest pain } = 2.9 \ \Rightarrow \chi_{P_1}(s_5) = 0.05e^{i\frac{2\pi}{3}}$$

$$\text{Persistent cough with mucus } = 1.5 \ \Rightarrow \chi_{P_1}(s_6) = 0.75e^{i\frac{2\pi}{3}}$$

$$\text{Frequent respiratory infections } = 3.5 \ \Rightarrow \chi_{P_1}(s_7) = 1e^{i\frac{4\pi}{3}}$$

$$\text{Fever, chills} = 2.5 \ \Rightarrow \chi_{P_1}(s_8) = 0.25e^{i\frac{2\pi}{3}}$$

$$\text{Weight loss } = 0.5 \ \Rightarrow \chi_{P_1}(s_9) = 0e^{i0}$$

$$\text{Rapid breathing and low oxygen levels} = 3 \ \Rightarrow \chi_{P_1}(s_{10}) = 1e^{i\frac{4\pi}{3}}$$

Then, the first row matrix for the patient P_1 is:

$$P_1\colon \begin{bmatrix} 1e^{i\frac{4\pi}{3}} & 0.5e^{i\frac{2\pi}{3}} & 0.75e^{i\frac{2\pi}{3}} & 0.25e^{i\frac{2\pi}{3}} & 0.05e^{i\frac{2\pi}{3}} & 0.75e^{i\frac{2\pi}{3}} & 1e^{i\frac{4\pi}{3}} & 0.25e^{i\frac{2\pi}{3}} & 0e^{i0} & 1e^{i\frac{4\pi}{3}} \end{bmatrix}$$

Here, for patient P_2 has breathing difficulty of scale 3.7, then the membership value become $\chi_{P_2}(s_1) = 0.75e^{i\frac{2\pi}{3}}$, obtained by substituting the value 1 in the corresponding membership function:

$$\text{Fatigue } = 1.5 \ \Rightarrow \chi_{P_2}(s_2) = 0.75e^{i\frac{2\pi}{3}}$$

$$\text{Wheezing} = 2 \ \Rightarrow \chi_{P_2}(s_3) = 0.5e^{i\frac{2\pi}{3}}$$

$$\text{Coughing (especially at night)} = 2.5 \ \Rightarrow \chi_{P_2}(s_4) = 0.25e^{i\frac{2\pi}{3}}$$

$$\text{Chest pain } = 1.5 \ \Rightarrow \chi_{P_2}(s_5) = 0.75e^{i\frac{2\pi}{3}}$$

$$\text{Persistent cough with mucus } = 2.9 \ \Rightarrow \chi_{P_2}(s_6) = 0.05e^{i\frac{2\pi}{3}}$$

$$\text{Frequent respiratory infections } = 1.2 \ \Rightarrow \chi_{P_2}(s_7) = 0.9e^{i\frac{2\pi}{3}}$$

$$\text{Fever, chills} = 3.1 \Rightarrow \chi_{P_2}(s_8) = 1e^{i\frac{4\pi}{3}}$$

$$\text{Weight loss } = 1 \ \Rightarrow \chi_{P_2}(s_9) = 0e^{i0}$$

$$\text{Rapid breathing and low oxygen levels} = 0 \ \Rightarrow \chi_{P_2}(s_{10}) = 0e^{i0}$$

Then, the first row matrix for the patient P_2 is:

$$P_2:\begin{bmatrix} 1e^{i\frac{4\pi}{3}} & 0.75e^{i\frac{2\pi}{3}} & 0.5e^{i\frac{2\pi}{3}} & 0.25e^{i\frac{2\pi}{3}} & 0.75e^{i\frac{2\pi}{3}} & 0.05e^{i\frac{2\pi}{3}} & 0.9e^{i\frac{2\pi}{3}} & 1e^{i\frac{4\pi}{3}} & 0e^{i0} & 0e^{i0} \end{bmatrix}$$

Similarly, we can find the 10 row matrices for each of the patients by substituting the range values in the corresponding membership functions:

$$P_3:\begin{bmatrix} 0.65e^{i\frac{2\pi}{3}} & 0e^{i0} & 0e^{i0} & 0.8e^{i\frac{2\pi}{3}} & 0e^{i0} & 0e^{i0} & 0.2e^{i\frac{2\pi}{3}} & 1e^{i\frac{4\pi}{3}} & 0.5e^{i\frac{2\pi}{3}} & 0.55e^{i\frac{2\pi}{3}} \end{bmatrix}$$

$$P_4:\begin{bmatrix} 0.25e^{i\frac{2\pi}{3}} & 0.85e^{i\frac{2\pi}{3}} & 0.15e^{i\frac{2\pi}{3}} & 0.1e^{i\frac{2\pi}{3}} & 0e^{i0} & 0e^{i0} & 1e^{i\frac{4\pi}{3}} & 0.2e^{i\frac{2\pi}{3}} & 1e^{i\frac{4\pi}{3}} & 0.55e^{i\frac{2\pi}{3}} \end{bmatrix}$$

$$P_5:\begin{bmatrix} 1e^{i\frac{4\pi}{3}} & 0.3e^{i\frac{2\pi}{3}} & 0.85e^{i\frac{2\pi}{3}} & 0e^{i0} & 0e^{i0} & 1e^{i\frac{4\pi}{3}} & 0.1e^{i\frac{2\pi}{3}} & 1e^{i\frac{4\pi}{3}} & 0.15e^{i\frac{2\pi}{3}} & 1e^{i\frac{4\pi}{3}} \end{bmatrix}$$

$$P_6:\begin{bmatrix} 0e^{i0} & 0.6e^{i\frac{2\pi}{3}} & 1e^{i\frac{4\pi}{3}} & 0e^{i0} & 0e^{i0} & 0e^{i0} & 0.2e^{i\frac{2\pi}{3}} & 0e^{i0} & 0e^{i0} & 0e^{i0} \end{bmatrix}$$

$$P_7:\begin{bmatrix} 0.5e^{i\frac{2\pi}{3}} & 0.5e^{i\frac{2\pi}{3}} & 0.5e^{i\frac{2\pi}{3}} & 0.1e^{i\frac{2\pi}{3}} & 1e^{i\frac{4\pi}{3}} & 0.05e^{i\frac{2\pi}{3}} & 0.9e^{i\frac{2\pi}{3}} & 1e^{i\frac{4\pi}{3}} & 0e^{i0} & 0.25e^{i\frac{2\pi}{3}} \end{bmatrix}$$

$$P_8:\begin{bmatrix} 0.8e^{i\frac{2\pi}{3}} & 1e^{i\frac{4\pi}{3}} & 0.5e^{i\frac{2\pi}{3}} & 1e^{i\frac{4\pi}{3}} & 1e^{i\frac{4\pi}{3}} & 0.3e^{i\frac{2\pi}{3}} & 1e^{i\frac{4\pi}{3}} & 0e^{i0} & 1e^{i\frac{4\pi}{3}} & 1e^{i\frac{4\pi}{3}} \end{bmatrix}$$

$$P_9:\begin{bmatrix} 0.75e^{i\frac{2\pi}{3}} & 0.8e^{i\frac{2\pi}{3}} & 0e^{i0} & 0e^{i0} & 0e^{i0} & 1e^{i\frac{4\pi}{3}} & 0.7e^{i\frac{2\pi}{3}} & 0e^{i0} & 0e^{i0} & 0.75e^{i\frac{2\pi}{3}} \end{bmatrix}$$

$$P_{10}:\begin{bmatrix} 1e^{i\frac{4\pi}{3}} & 1e^{i\frac{4\pi}{3}} & 0.25e^{i\frac{2\pi}{3}} & 1e^{i\frac{4\pi}{3}} & 0.9e^{i\frac{2\pi}{3}} & 0.5e^{i\frac{2\pi}{3}} & 1e^{i\frac{4\pi}{3}} & 0.05e^{i\frac{2\pi}{3}} & 0e^{i0} & 1e^{i\frac{4\pi}{3}} \end{bmatrix}$$

Then, the patient-causes of input variable complex fuzzy matrix obtained by writing each of the column matrices as rows are given by:

$$\begin{bmatrix}
1e^{i\frac{4\pi}{3}} & 0.5e^{i\frac{2\pi}{3}} & 0.75e^{i\frac{2\pi}{3}} & 0.25e^{i\frac{2\pi}{3}} & 0.05e^{i\frac{2\pi}{3}} & 0.75e^{i\frac{2\pi}{3}} & 1e^{i\frac{4\pi}{3}} & 0.25e^{i\frac{2\pi}{3}} & 0e^{i0} & 1e^{i\frac{4\pi}{3}} \\
1e^{i\frac{4\pi}{3}} & 0.75e^{i\frac{2\pi}{3}} & 0.5e^{i\frac{2\pi}{3}} & 0.25e^{i\frac{2\pi}{3}} & 0.75e^{i\frac{2\pi}{3}} & 0.05e^{i\frac{2\pi}{3}} & 0.9e^{i\frac{2\pi}{3}} & 1e^{i\frac{4\pi}{3}} & 0e^{i0} & 0e^{i0} \\
0.65e^{i\frac{2\pi}{3}} & 0e^{i0} & 0e^{i0} & 0.8e^{i\frac{2\pi}{3}} & 0e^{i0} & 0e^{i0} & 0.2e^{i\frac{2\pi}{3}} & 1e^{i\frac{4\pi}{3}} & 0.5e^{i\frac{2\pi}{3}} & 0.55e^{i\frac{2\pi}{3}} \\
0.25e^{i\frac{2\pi}{3}} & 0.85e^{i\frac{2\pi}{3}} & 0.15e^{i\frac{2\pi}{3}} & 0.1e^{i\frac{2\pi}{3}} & 0e^{i0} & 0e^{i0} & 1e^{i\frac{4\pi}{3}} & 0.2e^{i\frac{2\pi}{3}} & 1e^{i\frac{4\pi}{3}} & 0.55e^{i\frac{2\pi}{3}} \\
1e^{i\frac{4\pi}{3}} & 0.3e^{i\frac{2\pi}{3}} & 0.85e^{i\frac{2\pi}{3}} & 0e^{i0} & 0e^{i0} & 1e^{i\frac{4\pi}{3}} & 0.1e^{i\frac{2\pi}{3}} & 1e^{i\frac{4\pi}{3}} & 0.15e^{i\frac{2\pi}{3}} & 1e^{i\frac{4\pi}{3}} \\
0e^{i0} & 0.6e^{i\frac{2\pi}{3}} & 1e^{i\frac{4\pi}{3}} & 0e^{i0} & 0e^{i0} & 0e^{i0} & 0.2e^{i\frac{2\pi}{3}} & 0e^{i0} & 0e^{i0} & 0e^{i0} \\
0.5e^{i\frac{2\pi}{3}} & 0.5e^{i\frac{2\pi}{3}} & 0.5e^{i\frac{2\pi}{3}} & 0.1e^{i\frac{2\pi}{3}} & 1e^{i\frac{4\pi}{3}} & 0.05e^{i\frac{2\pi}{3}} & 0.9e^{i\frac{2\pi}{3}} & 1e^{i\frac{4\pi}{3}} & 0e^{i0} & 0.25e^{i\frac{2\pi}{3}} \\
0.8e^{i\frac{2\pi}{3}} & 1e^{i\frac{4\pi}{3}} & 0.5e^{i\frac{2\pi}{3}} & 1e^{i\frac{4\pi}{3}} & 1e^{i\frac{4\pi}{3}} & 0.3e^{i\frac{2\pi}{3}} & 1e^{i\frac{4\pi}{3}} & 0e^{i0} & 1e^{i\frac{4\pi}{3}} & 1e^{i\frac{4\pi}{3}} \\
0.75e^{i\frac{2\pi}{3}} & 0.8e^{i\frac{2\pi}{3}} & 0e^{i0} & 0e^{i0} & 0e^{i0} & 1e^{i\frac{4\pi}{3}} & 0.7e^{i\frac{2\pi}{3}} & 0e^{i0} & 0e^{i0} & 0.75e^{i\frac{2\pi}{3}} \\
1e^{i\frac{4\pi}{3}} & 1e^{i\frac{4\pi}{3}} & 0.25e^{i\frac{2\pi}{3}} & 1e^{i\frac{4\pi}{3}} & 0.9e^{i\frac{2\pi}{3}} & 0.5e^{i\frac{2\pi}{3}} & 1e^{i\frac{4\pi}{3}} & 0.05e^{i\frac{2\pi}{3}} & 0e^{i0} & 1e^{i\frac{4\pi}{3}}
\end{bmatrix}_{10\times 10}$$

Calculate the relative values $f\left(\frac{P_i}{s_j}\right)$ using the equation: $f\left(\frac{P_i}{s_j}\right)=\frac{\left|\chi_{s_i}(P_j)\right| - \left|\chi_{P_i}(s_j)\right|}{\max\left\{\left|\chi_{s_i}(P_j)\right|, \left|\chi_{P_i}(s_j)\right|\right\}}$:

$$r_{11} = f\left(\frac{P_1}{s_1}\right) = \frac{\left|\chi_{s_1}(P_1)\right| - \left|\chi_{P_1}(s_1)\right|}{\max\left\{\left|\chi_{s_1}(P_1)\right|, \left|\chi_{P_1}(s_1)\right|\right\}} = \frac{1-1}{\max\{1,1\}} = \frac{0}{1} = 0$$

$$r_{12} = f\left(\frac{P_1}{s_2}\right) = \frac{\left|\chi_{s_1}(P_1)\right| - \left|\chi_{P_1}(s_1)\right|}{\max\left\{\left|\chi_{s_1}(P_1)\right|, \left|\chi_{P_1}(s_1)\right|\right\}} = \frac{1-0.5}{\max\{1,0.5\}} = \frac{0.5}{1} = 0.5$$

Similarly, we calculate all the remaining relative values and can form the (10×10) comparison matrix $\left[r_{ij}\right]_{10\times10}$.

So, the comparison matrix, $R = \left[r_{ij}\right]_{10\times10}$:

$$= \begin{array}{c} \\ P_1 \\ P_2 \\ P_3 \\ P_4 \\ P_5 \\ P_6 \\ P_7 \\ P_8 \\ P_9 \\ P_{10} \end{array} \begin{array}{c} \begin{array}{cccccccccc} s_1 & s_2 & s_3 & s_4 & s_5 & s_6 & s_7 & s_8 & s_9 & s_{10} \end{array} \\ \begin{bmatrix} 0.000 & 0.500 & -0.33 & 0.000 & 0.950 & -1.00 & -0.50 & 0.687 & 1.000 & 0.000 \\ -0.50 & 0.000 & -1.00 & 0.706 & -0.60 & 0.912 & -0.44 & 0.000 & 1.000 & 1.000 \\ 0.133 & 1.000 & 0.000 & -0.813 & 1.000 & 1.000 & 0.600 & -0.50 & 0.000 & 1.000 \\ -0.61 & -0.70 & 0.813 & 0.000 & 0.000 & 0.000 & -0.90 & 0.800 & -1.00 & 0.450 \\ -0.95 & 0.600 & -1.00 & 0.000 & 0.000 & -1.00 & -0.90 & 0.000 & -0.15 & -0.10 \\ 1.000 & -0.92 & 1.00 & 0.000 & 1.000 & 0.000 & -0.75 & 1.000 & 1.000 & 1.000 \\ 0.500 & 0.444 & -0.60 & 0.900 & -0.90 & 0.750 & 0.000 & 0.000 & 1.000 & 0.750 \\ -0.688 & 0.000 & 0.500 & -0.80 & 0.000 & -1.00 & 0.000 & 0.000 & 1.000 & 0.950 \\ -1.00 & -1.00 & 1.000 & 1.000 & 1.000 & -1.00 & -1.00 & 1.000 & 0.000 & 1.000 \\ 0.000 & -1.00 & 0.545 & -0.45 & 0.100 & -1.00 & -0.75 & 0.950 & 1.000 & 0.000 \end{bmatrix}_{10\times10} \end{array}$$

From the ranking of the problem, by taking maximum in each row, we can find the major symptoms occurring in each patient. Using these findings, we can conclude what type of lung disease is occurring in that particular patient. For the patient P_1 is s_5, s_9, i.e., chest pain and weight loss. So, the possible type of lung diseases is lung cancer. For the patient P_2 is s_4, s_6, s_9, s_{10}, i.e., coughing (especially at night), persistent cough with mucus, weight loss, and rapid breathing and low oxygen levels. So, the possible type of lung diseases is TB, lung cancer, and ARDS. For the patient P_3 are s_2, s_5, s_6, s_{10}, i.e., fatigue, chest pain, persistent cough with mucus, and rapid breathing and low oxygen levels. So, the corresponding type of lung diseases are ARDS and pulmonary fibrosis. For the patient P_4 is s_3, s_8, i.e., wheezing and fever. So, the corresponding type of lung diseases are asthma and bronchitis. For the patient P_5 is s_2, i.e., fatigue. So, the corresponding type of lung diseases are interstitial lung disease, ARDS, and pulmonary fibrosis. For the patient P_6 is s_1, s_5, s_8, s_9, i.e., shortness of breath, chest pain, fever, weight loss, and rapid breathing and low oxygen levels. So, the corresponding type of lung diseases are asthma, pneumonia, pulmonary fibrosis, lung cancer, and ARDS. For the patient P_7 is s_4, s_9, i.e., coughing (especially at night) and weight loss. So, the corresponding types of gestational diseases is lung cancer. For the patient

P_8 is s_6, s_9, i.e., persistent cough with mucus and weight loss. So, the corresponding type of lung disease is lung cancer. For the patient P_9 is $s_3, s_4, s_5, s_8, s_{10}$, i.e., coughing (especially at night), chest pain, fever, weight loss, and rapid breathing and low oxygen levels. So, the corresponding chance of lung diseases are ARDS, lung cancer, and TB. For the patient P_{10} is s_9, i.e., weight loss. So, the corresponding types of lung disease is lung cancer and TB.

6 Conclusion

This study dealt with the symptom-based analysis of various lung diseases with the use of complex fuzzy matrices. This approach effectively addressed both ambiguity and uncertainty often associated with test results, symptom descriptions, patient histories, and other medical data. The proposed method offered a more systematic and comprehensive understanding of respiratory health problems by finding out the key symptom that occurs. The method demonstrates how complex fuzzy matrices can serve as powerful decision-support tools for medical diagnosis by increasing accuracy and reliability when dealing with imprecise clinical data. Beyond respiratory health, this concept has potential for broad applications in a variety of medical fields where uncertainty is common.

Bibliography

[1] Zadeh LA. 1965. Fuzzy sets. . Inf Control. 8:338–353.
[2] Ramot D et.al. 2002. Complex fuzzy sets. IEEE Trans Fuzzy Syst. 10.
[3] Kim KH, Roush FW. 1980. Generalised fuzzy matrices. Fuzzy Sets Syst. 4:293–315.
[4] Zhao ZQ, Ma SQ. 2015. Complex fuzzy matrices and its convergence problem research. Fuzzy Syst Oper Res. 367:157–162.
[5] Atanassov KT. 1986. Intuitionistic fuzzy sets. Fuzzy Sets Syst. 20(1):87–96.
[6] Geetha T, Usha A. 2017. Circulant fuzzy matrix in human body diseases. Int J Pure Appl Math. 117:99–106.
[7] Thangaraj B, Mallika M. 2017. Application of fuzzy matrices in medical diagnosis. Int J Fuzzy Math Arch. 14.
[8] Thomson MG. 1997. Convergence of powers of fuzzy matrix. J Math Anal Appl. 57:476–480.
[9] Ejegwa PA, Agba JA, Ugwuh F. 2024. Weighted intuitionistic fuzzy distance metrics in solving cases of pattern recognition and disease diagnosis. . Res Adv Math Data Sci. 1(1):35–52.
[10] Ejegwa PA, Kausar N, Agba JA, Ugwuh F, Özbilge E, Özbilge E. 2024. Determination of medical emergency via new intuitionistic fuzzy correlation measures based on Spearman's correlation coefficient. . AIMS Math. 9(6):15639–15670.
[11] Luo M, Zhao R, Inyang UG, Eyoh IJ, Ejegwa PA. 2023. Intuitionistic fuzzy approach for predicting maternal outcomes. In: Fuzzy Optimization, Decision Making, and Operations Research. Springer; pp. 1–18.

Abhilash Kangsha Banik, Alakananda Konwar, Palash Dutta*

A Novel Multi-criteria Decision-Making Method Based on Hyperbolic Fuzzy Sets

Abstract: The presence of vagueness and ambiguity has been an unavoidable characteristic in complex decision-making problems. The Zadeh's fuzzy set theory, along with the upgraded concepts such as the Atanassov's intuitionistic fuzzy sets, the Yager's Pythagorean fuzzy sets, and the Yager's q-rung orthopair fuzzy sets, was introduced to resolve decision-making problems. But, all these notions could not handle extreme judgments and values of form (1, 1). In this study, we introduce a novel distance measure (DM) based on hyperbolic fuzzy sets, which can handle extreme values of form (1, 1). Further, a new multi-criteria decision-making (MCDM) method is introduced using the DM. The method is used in ranking the alternatives established on multiple criteria in uncertain environments. This study also consists of a comparative analysis, which shows that the proposed approach based on hyperbolic fuzzy sets serves as a better alternative to established fuzzy-based notions. This study gives a better framework for addressing complex decision-making problems under uncertain environments.

Keywords: decision-making, uncertainty, fuzzy set theory, hyperbolic fuzzy sets

1 Introduction

In real-world circumstances, uncertainty usually arises from insufficient understanding, unclear thinking, an absence of practical information, and data shortcomings. To solve these issues, Zadeh [1] proposed the fuzzy set theory (FST) to portray human reasoning in decision-making.

FST is compatible with situations where binary logic fails to adequately represent the inherent complexity. The FST model uses uncertain information in various decision-making issues through a membership function that ranges from 0 to 1. Still, in numerous real-world situations, judgments are extended to membership degree (MD) and non-membership degree (NMD). To combine these aspects, Atanassov [2] developed intuitionistic fuzzy set (IFS) theory to address uncertainty through three parameters: MD, NMD, and hesitation degree, whose sum equals 1. IFS offers a broader framework for

***Corresponding author: Palash Dutta,** Department of Mathematics, Dibrugarh University, Dibrugarh 786004, Assam, India, email: palashdutta@dibru.ac.in
Abhilash Kangsha Banik, Department of Mathematics, Dibrugarh University, Dibrugarh 786004, Assam, India, e-mail: abhibanik99@gmail.com
Alakananda Konwar, Department of Mathematics, Dibrugarh University, Dibrugarh 786004, Assam, India, e-mail: alakananda.konwar@gmail.com

 | https://doi.org/10.1515/9783112222249-007

framing uncertainty than traditional FST. Yet, its restrictive nature limits its ability to illustrate complex, unclear circumstances where the totality of MD and NMD exceeds 1. Yager [3] in 2013 developed the Pythagorean fuzzy set (PFS) theory to handle such a limitation. PFS theory is categorised as the summation of squares of the MD and NMD not exceeding 1, thereby offering a higher flexibility and portraying complex uncertainty in diverse real-life applications in comparison to IFST. Subsequently, Yager [4] presented q-rung orthopair fuzzy set (q-ROFS) theory as a generalised version of the fuzzy set concept, such that the totality of the qth powers of the MD and NMD is restricted to one, and further Senapati and Yager [5] developed Fermatean fuzzy set (FFS) theory as a distinct case of q-ROFS with $q=3$. Higher values of q in q-ROFS deliver broader ranges of MDs, accordingly encouraging higher flexibility of fuzzy models that successfully promote effective portrayal of uncertainty in diverse real-world applications such as decision-making and pattern recognition. Despite the advancements of IFS, PFS, FFS, and q-ROFS, these fuzzy sets sometimes fail to address uncertain situations where the assignment of MD one and NMD (or vice versa) is intuitive. In such cases, the capability towards independently determine MD and NMD turns out to be an effective way in precise decision-making. For example, the treatments of the COVID-19 disease exhibited effective short-term outcomes. Yet, the long-term effects of the COVID-19 disease, such as fatigue and joint pain, sustained to remain ambiguous. Although vaccines helped control the pandemic, their lasting health effects were not fully known. In such a scenario, the short-term effectiveness necessitates an optimistic degree of one and the uncertain long-term effects require a pessimistic degree between 0 and 1. However, the established fuzzy models cannot effectively handle this kind of scenario. To manage such paradoxical and complex real-life situations, Dutta and Borah [6] introduced the hyperbolic fuzzy set (HyFS). An HyFS is structured with a degree of optimism and a degree of pessimism, which are independent of each other and both within the range [0, 1]. The advanced framework of HyFSs permits a more reliable and accurate representation of uncertainty and is subsequently advantageous over other fuzzy set extensions in contributing higher flexibility to decision-makers towards expressing uncertainty in complex and paradoxical decision-making issues. Figure 1 demonstrates the graphical illustration of MD and NMD for the IFS, the PFS, the FFS, the q-ROFS, and the HyFS, whereas Table 1 delivers a structural comparison of these fuzzy frameworks. Decision-making in sustainable supplier selection (SSS) is a complex task, comprising multiple diverse criteria from environmental, economic, and social areas.

SSS is extremely significant for economic and social development, as well as vital for the conservation of the environment. From economic aspects, the selection of a sustainable supplier encourages reduction in operational costs, reliable quality, and punctual delivery of products. Further, sustainable suppliers support to acceptance of ecologically safe practices like recycling materials, and the utilisation of renewable energy, thereby lowering in carbon footprint, thus resulting in the mitigation of climate change and conservation of resources. Socially sustainable suppliers support human rights, offer justifiable wages, and secure workplaces for employees. Thus, the environmental, economic, and social factors are ideal for meticulous evaluation of the optimal sustainable supplier. An effective decision-making process encourages an efficient supplier selection based on diverse sustainable factors. The main goal of this study is to develop a multi-criteria decision-making (MCDM) approach using the HyFS framework to determine the optimal sustainable supplier. For this purpose, we integrate the CRiteria Importance Through Intercriteria Correlation (CRITIC) and Technique for Order Preference by Similarity to Ideal Solution (TOPSIS) MCDM approach, supported by a novel distance measure (DM) based on the HyF framework. The novel MCDM model is efficiently utilised to address uncertain information in several SSS decision-making problems. The HyF-CRITIC approach determines the uncertain weight information of numerous complex environmental, technical, social, and economic factors, and the TOPSIS approach effectively selects and ranks the suppliers for SSS. Furthermore, the novel HyF DM employed in the proposed HyF-CRITIC-TOPSIS method disables the difficulties of the established DM and improves the decision-making process through delivering more reliable and consistent results in SSS.

The primary aim of decision-making in SSS involves identifying and choosing the optimal suppliers, keeping in concern the environmental factors like prevention of pollution, utilisation of resources, management of wastes, as well as adopting socio-economic factors like ethical labour practices, compliance, adoption of health and safety regulations, ideal balance in price and quality, dependability on product distribution, technological advancements, and commercial steadiness. SSS balances between quality of products and cost-effectiveness while promoting sustainability through boosting long-term competitiveness, reducing environmental effects, protecting moral standards, and benefiting community well-being. Further, the scope of SSS includes an MCDM process that assesses suppliers according to various sustainability criteria. MCDM techniques like TOPSIS, grey relational analysis (GRA), VIseKriterijumska Optimizacija I Kompromisno Resenje (VIKOR), and complex proportional assessment (COPRAS) are used in ranking and choosing ideal suppliers in numerous real-life decision-making applications. The SSS approach promotes sustainability and ethical supply chain practices across various industries such as health care, agriculture, manufacturing, and oil and gas.

The residual section of this study is designed as follows:

Section 2 explores the established literature related to this study. Section 3 comprises the preliminaries on fuzzy sets and their extensions. Section 4 introduces an innovative HyF DM and its fundamental properties in the HyF framework. Additionally, some established DMs on the IFS, the PFS, the FFS and the q-ROFS are delivered, and a superiority check of the defined DM with these established measures is conducted in this section. In Section 5, we present a novel HyF-CRITIC-TOPSIS MCDM method supported by the proposed HyF DM from the previous section. In Section 6, we utilise the novel HyF-CRITIC-TOPSIS MCDM in a case study on SSS and perform a comparative analysis of our proposed methodology with some established MCDM methods to present its reliability and consistency. As a final point, in Section 7, the conclusion and upcoming studies of this chapter are delivered.

Table 1: Graphical presentation of the IFS, the PFS, the FFS, the q-ROFS, and the HyFS.

IFS	PFS	FFS	q-ROFS	HyFS
$0 \le \alpha + \beta \le 1$	$0 \le \alpha^2 + \beta^2 \le 1$	$0 \le \alpha^3 + \beta^3 \le 1$	$0 \le \alpha^q + \beta^q \le 1$	$0 \le \alpha \le 1;\ 0 \le \beta \le 1$
$\pi = 1 - (\alpha + \beta)$	$\pi = \sqrt{1 - (\alpha^2 + \beta^2)}$	$\pi = \sqrt[3]{1 - (\alpha^3 + \beta^3)}$	$\pi = \sqrt[q]{1 - (\alpha^q + \beta^q)}$	No hesitancy
$\alpha + \beta + \pi = 1$	$\alpha^2 + \beta^2 + \pi^2 = 1$	$\alpha^3 + \beta^3 + \pi^3 = 1$	$\alpha^q + \beta^q + \pi^q = 1$	$\alpha\beta = 1$

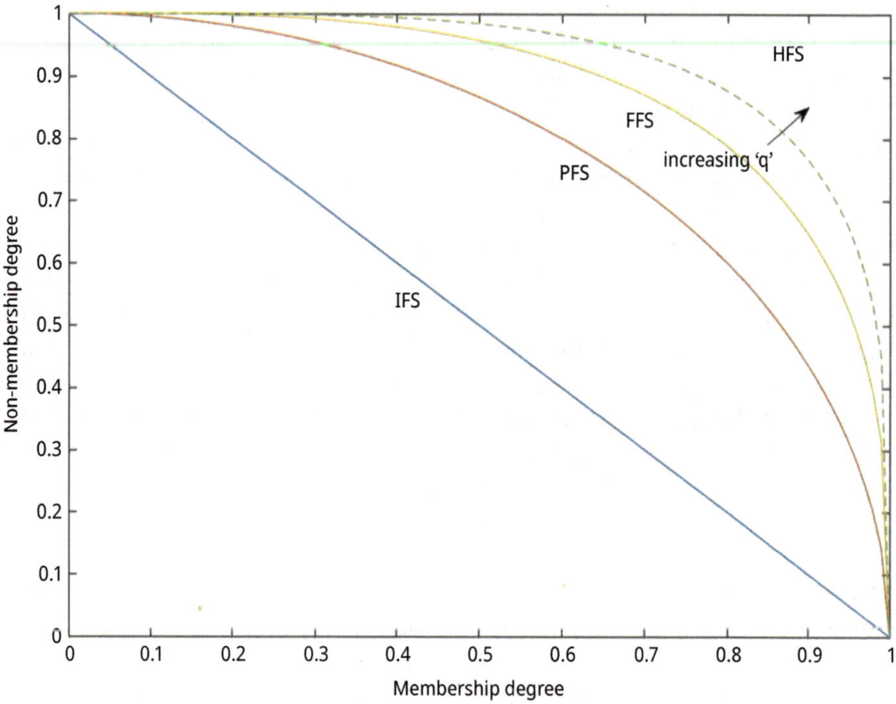

Figure 1: MD and NMD of IFSs, PFSs, FFSs, and HyFSs.

2 Literature Review

This study presents a concise literature review on DMs, CRITIC, TOPSIS MCDM method and some of the notable established decision-making procedures on the SSS problem.

In complex decision-making, DMs play a crucial role in enhancing decision-making processes. DMs compute the dissimilarity between two fuzzy sets. The literature includes notable DMs in IFS by Mahanta and Panda [7] for applications in pattern recognition and medical diagnostics, whereas Ngan et al. [8] developed the H-max DM in IFS for decision-making problems. The introduction of PFS by Yager [3], inspired by Xiao and Ding [9], introduced normalised Hamming and Euclidean DMs based on PFS for medical diagnosis, and Mahanta and Panda [10] introduced DM in PFS for applications in pattern identification and medical diagnosis. Besides, grounded on t-conorm, Ganie [11] created four DMs on FFS and established their application in pattern analysis; based on the triangular divergence and the Hellinger distance, Deng and Wang [12] proposed two novel DMs on FFS, respectively, along with applications in pattern recognition and medical diagnosis. Furthermore, Du [13] proposed the classical Hamming DM, while Hussain et al. [14] proposed Hausdorff metric-based DMs under q-ROFS, Kamacı and Petchimuthu [15] developed Soergel DMs for q-ROFS along with their applications, and Wang et al. [16] offered a Lance DM on q-ROFS with applications in decision-making. These advanced DMs are highly effective in decision-making for SSS issues, where evaluating uncertain information between suppliers and the criteria is a critical task.

The CRITIC approach, introduced by Diakoulaki et al. [17], is an objective weighting model of the attributes that utilises the standard deviation and the correlation coefficient to quantify the data of individual attributes to compute the attribute weights in diverse MCDM procedures [18]. Peng and Garg [19] evaluated content-centric networking cache placement strategy selection through merging the Combined Compromise Solution (CoCoSo) and CRITIC methods. Tripathi et al. [20] integrated CRITIC-Rank Sum (RS)-VIKOR approaches in IFSs for the evaluation of renewable energy. Ertemel et al. [21] developed Pythagorean fuzzy CRITIC-TOPSIS for the assessment of Smartphone addiction. Wang et al. [22] selected sustainable food suppliers using the CRITIC-Measurement of Alternatives and Ranking according to Compromise Solution (MARCOS) method in PFS. Mishra et al. [18] performed the selection of sustainable third-party reverse logistics providers by establishing the CRITIC-Evaluation based on the distance from average solution approach in FFSs. Yang et al. [23] developed the CRITIC-VIKOR method for q-ROFSs to address the Internet of Things challenges in sustainable supply chain management within the manufacturing sector.

Yoon and Hwang [24] initially developed the TOPSIS method for ranking alternatives in decision-making problems, including multiple and conflicting criteria. According to Dutta and Konwar [25], the TOPSIS method operates by determining the positive ideal solution with a highest value for the benefit and a least value for the cost criteria, while a negative ideal solution with a least value for the benefit and a highest value for the cost criteria, thereby computing the least and greatest distance of the alternatives from the positive and the negative ideal solution, respectively. Mukherjee et al. [26] examined supply chain risk prioritisation based on IFSs. Hajiaghaei-Keshteli et al. [27] utilised PFSs for the evaluation of green supplier selection in the food industry. Majd et al. [28] developed the FFS-based TOPSIS method with its application in the ranking of business intelligence-based strategies from a smart city perspective. Zulqarnain et al. [29] applied supplier selection in green supply chain management using correlation-based TOPSIS in a q-ROFS.

The main objectives of the study are given below:

- Introduction of a novel DM in the HyF framework
- A comparative analysis of the proposed HyF DM with the established DMs is made to test the reliability and dominance of the proposed DM
- Introduction of the HyF-CRITIC-TOPSIS approach supported by our proposed measure
- Application of the introduced HyF-CRITIC-TOPSIS approach for selection of sustainable suppliers in the HyFS environment
- Comparative analysis of the extended HyF-CRITIC-TOPSIS method with established MCDM methods to validate the efficiency and robustness of the suggested method

3 Preliminaries

An outline of the FS and its variants like the IFS, the PFS, the q-ROFS, the FFS, and the HyFS and a description of the fundamental operating rules of HyFS are presented in this section.

Definition 3.1 ([1]). An FS F on a finite universe of discourse (FUD) $\underset{.}{W}$, is characterised by a membership function (MF) of the element $\underset{.}{w}_i \in \underset{.}{W}$ toward F such that, $\alpha_F: \underset{.}{W} \to [0,1]$ and represented as $F = \{\langle \underset{.}{w}_i, \alpha_F(\underset{.}{w}_i): \underset{.}{w}_i \in \underset{.}{W}\rangle\}$.

Definition 3.2 ([2]). An IFS on an FUD $\underset{.}{W}$, is characterised by an MF $\alpha_{\check{I}}: \underset{.}{W} \to [0,1]$ and a non-membership function (NMF) $\beta_{\check{I}}: \underset{.}{W} \to [0,1]$ of the element $\underset{.}{w}_i \in \underset{.}{W}$ to the set $\check{I}$, such that $0 \le \alpha_{\check{I}}(\underset{.}{w}_i) + \beta_{\check{I}}(\underset{.}{w}_i) \le 1$. An IFS, $\check{I}$ is represented as $\check{I} = \{\langle \underset{.}{w}_i, \alpha_{\check{I}}(\underset{.}{w}_i), \beta_{\check{I}}(\underset{.}{w}_i): \underset{.}{w}_i \in \underset{.}{W}\rangle\}$ and the hesitancy degree (HD) of the element $\underset{.}{w}_i \in \underset{.}{W}$ toward the set $\check{I}$ is indicated as $\pi_I(\underset{.}{w}_i) = 1 - \alpha_{\check{I}}(\underset{.}{w}_i) - \beta_{\check{I}}(\underset{.}{w}_i)$.

Definition 3.3 ([3]). A PFS, on an FUD $\underset{.}{W}$, is characterised by an MF $\alpha_p: \underset{.}{W} \to [0,1]$ and an NMF $\beta_{\dot{p}}: \underset{.}{W} \to [0,1]$ of the element $x \in \underset{.}{W}$ to the set $\dot{P}$, such that $0 \le \alpha_{\dot{p}}^2(\underset{.}{w}_i) + \beta_{\dot{p}}^2(\underset{.}{w}_i) \le 1$. A PFS, $\dot{P}$ is represented as $\dot{P} = \{\langle \underset{.}{w}_i, \alpha_{\dot{P}}(\underset{.}{w}_i), \beta_{\dot{P}}(\underset{.}{w}_i): \underset{.}{w}_i \in \underset{.}{W}\rangle\}$ and the HD of the element $\underset{.}{w}_i \in \underset{.}{W}$ toward the set $\dot{P}$ is indicated as $\pi_I(\underset{.}{w}_i) = 1 - \alpha_{\dot{p}}^2(\underset{.}{w}_i) - \beta_{\dot{P}}^2(\underset{.}{w}_i)$.

Definition 3.4 ([4]). A q-ROFS, on an FUD $\underset{.}{W}$, is characterised by an MF $\alpha_Q: \underset{.}{W} \to [0,1]$ and an NMF $\beta_Q: \underset{.}{W} \to [0,1]$ of the element $x \in \underset{.}{W}$ to the set Q, such that $0 \le \alpha_Q^q(\underset{.}{w}_i) + \beta_Q^q(\underset{.}{w}_i) \le 1$. A q-ROFS, Q is represented as $Q = \{\langle \underset{.}{w}_i, \alpha_Q(\underset{.}{w}_i), \beta_Q(\underset{.}{w}_i): \underset{.}{w}_i \in \underset{.}{W}\rangle\}$ and the degree of hesitancy of the element $\underset{.}{w}_i \in \underset{.}{W}$ to the set Q is indicated as $\pi_Q(\underset{.}{w}_i) = 1 - \alpha_Q^q(\underset{.}{w}_i) - \beta_Q^q(\underset{.}{w}_i)$.

Definition 3.5 ([5]). On an FUD $\underset{.}{W}$, a Fermatean fuzzy set (FFS) is characterised by an MF $\alpha_{\dot{F}}: \underset{.}{W} \rightarrow [0,1]$ and an NMF $\beta_{\dot{F}}: \underset{.}{W} \rightarrow [0,1]$ of the element $x \in \underset{.}{W}$ to the set $\dot{F}$, such that $0 \leq \alpha_{\dot{F}}^3(\underset{.}{w}_i) + \beta_{\dot{F}}^3(\underset{.}{w}_i) \leq 1$. An FFS, F is represented as $\dot{F} = \{\langle \underset{.}{w}_i, \alpha_{\dot{F}}(\underset{.}{w}_i), \beta_{\dot{F}}(\underset{.}{w}_i): \underset{.}{w}_i \in \underset{.}{W}\rangle\}$ and the degree of hesitancy of the element $\underset{.}{w}_i \in \underset{.}{W}$ toward the set F is indicated as $\pi_{\dot{F}}(\underset{.}{w}_i) = 1 - \alpha_{\dot{F}}^3(\underset{.}{w}_i) - \beta_{\dot{F}}^3(\underset{.}{w}_i)$.

Definition 3.6 ([6]). A hyperbolic fuzzy set (HyFS) on an FUD $\underset{.}{W}$, is characterised by an optimistic degree $\alpha_H: \underset{.}{W} \rightarrow [0,1]$ and a pessimistic degree: $\underset{.}{W} \rightarrow [0,1]$ of $\underset{.}{w}_i \in \underset{.}{W}$ such that the optimistic and pessimistic degrees are independent of each other, with the property $0 \leq (\alpha_H(\underset{.}{w}_i))(\beta_H(\underset{.}{w}_i)) \leq 1$ for all $\underset{.}{w}_i \in \underset{.}{W}$. We signify a hyperbolic fuzzy set on a finite universe of discourse, $\underset{.}{W}$ as $HyFS(\underset{.}{W})$

Definition 3.7 ([6]). The operational laws on HyFS

For any three $H = \{(\underset{.}{w}_i, \alpha_H(\underset{.}{w}_i), \beta_H(\underset{.}{w}_i),): \underset{.}{w}_i \in \underset{.}{W}\}, H_1 = \{(\underset{.}{w}_i, \alpha_{H_1}(\underset{.}{w}_i), \beta_{H_1}(\underset{.}{w}_i)): \underset{.}{w}_i \in \underset{.}{W}\}$, $H_2 = \{(\underset{.}{w}_i, \alpha_{H_2}(\underset{.}{w}_i), \beta_{H_2}(\underset{.}{w}_i)): \underset{.}{w}_i \in \underset{.}{W}\} \in HyFS(\underset{.}{W})$, the fundamental set operations are defined as follows:

i. $H^c = \{(x, 1 - \alpha_H(\underset{.}{w}_i), 1 - \beta_H(\underset{.}{w}_i)): \underset{.}{w}_i \in \underset{.}{W}\}$
ii. $(H^c)^c = H^c$
iii. $H_1 \subseteq H_2$ if and only if $\alpha_{H_1}(\underset{.}{w}_i) \leq \alpha_{H_2}(\underset{.}{w}_i)$ and $\beta_{H_1}(\underset{.}{w}_i) \geq \beta_{H_2}(\underset{.}{w}_i)$
iv. $H_1 = H_2$ if and only if $\alpha_{H_1}(\underset{.}{w}_i) \subseteq \alpha_{H_2}(\underset{.}{w}_i)$ and $\beta_{H_1}(\underset{.}{w}_i) \supseteq \beta_{H_2}(\underset{.}{w}_i)$
v. $H_1 \wedge H_2 = \{(\underset{.}{w}_i, \min(\alpha_{H_1}(x_i), \alpha_{H_2}(\underset{.}{w}_i))), \max((\beta_{H_1}(\underset{.}{w}_i), \beta_{H_2}(\underset{.}{w}_i))): \underset{.}{w}_i \in \underset{.}{W}\}$
vi. $H_1 \vee H_2 = \{(\underset{.}{w}_i, \max(\alpha_{H_1}(\underset{.}{w}_i), \alpha_{H_2}(\underset{.}{w}_i)), \min(\beta_{H_1}(\underset{.}{w}_i), \beta_{H_2}(\underset{.}{w}_i))): \underset{.}{w}_i \in \underset{.}{W}\}$

Definition 3.8 ([6]). For $H \in HyFS(\underset{.}{W})$, the score $S(H)$ and the accuracy function $A(H)$ is defined, respectively, as

$$S(H) = 2\alpha_H(\underset{.}{w}_i) - \alpha_H(\underset{.}{w}_i)\beta_H(\underset{.}{w}_i), \text{where } S(H) \in [0,2]$$

$$A(H) = 2\alpha_H(\underset{.}{w}_i) - \alpha_H(\underset{.}{w}_i)\beta_H(\underset{.}{w}_i), \text{where } A(H) \in [0,2]$$

4 A Novel HyF DM

DM shows a substantial role in handling the degree of alteration between uncertain information in diverse decision-making environments. This section comprises the previously established DMs, as well as creates an effort to put forward a novel DM based on HyFS that satisfies the basic properties and the related theorems on a distance metric. Further, this proposed DM is compared with the earlier DMs to present its dominance, reliability and to conquer the limitations of the established DMs.

4.1 Established DMs

In the literature, researchers across the world have developed various DMs to enumerate unclear information. In this sub-section, we discuss some previously established DMs based on the IFS, the PFS, the FFS and the q-ROFS from the literature. Consider, $H_1 = \{(w_i, \alpha_{H_1}(w_i), \beta_{H_1}(w_i)) : w_i \in W\}, H_2 = \{(w_i, \alpha_{H_2}(w_i), \beta_{H_2}(w_i)) : w_i \in W\} \in HyFS(W)$, where $W = \{w_i : i = 1, 2, 3 \ldots, n\}$. Evidently, H_1 *and* H_2 also behave as IFSs, PFSs, FFSs, and q-ROFSs. Followed by, Table 2 shows the established DMs for H_1 and H_2.

4.2 Proposed Novel DM on HyFSs

In this subsection, a novel DM on HyFSs is constructed. Following, the fundamental properties and propositions for the presented DM are also discussed in this subsection.

Definition 4.1. Given an FUD $W = \{w_i : i = 1, 2, 3 \ldots, n\}$, let $H_1 = \{(w_i, \alpha_{H_1}(w_i), \beta_{H_1}(w_i)) : w_i \in W\}, H_2 = \{(w_i, \alpha_{H_2}(w_i), \beta_{H_2}(w_i)) : w_i \in W\}$, be two HyFSs, then we define the novel DM between H_1 and H_2 as

$$\begin{aligned} D_{\text{HyFKD}}(H_1, H_2) = \frac{1}{2n} \sum_{i=1}^{n} \Bigg[& \left\{ \frac{|\alpha_{H_1}(w_i) - |\alpha_{H_2}(w_i)| + |\beta_{H_1}(w_i) - \beta_{H_2}(w_i)|}{2} \right\} \\ & \times \{1 + |\alpha_{H_1}(w_i)\beta_{H_2}(w_i) - \beta_{H_1}(w_i)\alpha_{H_2}(w_i)| \\ & + |\max(\alpha_{H_1}(w_i), \beta_{H_2}(w_i)) - \max(\beta_{H_1}(w_i)\alpha_{H_2}(w_i))| \\ & + |\min(\alpha_{H_1}(w_i), \beta_{H_2}(w_i)) - \min(\beta_{H_1}(w_i)\alpha_{H_2}(w_i))|\}^{\frac{1}{2}} \Bigg] \end{aligned}$$

and n is the number of pairs of HyFSs.

Definition 4.2 ([6]). Let $H_1 = \{(w_i, \alpha_{H_1}(w_i), \beta_{H_1}(w_i)) : w_i \in W\}, H_2 = \{(x_i, \alpha_{H_2}(x_i), \beta_{H_2}(x_i)) : w_i \in W\}$ and $H_3 = \{(w_i, \alpha_{H_3}(w_i), \beta_{H_3}(w_i)) : w_i \in W\}$ be three HyFSs on an FUD $W = \{w_i : i = 1, 2, 3 \ldots, n\}$. Then the novel HyF DM D_{HyFKD}, satisfies the following properties:

P_1: $0 \le D_{\text{HyFKD}}(H_1, H_2) \le 1$
P_2: $D_{\text{HyFKD}} = 0$ iff $H_1 = H_2$
P_3: $D_{\text{HyFKD}}(H_1, H_2) = D_{\text{HyFKD}}(H_2, H_1)$
P_4: For $H_1 \subseteq H_2 \subseteq H_3$, then $D_{\text{HyFKD}}(H_1, H_3) \ge D_{\text{HyFKD}}(H_1, H_2)$ and $D_{\text{HyFKD}}(H_1, H_3) \ge D_{\text{HyFKD}}(H_2, H_3)$

Proof (P_1): Since, H_1, H_2, and H_3 are HyFSs, then clearly $0 \le D_{\text{HyFKD}}(H_1, H_2) \le 1$
Proof (P_2): Let $D_{\text{HyFKD}}(H_1, H_2) = 0$, then for all $i = 1, 2, \ldots, n$

Table 2: Established DMs.

Authors	DM
Ngan et al. [8]	$dis^1_{ING}(H_1,H_2)=\frac{1}{3n}\sum_{i=1}^{n}\left(\lvert\alpha_{H_1}(w_i)-\alpha_{H_2}(w_i)\rvert+\lvert\beta_{H_1}(w_i)-\beta_{H_2}(w_i)\rvert+\lvert\max\left(\alpha_{H_1}(w_i),\beta_{H_2}(w_i)\right)-\max\left(\alpha_{H_1}(w_i),\beta_{H_2}(w_i)\right)\rvert\right)$ $dis^2_{ING}(H_1,H_2)=\frac{1}{n}\sum_{i=1}^{n}\left(\frac{\lvert\alpha_{H_1}(w_i)-\alpha_{H_2}(w_i)\rvert+\lvert\beta_{H_1}(w_i)-\beta_{H_2}(w_i)\rvert}{4}+\frac{\lvert\max\left(\alpha_{H_1}(w_i)-\beta_{H_2}(w_i)\right)-\max\left(\alpha_{H_1}(w_i)-\beta_{H_2}(w_i)\right)\rvert}{2}\right)$
Mahanta and Panda [7]	$dis_{IMP}(H_1,H_2)=\frac{1}{n}\sum_{i=1}^{n}\left(\frac{\lvert\alpha_{H_1}(w_i)-\alpha_{H_2}(w_i)\rvert+\lvert\beta_{H_1}(w_i)-\beta_{H_2}(w_i)\rvert}{\alpha_{H_1}(w_i)+\alpha_{H_2}(w_i)+\beta_{H_1}(w_i)+\beta_{H_2}(w_i)}\right)$
Xiao and Ding [9]	$dis^1_{PXD}(H_1,H_2)=\sqrt{\frac{1}{2n}\sum_{i=1}^{n}\left[\left(\alpha^2_{H_1}(w_i)-\alpha^2_{H_2}(w_i)\right)^2+\left(\beta^2_{H_1}(w_i)-\beta^2_{H_2}(w_i)\right)^2+\left(\pi^2_1(w_i)-\pi^2_2(w_i)\right)^2\right]}$ $dis^2_{PXD}(H_1,H_2)=\frac{1}{2n}\sum_{i=1}^{n}\left(\lvert(\alpha^2_{H_1}(w_i)-\alpha^2_{H_2}(w_i))\rvert+\lvert(\beta^2_{H_1}(w_i)-\beta^2_{H_2}(w_i))\rvert+\lvert(\pi^2_1(w_i)-\pi^2_2(w_i))\rvert\right)$
Mahanta and Panda [10]	$dis_{PMP}(H_1,H_2)=\frac{1}{n}\sum_{i=1}^{n}\frac{\lvert\alpha^2_{H_1}(w_i)-\alpha^2_{H_2}(w_i)\rvert+\lvert\beta^2_{H_1}(w_i)-\beta^2_{H_2}(w_i)\rvert}{\alpha^2_{H_1}(w_i)+\alpha^2_{H_2}(w_i)+\beta^2_{H_1}(w_i)+\beta^2_{H_2}(w_i)}$
Senapati and Yager [5]	$dis^1_{FSY}(H_1,H_2)=\sqrt{\frac{1}{2n}\sum_{i=1}^{n}\left[\left(\lvert\alpha^3_{H_1}(w_i)-\alpha^3_{H_2}(w_i)\right)^2+\left(\beta^3_{H_1}(w_i)-\beta^3_{H_2}(w_i)\right)^2+\left(\pi^3_1(w_i)-\pi^3_2(w_i)\right)^2\right]}$ $dis^2_{FSY}(H_1,H_2)=\frac{1}{2n}\sum_{i=1}^{n}\left(\left\lvert(\alpha^3_{H_1}(w_i)-\alpha^3_{H_2}(w_i))\right\rvert+\left\lvert\beta^3_{H_1}(w_i)-\beta^3_{H_2}(w_i)\right\rvert+\lvert(\pi^3_1(w_i)-\pi^3_2(w_i))\rvert\right)$

(continued)

Table 2 (continued)

Authors	DM
Ganie [11]	$dis_{FG}^1(H_1,H_2)=\frac{1}{n}\sum_{i=1}^{n}\left[\frac{\lvert\alpha^3_{H_1}(w_i)-\alpha^3_{H_2}(w_i)\rvert+\lvert\left(\beta^3_{H_1}(w_i)-\beta^3_{H_2}(w_i)\right)\rvert-2\lvert\left(\alpha^3_{H_1}(w_i)-\alpha^3_{H_2}(w_i)\right)\rvert\,\lvert\left(\beta^3_{H_1}(w_i)-\beta^3_{H_2}(w_i)\right)\rvert}{1-\lvert\left(\alpha^3_{H_1}(w_i)+\alpha^3_{H_2}(w_i)\right)\rvert\,\lvert\left(\beta^3_{H_1}(w_i)+\beta^3_{H_2}(w_i)\right)\rvert}\right]$ $dis_{FG}^2(H_1,H_2)=\frac{1}{n}\sum_{i=1}^{n}\left[\left(\lvert\alpha^3_{H_1}(w_i)-\alpha^3_{H_2}(w_i)\right)\rvert+\lvert\beta^3_{H_1}(w_i)-\beta^3_{H_2}(w_i)\rvert-\lvert\left(\alpha^3_{H_1}(w_i)-\alpha^3_{H_2}(w_i)\right)\rvert\,\lvert\beta^3_{H_1}(w_i)-\beta^3_{H_2}(w_i)\rvert\right]$ $dis_{FG}^3(H_1,H_2)=\frac{1}{n}\sum_{i=1}^{n}\left[\min\left(1,\lvert\alpha^3_{H_1}(w_i)-\alpha^3_{H_2}(w_i))\rvert+\lvert\beta^3_{H_1}(w_i)-\beta^3_{H_2}(w_i)\rvert\right)\right]$ $dis_{FG}^4(H_1,H_2)=\frac{1}{n}\sum_{i=1}^{n}\left[\frac{\lvert\alpha^3_{H_1}(w_i)-\alpha^3_{H_2}(w_i)\rvert+\lvert\beta^3_{H_1}(w_i)-\beta^3_{H_2}(w_i)\rvert}{1+\lvert\left(\alpha^3_{H_1}(w_i)-\alpha^3_{H_2}(w_i)\right)\rvert\,\lvert\beta^3_{H_1}(w_i)-\beta^3_{H_2}(w_i)\rvert}\right]$
Deng and Wang [12]	$dis_{FDW}^1(H_1,H_2)=\sqrt{\frac{1}{2n}\sum_{i=1}^{n}\left[\left(\sqrt{\alpha^3_{H_1}(w_i)}-\sqrt{\alpha^3_{H_2}(w_i)}\right)^2+\left(\sqrt{\beta^3_{H_1}(w_i)}-\sqrt{\beta^3_{H_2}(w_i)}\right)^2\right]}$ $dis_{FDW}^2(H_1,H_2)=\sqrt{\frac{1}{2n}\sum_{i=1}^{n}\left[\frac{\left(\sqrt{\alpha^3_{H_1}(w_i)}-\sqrt{\alpha^3_{H_2}(w_i)}\right)^2}{\alpha^3_{H_1}(w_i)+\alpha^3_{H_2}(w_i)}+\frac{\left(\sqrt{\beta^3_{H_1}(w_i)}-\sqrt{\beta^3_{H_2}(w_i)}\right)^2}{\beta^3_{H_1}(w_i)+\beta^3_{H_2}(w_i)}\right]}$
Du [13]	$dis_{QD}^1(H_1,H_2)=\frac{1}{2n}\sum_{i=1}^{n}\left(\lvert\zeta_{H_1}(w_i)^q-\left(\zeta_{H_2}(w_i)\right)^q\rvert+\lvert\left(\beta^3_{H_1}(w_i)\right)^q-\left(\beta^3_{H_2}(w_i)^q\rvert\right)\right)$ $dis_{QD}^2(H_1,H_2)=\left[\frac{1}{2n}\sum_{i=1}^{n}\left\{\left(\left(\alpha_{H_1}\right)\right)^q-\left(\left(\alpha_{H_1}\right)^q\right)^2+\left(\beta^3_{H_1}(w_i)\right)^q-\left(\beta^3_{H_2}(w_i)^q\right)^2\right\}\right]^{\frac{1}{2}}$ $dis_{QD}^3(H_1,H_2)=\frac{1}{n}\sum_{i=1}^{n}\max\left(\lvert\left(\alpha_{H_1}(w_i)\right)^q-\left(\alpha_{H_2}(w_i)\right)^q\rvert,\lvert\left(\beta_{H_1}(w_i)\right)^q-\left(\beta^3_{H_2}(w_i)^q\rvert\right)\right)$

Hussain et al. [14]	$dis_{QH}(H_1,H_2)=\frac{1}{n}\sum_{i=1}^{n}\max\left(\begin{array}{c}\lvert(\alpha'_{H_1}(w_i))^q-(\alpha'_{H_2}(w_i))^q\rvert,\\ \lvert\left(1-\left(\eta'_{H_1}(w_i)\right)^q\right)-\left(1-\left(\eta'_{H_2}(w_i)\right)^q\right)\rvert\end{array}\right)$
Kamacı and Petchimuthu [15]	$dis_{QKF1}(H_1,H_2)=\frac{1}{n}\sum_{i=1}^{n}\frac{\lvert\left(\alpha'_{H_1}(w_i)\right)^q-\left(\alpha'_{H_2}(w_i)\right)^q\rvert+\lvert\left(\beta'_{H_1}(w_i)\right)^q-\left(\beta'_{H_2}(w_i)\right)^q\rvert}{\max\left\{\left(\alpha'_{H_1}(w_i)\right)^q,\left(\alpha'_{H_2}(w_i)\right)^q\right\}+\max\left\{\left(\beta'_{H_1}(w_i)\right)^q,\left(\beta'_{H_2}(w_i)\right)^q\right\}}$ $dis_{QKF2}(H_1,H_2)=\frac{1}{n}\sum_{i=1}^{n}\frac{\lvert\left(\alpha'_{H_1}(w_i)\right)^q-\left(\alpha'_{H_2}(w_i)\right)^q\rvert+\lvert\left(\beta'_{H_1}(w_i)\right)^q-\left(\beta'_{H_2}(w_i)\right)^q\rvert+\lvert(\pi_{H_1}(w_i))^q-(\pi_{H_2}(w_i))^q\rvert}{\max\left\{\left(\alpha'_{H_1}(w_i)\right)^q,\left(\alpha'_{H_2}(w_i)\right)^q\right\}+\max\left\{\left(\beta'_{H_1}(w_i)\right)^q,\left(\beta'_{H_2}(w_i)\right)^q\right\}+\max\{(\pi_{H_1}(w_i))^q,(\pi_{H_2}(w_i))^q\}}$ $dis_{QKF3}(H_1,H_2)=\frac{\sum_{i=1}^{n}\lvert\left(\alpha'_{H_1}(w_i)\right)^q-\left(\alpha'_{H_2}(w_i)\right)^q\rvert+\lvert\left(\beta'_{H_1}(w_i)\right)^q-\left(\beta'_{H_2}(w_i)\right)^q\rvert+\lvert(\pi_{H_1}(w_i))^q-(\pi_{H_2}(w_i))^q\rvert}{\sum_{i=1}^{n}\max\left\{\left(\alpha'_{H_1}(w_i)\right)^q,\left(\alpha'_{H_2}(w_i)\right)^q\right\}+\max\left\{\left(\beta'_{H_1}(w_i)\right)^q,\left(\beta'_{H_2}(w_i)\right)^q\right\}+\max\{(\pi_{H_1}(w_i))^q,(\pi_{H_2}(w_i))^q\}}$
Wang et al. [16]	$dis_{QW}(H_1,H_2)=\frac{1}{3n}\sum_{i=1}^{n}\left(\frac{\lvert\left(\alpha'_{H_1}(w_i)\right)^q-\left(\alpha'_{H_2}(w_i)\right)^q\rvert}{1+\left(\alpha'_{H_1}(w_i)\right)^q+\left(\alpha'_{H_2}(w_i)\right)^q}+\frac{\lvert\left(\beta'_{H_1}(w_i)\right)^q-\left(\beta'_{H_2}(w_i)\right)^q\rvert}{1+\left(\beta'_{H_1}(w_i)\right)^q+\left(\beta'_{H_2}(w_i)\right)^q}+\frac{\lvert\left(\pi'_{H_1}(w_i)\right)^q-\left(\pi'_{H_2}(w_i)\right)^q\rvert}{1+\left(\pi'_{H_1}(w_i)\right)^q+\left(\pi'_{H_2}(w_i)\right)^q}\right)$

we have

$$\frac{1}{2}\left[\left\{\frac{|\alpha_{H_1}(w_i)-\alpha_{H_2}(w_i)|+\left|\beta_{H_1}(w_i)-\beta_{H_2}(w_i)\right|}{2}\right\}\times\left\{1+\left|\alpha_{H_1}(w_i)\beta_{H_2}(w_i)-\beta_{H_1}(w_i)\alpha_{H_2}(w_i)\right|\right.\right.$$

$$+\left|\max\left(\alpha_{H_1}(w_i),\beta_{H_2}(w_i)\right)-\max\left(\beta_{H_1}(w_i)\alpha_{H_2}(w_i)\right)\right|$$

$$\left.\left.+\left|\min\left(\alpha_{H_1}(w_i),\beta_{H_2}(w_i)\right)-\min\left(\beta_{H_1}(w_i)\alpha_{H_2}(w_i)\right)\right|\right\}^{\frac{1}{2}}\right]$$

$$=0\Leftrightarrow\left|\alpha_{H_1}(w_i)-\alpha_{H_2}(w_i)\right|+\left|\beta_{H_1}(w_i)-\beta_{H_2}(w_i)\right|=0$$

$$\Leftrightarrow\alpha_{H_1}(w_i)=\alpha_{H_2}(w_i)\text{ and }\beta_{H_1}(w_i)-\beta_{H_2}(w_i)$$

$$\Leftrightarrow H_1=H_2$$

Proof (P_3): Clearly, it is obvious that $D_{HyFKD}(H_1,H_2)=D_{HyFKD}(H_2,H_1)$
Proof (P_4): Consider, $H_1\subseteq H_2\subseteq H_3$

Then, for all $i=1,2,\ldots,n$ we have:

$$0\le\alpha_{H_1}(w_i)\le\alpha_{H_2}(w_i)\le\alpha_{H_3}(w_i)\le1\text{ and }0\ge\beta_{H_1}(w_i)\ge\beta_{H_2}(w_i)\ge\beta_{H_3}(w_i)\ge1$$

It follows that:

$$|\alpha_{H_1}(w_i)-\alpha_{H_2}(w_i)|\le|\alpha_{H_1}(w_i)-\alpha_{H_3}(w_i)|$$

Also, $\alpha_{H_1}(w_i)\beta_{H_3}(w_i)\le\alpha_{H_1}(w_i)\beta_{H_2}(w_i)\le\alpha_{H_2}(w_i)\beta_{H_1}(w_i)\le\alpha_{H_3}(w_i)\beta_{H_1}(w_i)$

$$\Rightarrow\left|(\alpha_{H_2}(w_i)\beta_1(w_i))-\left(\alpha_{H_1}(w_i)\beta_{H_2}(w_i)\right)\right|\le\left|\left(\alpha_{H_3}(w_i)\beta_{H_1}(w_i)\right)-\left(\alpha_{H_1}(w_i)\beta_{H_3}(w_i)\right)\right|$$

Further, $\left|\max\left(\alpha_{H_1}(w_i),\beta_{H_2}(w_i)\right)-\max\left(\beta_{H_1}(w_i)\alpha_{H_2}(w_i)\right)\right|\le\left|\max\left(\alpha_{H_1}(w_i),\beta_{H_3}(w_i)\right)-\max\left(\beta_{H_1}(w_i)\alpha_{H_3}(w_i)\right)\right|$

And $\left|\min\left(\alpha_{H_1}(w_i),\beta_{H_2}(w_i)\right)-\max\left(\beta_{H_1}(w_i)\alpha_{H_2}(w_i)\right)\right|\le\left|\min\left(\alpha_{H_1}(w_i),\beta_{H_3}(w_i)\right)-\min\left(\beta_{H_1}(w_i)\alpha_{H_3}(w_i)\right)\right|$

So, we have

$$\frac{1}{2}\left[\left\{\frac{|\alpha_{H_1}(w_i)-\alpha_{H_2}(w_i)|+|\beta_{H_1}(w_i)-\beta_{H_2}(w_i)|}{2}\right\}\times\left\{1+\left|\alpha_{H_1}(w_i)\beta_{H_2}(w_i)-\beta_{H_1}(w_i)\alpha_{H_2}(w_i)\right|\right.\right.$$
$$+\left|\max\left(\alpha_{H_1}(w_i),\beta_{H_2}(w_i)\right)-\max\left(\beta_{H_1}(w_i)\alpha_{H_2}(w_i)\right)\right|$$
$$\left.\left.+\left|\min\left(\alpha_{H_1}(w_i),\beta_{H_2}(w_i)\right)-\min\left(\beta_{H_1}(w_i)\alpha_{H_2}(w_i)\right)\right|\right\}^{\frac{1}{2}}\right]$$
$$\leq\frac{1}{2}\left[\left\{\frac{|\alpha_{H_1}(w_i)-\alpha_{H_3}(w_i)|+|\beta_{H_1}(w_i)-H_3(w_i)|}{2}\right\}\times\left\{1+\left|\alpha_{H_1}(w_i)\beta_{H_3}(w_i)-\beta_{H_1}(w_i)\alpha_{H_3}(w_i)\right|\right.\right.$$
$$+\left|\max\left(\alpha_{H_1}(w_i),\beta_{H_3}(w_i)\right)-\max\left(\beta_{H_1}(w_i)\alpha_{H_3}(w_i)\right)\right|$$
$$\left.\left.+\left|\min\left(\alpha_{H_1}(w_i),\beta_{H_3}(w_i)\right)-\min\left(\beta_{H_1}(w_i)\alpha_{H_3}(w_i)\right)\right|\right\}^{\frac{1}{2}}\right]$$

Thus, we obtain $D_{\text{HyFKD}}(H_1,H_3)\geq D_{\text{HyFKD}}(H_1,H_2)$

Similarly, we have $D_{\text{HyFKD}}(H_1,H_3)\geq D_{\text{HyFKD}}(H_2,H_3)$

So, the DM fulfils all the properties of a DM.

Next, a few propositions grounded on the DM are given as follows:

Proposition 4.1. For $A=<\alpha,\beta>, B=<1-\alpha,1-\beta>$, then

$$D_{\text{HyFKD}}(A,B)=\frac{1}{4}\left[\{|2\alpha-1|+|2\beta-1|\}\{1+3|\alpha-\beta|\}^{\frac{1}{2}}\right]$$

Graphically the distance of $D_{\text{HyFKD}}(A,B)$ is shown in Figure 2.

Proposition 4.2. For $A=<\alpha,\alpha>, B=<\beta,1-\beta>, C=<1-\beta,\beta>$, then

$$D_{\text{HyFKD}}(A,B)=\frac{1}{4}\left[\{|\alpha-\beta|+|\alpha+\beta-1|\}\{1+|1-2\beta|+|\alpha(1-2\beta)|\}^{\frac{1}{2}}\right]-D_{\text{HyFKD}}(A,C)$$

Graphically, the distance of $D_{\text{HyFKD}}(A,B)=D_{\text{HyFKD}}(A,C)$ is shown in Figure 3.

Proposition 4.3. For $A=<\alpha,1-\alpha>, B=<\beta,1-\beta>$ then,

$$D_{\text{HyFKD}}(A,B)=\frac{1}{2}\left[\{|\alpha-\beta|\}\{1+3|\alpha-\beta|\}^{\frac{1}{2}}\right]$$

Graphically, the distance of $D_{\text{HyFKD}}(A,B)$ is shown in Figure 4.

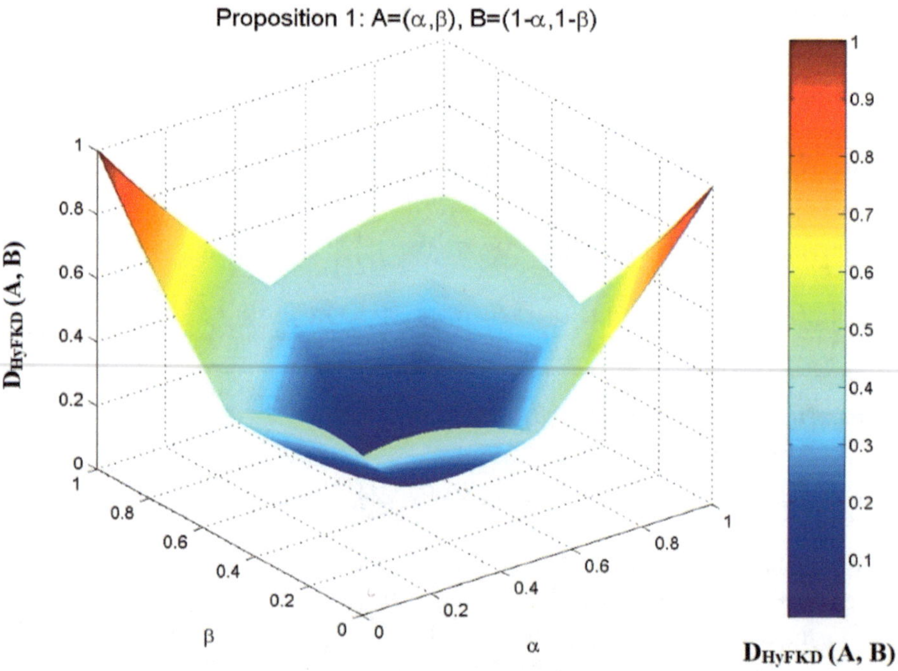

Figure 2: Graphical representation of the distance of for Proposition 1.

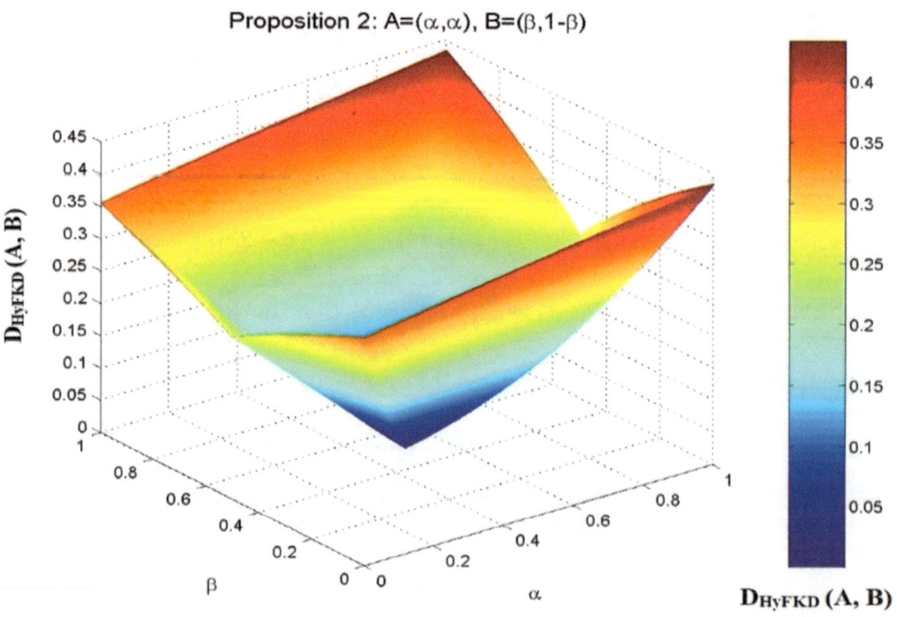

Figure 3: Graphical representation of the distance of $D_{\mathbf{HyFKD}}(A, B) = D_{\mathbf{HyFKD}}(A, C)$.

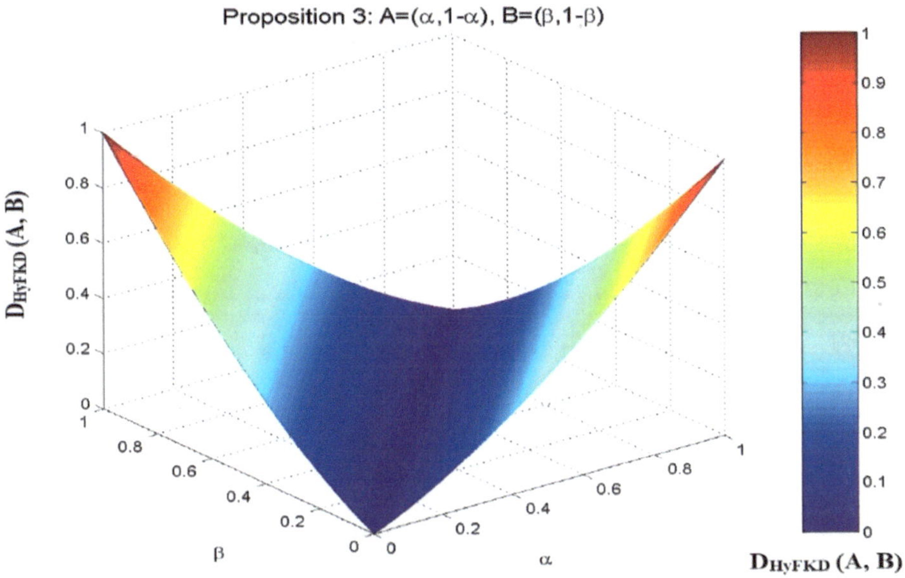

Figure 4: Graphical representation of the distance of $D_{\text{HyFKD}}(A, B)$ for Proposition 3.

Proposition 4.4. Consider the following three pairs of HyFSs:
1. $A = (1, 0), B = (\beta, 1 - \beta)$
2. $A = (0.5, 0.5), B = (\beta, 1 - \beta)$
3. $A = (0.3, 0.7), B = (\beta, 1 - \beta)$

Then, the non-linear characteristics of the above three pairs of HyFSs are shown in Figures 5–7.

4.3 Established DMs

The practice of DMs in decision-making problems is extremely beneficial for comparing two data sets by determining the distance between them. A comparative study of our suggested DM and the various established DMs has been discussed below, where we considered $q = 1$ for the q-ROFS-based measure. The comparison is shown in Table 3.

In Table 3, we see the DM values of the established measures along with our proposed DM. We have considered $q = 1$ for q-ROFS-based measures. In Table 3, several shortcomings of established DMs can be seen. Some of the DMs like $\text{dis}^1_{\text{PXD}}$, $\text{dis}^2_{\text{PXD}}$, $\text{dis}^1_{\text{FSY}}$, $\text{dis}^2_{\text{FSY}}$, dis^1_{FG}, dis^2_{FG}, dis^3_{FG}, dis^4_{FG}, dis^3_{QD}, dis_{QH} give a DM of 1 for Profile 3, which is not logical, as the values taken in Profile 3 are not complementary to each other. Again, DMs like dis_{IMP}, dis_{PMP}, dis_{QKP1}, dis_{QKP2}, and dis_{QKP3} give DM value 1 for Profiles 3 and 4, showing its limitation as the values taken in Profiles 3 and 4 are not comple-

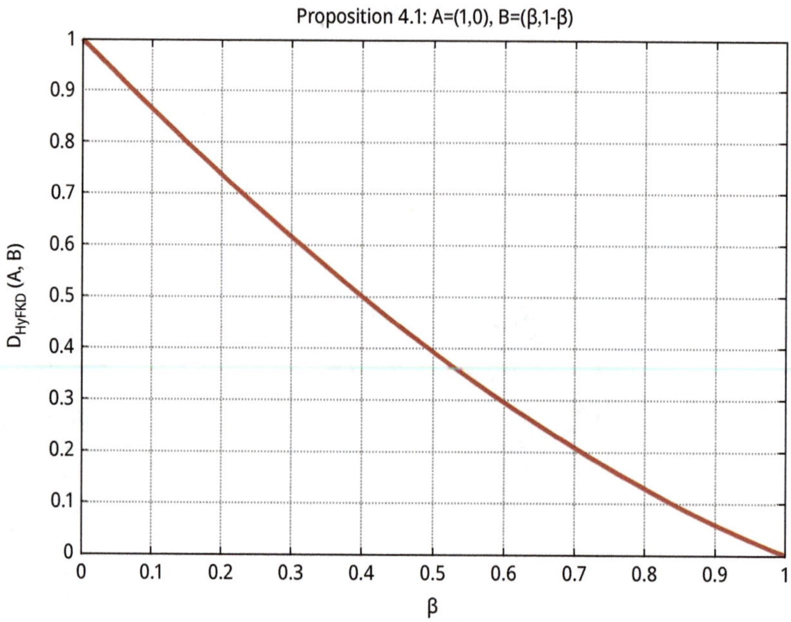

Figure 5: Proposed DM between $A=(1,0)$ and $B=(\beta,1-\beta)$.

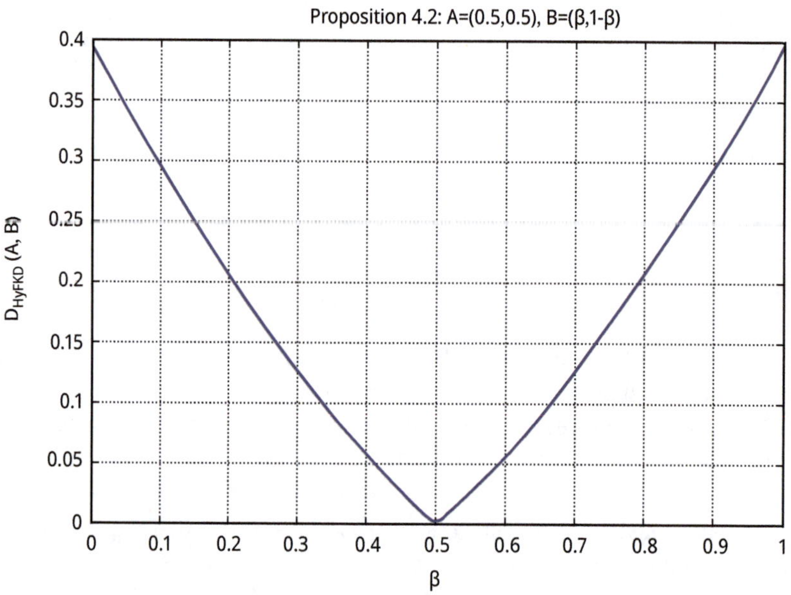

Figure 6: Proposed DM between $A=(0.5,0.5)$ and $B=(\beta,1-\beta)$.

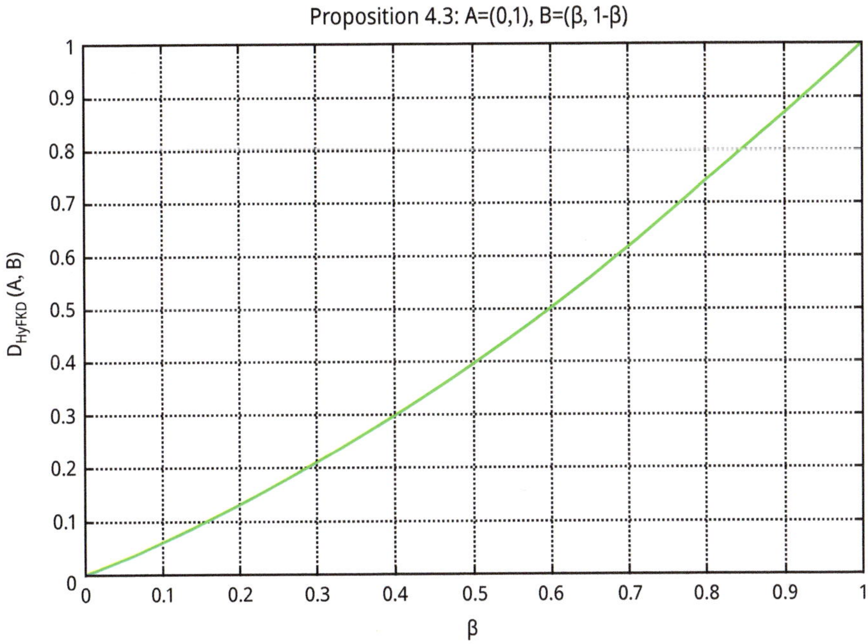

Figure 7: Proposed DM between $A = (0,1)$ and $B = (\beta, 1-\beta)$.

Table 3: Comparison table.

Table	**Profile 1**	**Profile 2**	**Profile 3**	**Profile 4**	**Profile 5**	**Profile 6**
$'H_1$	(0.3, 0.3)	(0.3, 0.4)	(1, 0)	(0.5, 0.5)	(0.4, 0.2)	(0.4, 0.2)
$'H_2$	(0.4, 0.4)	(0.4, 0.3)	(0, 0)	(0, 0)	(0.5, 0.3)	(0.5, 0.2)
$dis^1_{ING}('H_1, 'H_2)$ Ngan et al. [8]	**0.0667**	**0.1**	0.6667	0.3333	**0.1**	**0.0667**
$dis^2_{ING}('H_1, 'H_2)$ Ngan et al. [8]	0.05	**0.1**	**0.75**	0.25	**0.1**	**0.075**
$dis_{IMP}('H_1, 'H_2)$ Mahanta and Panda [7]	**0.1429**	**0.1429**	**1**	**1**	**0.1429**	0.0769
$dis^1_{PXD}('H_1, 'H_2)$ Xiao and Ding [9]	0.1212	0.07	**1**	0.433	0.1229	0.09
$dis^2_{PXD}('H_1, 'H_2)$ Xiao and Ding [9]	**0.14**	0.07	**1**	0.5	**0.14**	0.09
$dis_{PMP}('H_1, 'H_2)$ Mahanta and Panda [10]	**0.28**	**0.28**	**1**	**1**	0.2593	0.1837
$dis^1_{FSY}('H_1, 'H_2)$ Senapati and Yager [5]	0.0641	0.037	**1**	0.2165	0.0724	0.061
$dis^2_{FSY}('H_1, 'H_2)$ Senapati and Yager [5]	0.074	0.037	**1**	0.25	0.08	0.061
$dis^1_{FG}('H_1, 'H_2)$ Ganie [11]	**0.0714**	**0.0714**	**1**	0.2222	0.0778	0.061
$dis^2_{FG}('H_1, 'H_2)$ Ganie [11]	**0.0726**	**0.0726**	**1**	0.2344	0.0788	0.061
$dis^3_{FG}('H_1, 'H_2)$ Ganie [11]	**0.074**	**0.074**	**1**	0.25	0.08	0.061
$dis^4_{FG}('H_1, 'H_2)$ Ganie [11]	**0.0739**	**0.0739**	**1**	0.2462	0.0799	0.061
$dis^1_{FDW}('H_1, 'H_2)$ Deng and Wang (2022)	**0.0887**	**0.0887**	0.7071	0.3536	0.0887	0.0711
$dis^2_{FDW}('H_1, 'H_2)$ Deng and Wang [12]	**0.4157**	**0.4157**	**N/A**	**1.4142**	0.4623	0.2313

Table 3 (continued)

Table	Profile 1	Profile 2	Profile 3	Profile 4	Profile 5	Profile 6
$dis^1_{QD}(\acute{H}_1, \acute{H}_2)$ Du [13]	**0.1**	**0.1**	**0.5**	**0.5**	**0.1**	0.05
$dis^2_{QD}(\acute{H}_1, \acute{H}_2)$ Du [13]	**0.1**	**0.1**	0.7071	0.5	**0.1**	0.0707
$dis^3_{QD}(\acute{H}_1, \acute{H}_2)$ Du [13]	**0.1**	**0.1**	**1**	0.5	**0.1**	**0.1**
$dis_{QH}(\acute{H}_1, \acute{H}_2)$ Hussain et al. [14]	**0.1**	**0.1**	**1**	0.5	**0.1**	**0.1**
$dis_{QKP1}(\acute{H}_1, \acute{H}_2)$ Kamaci and Petchimuthu [15]	**0.25**	**0.25**	**1**	**1**	**0.25**	0.1429
$dis_{QKP2}(\acute{H}_1, \acute{H}_2)$ Kamaci and Petchimuthu [15]	**0.3333**	**0.1818**	**1**	**1**	**0.3333**	**0.1818**
$dis_{QKP3}(\acute{H}_1, \acute{H}_2)$ Kamaci and Petchimuthu [15]	**0.3333**	**0.1818**	**1**	**1**	**0.3333**	**0.1818**
$dis_{QW}(\acute{H}_1, \acute{H}_2)$ Wang et al. [16]	0.0809	0.0392	0.3333	0.3889	0.0814	0.0372
$D_{HyFKD}(H_1, H_2)$ (proposed)	0.05	0.0563	0.3536	0.25	0.0552	0.0265

Bold signifies unreasonable results; N/A denotes division by zero; $q = 1$ for established measures on q-ROFSs.

mentary to each other. Again, DMs, such as dis_{IMP}, dis^2_{QD}, and dis_{QKP1}, give the same DM values for Profiles 1, 2, and 5. DMs dis^1_{ING} cannot distinguish the values in Profiles 1 and 6, and 2 and 5; dis^2_{ING} cannot distinguish the values in Profiles 3 and 6, and 2 and 5; and dis^1_{QD} cannot distinguish the values in Profiles 1, 2 and 5, and 3 and 4. Similarly, dis_{QKP2}, and dis_{QKP3} cannot distinguish the values in Profiles 1 and 5, and 2 and 6, respectively. In addition, dis^3_{QD}, dis_{QH} do not distinguish between the values in Profiles 1, 2, 5, and 6. Table 3 shows that the established measures could not correctly distinguish the distance values in the six profiles. Whereas our proposed DM correctly gives different distance values for all the profiles, showing its advantage over established methods.

We see that the values considered in the Profile 1 are similar type of form (a, a), but the values considered in the Profile 2 are compliment of each other. As a result, it is logical that the DM value in Profile 2 must be more than Profile 1. In Table 3, DMs, such as dis_{PMP}, dis^1_{FG}, dis^2_{FG}, dis^3_{FG}, dis^4_{FG}, dis^1_{FDW}, dis^2_{FDW}, dis^1_{QD}, dis^2_{QD}, dis^3_{QD}, dis_{QH}, and dis_{QKP1}, give equal DM values and DMs, such as dis_{QKP2}, dis_{QKP3}, and dis_{QW}, give the DM value for Profile 1 more than Profile 2. But our proposed measure correctly gives the DM values in Profile 2 more than the values in Profile 1.

Now, if we consider Profiles 3 and 4 from Table 3, the DM values obtained in Profile 3 should be more than the values obtained in Profile 4. DMs like dis_{QW} give the DM values of the Profile 3 less than those of the Profile 4, and dis^1_{QD} give equal DM values in Profiles 3 and 4. Again, the values considered in Profiles 3 and 4 are not compliment to each other. Thus, it is logical that the DM values should not be 1. In Table 3, we see that some of the DMs, such as dis^1_{PXD}, dis^2_{PXD}, dis^1_{FSY}, dis^2_{FSY}, dis^1_{FG}, dis^2_{FG}, dis^3_{FG}, dis^4_{FG}, dis^3_{QD}, and dis_{QH}, give DM value 1 for Profile 3, and DMs like dis_{IMP}, dis_{PMP}, dis_{QKP1}, dis_{QKP2}, and

dis_{QKP3} give DM value 1 for both Profiles 3 and 4, which is not logical. Further, $\text{dis}^2_{\text{FDW}}$ gives a division by zero problem for Profile 3, whereas for Profile 4, the distance value is greater than 1. But our proposed measure satisfies all the distance properties and correctly distinguishes distance values of Profiles 3 and 4, such that the values in Profile 3 are more than the values in Profile 4.

Finally, we can see that the MD (or optimism) of the values of Profiles 5 and 6 are of a similar pattern, showing resemblance in value. But, we see the dissimilarity in the NMD (or pessimism), which shows values in Profile 6 are closer than those in Profile 5. So, the DM value of Profile 5 must be more than that of Profile 6. In the established measures, such as dis^3_{QD}, dis_{QH}, it can be seen that the DM values are the same in Profiles 5 and 6. However, in our proposed measure, the distance value of Profile 5 is more than that of Profile 6.

So, based on the results in Table 3, we can say that our proposed DM does not show the drawbacks of the established DMs, which shows the superiority of our proposed measure.

5 Proposed HyF-CRITIC-TOPSIS Approach

In this section, we expand the traditional CRITIC and TOPSIS methods constructed on the proposed DM in the HyF environment. The HyF-CRITIC is an objective weight-determining tool, whereas the TOPSIS method ranks alternatives in numerous MCDM applications. The procedural steps of the proposed HyF-CRITIC-TOPSIS approach, supported by the proposed DM under the HyF framework, are presented as follows:

Step 1. Problem description
Suppose a set of decision-makers $\{e_1, e_2, \ldots, e_k\}$ assesses a set of alternatives $\{A_1, A, \ldots, A_m\}$ grounded on the criteria set $\{C_1, C_2, \ldots, C_n\}$. Let every decision-maker $e_r, r = 1, 2, \ldots, k$ defines a decision matrix $Y^r = [y^r_{ij}], r = 1, 2, \ldots, k, i = 1, 2, \ldots, m, j = 1, 2, \ldots, n$ for each alternative $A_p, p = 1, 2, \ldots, m$ based on every criterion C_q in the HyF environment. Here, y^r_{ij} of the rth individual matrix depicts the judgment of the rth decision-maker to the alternative a_i connected to the criteria C_j.

Step 2. Calculate the aggregated HyF decision matrix
Determine the aggregated HyF decision matrix (AHyFDM) $Y = (y_{ij})_{m \times n}$ as follows, where

$$y_{ij} = \left(\alpha^*_{ij}, \beta^*_{ij}\right) = \left(\prod_{r=1}^{k} u_r \alpha_{ij}, \prod_{r=1}^{k} u_r \beta_{ij}\right), i = 1, 2, \ldots, m, j = 1, 2, \ldots, n$$

and $u_r > 0, r = 1, 2, \ldots, k$ represents the weight of each decision expert such that $\sum_{r=1}^{k} u_r = 1$ is fulfilled.

Step 3. Compute the criteria weights by the CRITIC method
The following are the steps to estimate the objective weights of the criteria using the CRITIC tool.

Step 3.1 Determine the score values of the AHyDM by employing the score function of Dutta and Borah [6] to obtain the score matrix as follows:

$$S=(s_{ij})_{m\times n}, \text{where } s_{ij}=2\alpha_{ij}^{*}\left(1-\beta_{ij}^{*}\right), p=1,2,\ldots,m, q=1,2,\ldots,n$$

Step 3.2 Convert the score matrix $S=(s_{ij})_{m\times n}$, into a standard HyF matrix

$$\bar{S}=(\overline{s_{ij}})_{m\times n}, \text{ where } \overline{s_{ij}}=\begin{cases}\dfrac{s_{ij}-s_j^{+}}{s_j^{+}-s_j^{-}}, \text{for benefit criteria}\\ \dfrac{s_j^{-}-s_{ij}}{s_j^{+}-s_j^{-}}, \text{for cost criteria}\end{cases}$$

where $s_j^{+}=\max\limits_{i} s_{ij}$ and $s_j^{-}=\min\limits_{i} s_{ij}$

Step 3.3 Next, we compute the standard deviation of each criterion as follows:

$$\sigma_j=\sqrt{\frac{\sum\limits_{i=1}^{m}\left(\overline{s_{ij}}-\overline{s_j}\right)^2}{m}}, \text{ where } \overline{s_j}=\sum_{i=1}^{m}\frac{\overline{s_{ij}}}{m}.$$

Step 3.4 Compute the correlation among the criteria C_j and C_t by utilising Karl-Pearson's correlation in the following manner:

$$r_{jk}=\frac{\sum\limits_{i=1}^{t}\left(\overline{s_{ij}}-\overline{s_j}\right)\left(\overline{s_{ij}}-\overline{s_k}\right)}{\sqrt{\sum\limits_{i=1}^{t}\left(\overline{s_{ij}}-\overline{s_j}\right)^2\sum\limits_{i=1}^{t}\left(\overline{s_{ij}}-\overline{s_k}\right)^2}}$$

Step 3.5 Next, evaluate the quantity of information for each criterion as follows:

$$p_j=\sigma_j\sum_{k=1}^{n}(1-r_{jk})$$

Step 3.6 Finally, determine the objective criteria weights as follows:

$$w_j=\frac{p_j}{\sum\limits_{j=1}^{n}p_j}$$

Step 4. Execution of the TOPSIS method
The next steps implement the TOPSIS approach to rank the alternatives.

Step 4.1: Calculate the HyF positive ideal solution, HyFPIS (P^+) and the HyF negative ideal solution, HyFNIS (P^-), respectively, from the AHyFDM based on the evaluated score matrix in step 3.1 as follows:

$$P^+ = \left(\gamma_j^+\right)_{1\times n} = \left\{\left(\alpha_j^+, \beta_j^+\right)\right\} = \begin{cases} \max\limits_j \left\langle S\left(\alpha_{ij}^*, \beta_{ij}^*\right), j = 1, 2, \ldots, n, C_j \text{ is a benefit criteria} \right\rangle \\ \min\limits_j \left\langle S\left(\alpha_{ij}^*, \beta_{ij}^*\right), j = 1, 2, \ldots, n, C_j \text{ is a cost criteria} \right\rangle \end{cases}$$

$$P^- = \left(\gamma_j^-\right)_{1\times n} = \left\{\left(\alpha_j^-, \beta_j^-\right)\right\} = \begin{cases} \min\limits_j \left\langle S\left(\alpha_{ij}^*, \beta_{ij}^*\right), j = 1, 2, \ldots, n, C_j \text{ is a benefit criteria} \right\rangle \\ \max\limits_j \left\langle S\left(\alpha_{ij}^*, \beta_{ij}^*\right), j = 1, 2, \ldots, n, C_j \text{ is a cost criteria} \right\rangle \end{cases}$$

Step 4.2: Determine the distance of each alternative A_i from the hyperbolic fuzzy positive ideal solution (HyFPIS, P^+) and from hyperbolic fuzzy negative ideal solution (HyFNIS, P^-) referring to the following equations:

$$D\left(\gamma_{ij}, \gamma_j^+\right) = \sum_{i=1}^{n} w_j d_{\mathrm{BAP}}\left(\gamma_{ij}, \gamma_j^+\right)$$

and

$$D\left(\gamma_{ij}, \gamma_j^-\right) = \sum_{j=1}^{n} w_j d_{\mathrm{BAP}}\left(\gamma_{ij}, \gamma_j^-\right)$$

where

$$d_{\mathrm{BAP}}\left(\gamma_{ij}, \gamma_j^+\right) = \frac{1}{2n}\sum_{j=1}^{n}\left[\begin{array}{c} \left\{\dfrac{\left|\alpha_{ij}^* - \alpha_j^+\right| + \left|\beta_{ij}^* - \beta_j^+\right|}{2}\right\} \\ \times\left\{\begin{array}{c} 1 + \left|\left(\alpha_{ij}^*\right)\left(\beta_j^+\right) - \left(\alpha_j^+\right)\left(\beta_{ij}^*\right)\right| \\ + \left|\max\left\{\left(\alpha_{ij}^*\right), \left(\beta_j^+\right)\right\} - \max\left\{\left(\alpha_j^+\right), \left(\beta_{ij}^*\right)\right\}\right| \\ + \left|\max\left\{\left(\alpha_{ij}^*\right), \left(\beta_j^+\right)\right\} - \max\left\{\left(\alpha_j^+\right), \left(\beta_{ij}^*\right)\right\}\right| \end{array}\right\}^{\frac{1}{2}} \end{array}\right]$$

and

$$d_{\text{BAP}}\left(\gamma_{ij}, \gamma_j^-\right) = \frac{1}{2n}\sum_{j=1}^{n}\left[\begin{array}{c}\left\{\dfrac{\left|\alpha_{ij}^{*} - \alpha_j^-\right| + \left|\beta_{ij}^{*} - \beta_j^-\right|}{2}\right\} \\ \times\left\{\begin{array}{c} 1 + \left|\left(\alpha_{ij}^{*}\right)\left(\beta_j^-\right) - \left(\alpha_j^-\right)\left(\beta_{ij}^{*}\right)\right| \\ + \left|\max\left\{\left(\alpha_{ij}^{*}\right), \left(\beta_j^-\right)\right\} - \max\left\{\left(\alpha_j^-\right), \left(\beta_{ij}^{*}\right)\right\}\right| \\ + \left|\max\left\{\left(\alpha_{ij}^{*}\right), \left(\beta_j^-\right)\right\} - \max\left\{\left(\alpha_j^-\right), \left(\beta_{ij}^{*}\right)\right\}\right| \end{array}\right\}\end{array}\right]^{\frac{1}{2}}$$

Here, $d_{\text{BAP}}\left(\gamma_{ij}, \gamma_j^+\right)$ denotes the HyF distance computed between the aggregated value of each alternative $A_i, i = 1, 2, \ldots, m$ with respect to all $C_j, j = 1, 2, \ldots, n,$ and the HyFPIS (P^+) and $d_{\text{BAP}}\left(\gamma_{ij}, \gamma_j^-\right)$ denotes the HyF distance computed between the aggregated value of each alternative $A_i, i = 1, 2, \ldots, m$ with respect to all $C_j, j = 1, 2, \ldots, n,$ and the HyFPIS (P^-), respectively.

Step 4.3. Compute the relative closeness index $\xi(RC_i)$ for individual alternative, defined as follows:

$$\xi(RC_i) = \frac{D\left(\gamma_{ij}, \gamma_j^-\right)}{D_{\max}\left(\gamma_{ij}, \gamma_j^-\right)} - \frac{D\left(\gamma_{ij}, \gamma_j^+\right)}{D_{\min}\left(\gamma_{ij}, \gamma_j^-\right)}, \quad i = 1, 2, \ldots, m$$

Step 5. Evaluate the revised closeness index (N^*) in accordance with the equation below, and accordingly, govern the ranking of the alternatives such that the alternative with the highest $\xi(RC_i)$ value is regarded to be optimal.

$$N^* = \left\{N_i: \left(i: \xi(N_i) = \max_{1 \leq i \leq m} \xi(RC_i)\right)\right\}$$

The graphical flowchart of the HyF-CRITIC-TOPSIS approach based on DM is shown in Figure 8.

6 Application of the Proposed MCDM Approach to an SSS Problem

SSS mitigates climate alteration, increases economic and technical benefits, while promoting social growth, as well as protecting the quality of products. Selection of a sustainable supplier is a complex task that comprises high uncertainty and involves numerous multi-criteria in social, economic, technical, and environmental aspects. Here, we adopt a numerical example on SSS from Zulqarnain et al. [29]. The authors considered four sustainable suppliers that are examined based on the criteria set Csuch that

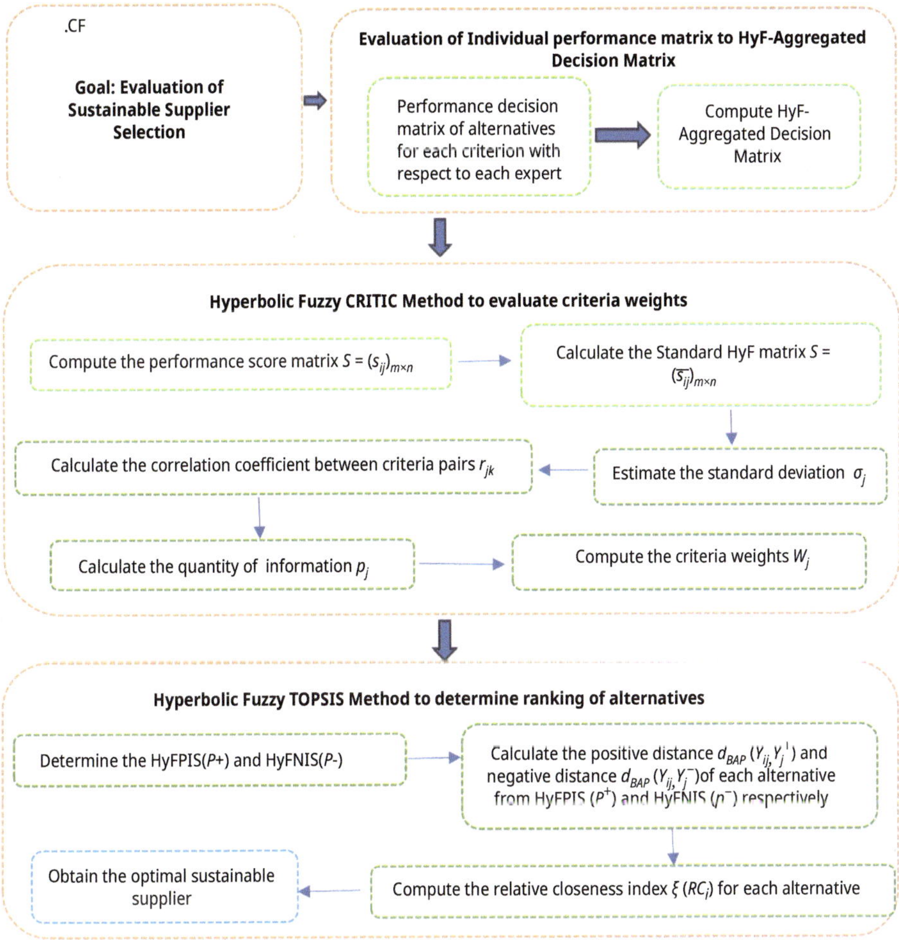

Figure 8: Graphical flowchart of the HyF-CRITIC-TOPSIS approach based on DM.

$$C = \left\{ \begin{array}{l} C_1\text{: Quality and reliability, } C_2\text{: Capacity and scalability,} \\ C_3\text{: Compliance and certifications, } C_4\text{: Sustainability practices} \end{array} \right\}$$

by four decision-makers with weights $\{0.2, 0.23, 0.27, 0.3\}$.

Here, all the criteria are considered as benefit criteria. The individual performance matrix for the evaluation of SSS is presented in Table 4. Next, utilising step 2 from the previous algorithm and taking the given decision experts' weights, we obtain the HyF aggregated decision matrix for SSS in Table 5. Next, applying the procedural steps of the CRITIC tool from step 3, we compute the standard HyF matrix, standard deviation, quantity of information, and finally, the criteria weights as $w_1 = 0.278, w_2 = 0.259, w_3 = 0.234$, and $w_4 = 0.228$, which are displayed in Table 6.

Following the steps of the TOPSIS approach from step 4, we compute the score performance matrix as shown in Table 7, and subsequently we compute the HyFPIS and HyFNIS, and finally determine the rank of the alternatives as shown in Table 8. Clearly, Table 8 depicts that A_2 is the optimal SSS alternative.

Table 4: Individual decision matrices for SSS evaluation.

Decision-maker	SSS alternatives	C_1	C_2	C_3	C_4
e_1	A_1	(0.2, 0.9)	(0.8, 0.9)	(0.1, 0.6)	(0.4, 0.5)
	A_2	0.228	(0.5, 0.9)	(0.3, 0.5)	(0.9, 0.6)
	A_3	(0.6, 0.4)	(0.4, 0.4)	(0.6, 0.5)	(0.5, 0.9)
	A_4	(0.5, 0.9)	(0.2, 0.9)	(0.4, 0.9)	(0.3, 0.9)
e_2	A_1	(0.5, 0.6)	(0.8, 0.7)	(0.2, 0.5)	(0.9, 0.9)
	A_2	(0.9, 0.8)	(0.6, 0.4)	(0.5, 0.6)	(0.2, 0.7)
	A_3	(0.7, 0.5)	(0.6, 0.9)	(0.5, 0.6)	(0.3, 0.7)
	A_4	(0.7, 0.3)	(0.4, 0.8)	(0.7, 0.6)	(0.8, 0.8)
e_3	A_1	(0.7, 0.7)	(0.7, 0.8)	(0.9, 0.9)	(0.6, 0.6)
	A_2	(0.8, 0.7)	(0.8, 0.9)	(0.2, 0.3)	(0.8, 0.3)
	A_3	(0.9, 0.7)	(0.8, 0.5)	(0.8, 0.6)	(0.9, 0.5)
	A_4	(0.9, 0.5)	(0.9, 0.8)	(0.5, 0.3)	(0.7, 0.6)
e_4	A_1	(0.3, 0.7)	(0.9, 0.7)	(0.8, 0.6)	(0.3, 0.8)
	A_2	(0.7, 0.6)	(0.7, 0.6)	(0.2, 0.6)	(0.5, 0.9)
	A_3	(0.8, 0.6)	(0.7, 0.6)	(0.7, 0.2)	(0.4, 0.8)
	A_4	(0.8, 0.9)	(0.9, 0.2)	(0.3, 0.7)	(0.2, 0.5)

Table 5: Aggregated decision matrix for SSS evaluation.

SSS alternatives	C_1	C_2	C_3	C_4
A_1	(0.729, 0.596)	(0.553, 0.540)	(0.487, 0.699)	(0.418, 0.658)
A_2	(0.756, 0.553)	(0.626, 0.578)	(0.651, 0.416)	(0.487, 0.699)
A_3	(0.392, 0.710)	(0.799, 0.763)	(0.396, 0.642)	(0.493, 0.692)
A_4	(0.768, 0.725)	(0.655, 0.661)	(0.267, 0.479)	(0.517, 0.582)

Table 6: The standard HyF matrix, standard deviation, quantity of information, and criteria weights.

SSS alternatives	C_1	C_2	C_3	C_4	σ_j	p_j	w_j
A_1	0	1	0.209	0.487	0.394	1.372	0.278
A_2	0.807	0.383	0	1	0.356	1.275	0.259
A_3	1	0.459	1	0.421	0.373	1.152	0.234
A_4	0.879	0	0.363	0	0.355	1.122	0.228

Table 7: Score performance matrix for SSS evaluation.

SSS alternatives	C_1	C_2	C_3	C_4
A_1	1.0232	0.8069	0.6336	0.5614
A_2	1.0939	0.8904	1.0315	0.6336
A_3	0.5043	0.9888	0.5379	0.6451
A_4	0.9805	0.8766	0.4071	0.7332

Table 8: Decision achieved by the HyF-TOPSIS method for SSS evaluation.

SSS alternatives	$D\left(\gamma_{ij}, \gamma_j^+\right)$	$D\left(\gamma_{ij}, \gamma_j^-\right)$	$\xi(RC_i)$	Rank
A_1	0.0818	0.0726	−1.6869	3
A_2	0.0336	0.0972	0	1
A_3	0.0948	0.0573	−2.2323	4
A_4	0.0648	0.0607	−1.3032	2

Furthermore, a strength and uniformity check of our suggested MCDM approach is performed by a comparative analysis with a few of the established notable MCDM approaches, such as FF-TOPSIS [5], FF-ARAS (Fermatean fuzzy additive ratio assessment) [30], and FF-SAW (Fermatean fuzzy simple additive weighting) [30]. Table 9 presents the ranking of the SSS alternatives with respect to our proposed HyF-CRITIC-TOPSIS approach and the other three established methods. Clearly, Table 9 displays A_2 as the best sustainable supplier among all the considered established MCDM methods. Figure 9 contributes a graphical representation of the rankings of the sustainable suppliers.

Table 9: Comparison of the proposed HyF technique with the established fuzzy MCDM approaches.

MCDM methods	Ranking order
FF-TOPSIS [5]	$A_2 \geq A_4 \geq A_1 \geq A_3$
FF-ARAS [30]	$A_2 \geq A_4 \geq A_1 \geq A_3$
FF-SAW [30]	$A_2 \geq A_4 \geq A_1 \geq A_3$
q-ROFS-TOPSIS [29]	$A_2 \geq A_1 \geq A_3 \geq A_4$
HyF-CRITIC-TOPSIS (proposed)	$A_2 \geq A_4 \geq A_1 \geq A_3$

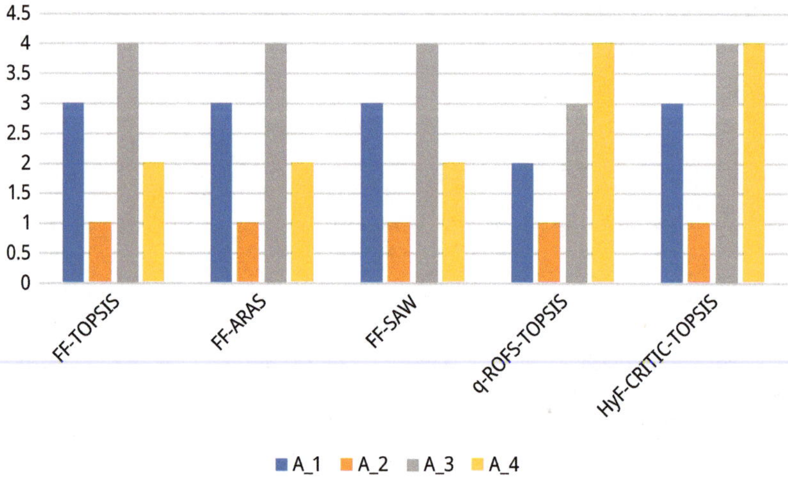

Figure 9: Graphical comparison of the ranking of SSS alternatives.

7 Conclusion

The increasing impacts of environmental degradation, social inequality, demand for economic growth, and advanced technological requirements, shorter delivery time while improving effectiveness and quality of products, and resilience, have significantly accelerated the requirement for SSS. The chief limitations of FSs and their various extensions, such as the IFS, the PFS, the q-ROFS, and the FFS, are successfully conquered by HyFS in this study. Correspondingly, the proposed HyF DM successfully overpowers the restrictions of a few notable established DMs from the literature. Moreover, decision making in the HyF framework is more flexible than IFS, PFS, FFS, and q-ROFS, thus representing uncertainty and reducing information loss in a wide range of complex real-life applications. Furthermore, the extended CRITIC tool in our developed hybrid MCDM approach produces meticulous criteria weights, thereby enhancing the effectiveness of the TOPSIS MCDM approach under the HyF framework to obtain a precise and consistent ranking of SSS alternatives in MCDM scenarios. In addition, the developed HyF-CRITIC-TOPSIS model is also supported by the novel DM in the HyF framework, which efficiently portrays uncertain scenarios and thus enhances the decision outcomes by preventing information loss. Correspondingly, the comparative analysis of our proposed HyF-CRITIC-TOPSIS method grounded on the novel DM with the notable established methods such as FF-TOPSIS [5], FF-ARAS [30], FF-SAW [30] authenticates the reliability and consistency of our suggested methodology. In the future, we will expand the traditional MCDM methods like VIKOR [31], AHP [32], ARAS [33] with our DM under the HyF framework for handling complex MCDM problems, and also develop tools to examine the subjective criteria weights in diverse real-life MCDM problems.

References

[1] L. A. Zadeh, I. Introduction, and U. S. Navy. Fuzzy Sets Syst. 1965;353:338–353.

[2]. K. T. Atanassov. Intuitionistic fuzzy sets. *Fuzzy Sets Syst.* 1986;20(1):87–96. 10.1016/S0165-0114(86)80034-3.

[3] Yager, R. R. June. Pythagorean fuzzy subsets. *2013 joint IFSA world congress and NAFIPS annual meeting (IFSA/NAFIPS).* IEEE; 2013 pp. 57–61.

[4] R. R. Yager and F. Ieee. Generalized Orthopair Fuzzy Sets. 2016;6706(x). 10.1109/TFUZZ.2016.2604005.

[5] T. Senapati and R. R Yager. Fermatean fuzzy sets. *J Ambient Intell Human Comput.* 2019;(2015). 10.1007/s12652-019-01377-0.

[6] Dutta, P., & Borah, G. Construction of hyperbolic fuzzy set and its applications in diverse COVID-19 associated problems. *New Math Nat Comput.* 2023;*19*(01):217–288.

[7] Mahanta, J., & Panda, S. A novel distance measure for intuitionistic fuzzy sets with diverse applications. *Int J Intell Syst.* 2021;*36*(2):615–627.

[8] Ngan, R. T., Cuong, B. C., & Ali, M. H-max distance measure of intuitionistic fuzzy sets in decision making. *Appl Soft Comput.* 2018;*69*:393–425.

[9] Xiao, F., & Ding, W. Divergence measure of Pythagorean fuzzy sets and its application in medical diagnosis. *Appl Soft Comput.* 2019;*79*:254–267.

[10] Mahanta, J., & Panda, S. Distance measure for Pythagorean fuzzy sets with varied applications. *Neural Comput Appl.* 2021;*33*(24):17161–17171.

[11] A. H. Ganie. Multicriteria decision-making based on DMs and knowledge measures of Fermatean fuzzy sets. *Granul Comput.* 2022;8(2016). 10.1007/s41066-021-00309-8.

[12] Deng, Z., & Wang, J. 2022. New DM for Fermatean fuzzy sets and its application. *Int J Intell Syst.* 1903–1930;*37*(3).

[13] Du, W. S. Minkowski-type DMs for generalized orthopair fuzzy sets. Int J Intell Syst. 2018;33:802–817. 47.

[14] Hussain, Z., Abbas, S., & Yang, M. S. Distances and similarity measures of q-rung orthopair fuzzy sets based on the Hausdorff metric with the construction of orthopair fuzzy TODIM. Symmetry. 2022;14(11):2467.

[15] Kamacı, H., & Petchimuthu, S. Soergel distance measures for q-rung orthopair fuzzy sets and their applications. q-rung orthopair fuzzy sets: theory and applications. Springer Nature Singapore. Singapore;2022. pp. 67–107.

[16] Wang, H., Xu, T., Pamucar, D., Li, X., & Feng, L. A Lance distance-based MAIRCA method for q-rung orthopair fuzzy MCDM with completely unknown weight information. *Informatica.* 2024;*35*(1):179–202.

[17] Diakoulaki, D., Mavrotas, G., & Papayannakis, L. Determining objective weights in multiple criteria problems: The critic method. *Comput Oper Res.* 1995;*22*(7):763–770.

[18] Mishra, A. R., Rani, P., & Pandey, K. Fermatean fuzzy CRITIC-EDAS approach for the selection of sustainable third-party reverse logistics providers using improved generalized score function. *J Ambient Intell Human Comput.* 2022;*13*(1):295–311.

[19] Peng, X., & Garg, H. Intuitionistic fuzzy soft decision making method based on CoCoSo and CRITIC for CCN cache placement strategy selection. *Artif Intell Rev.* 2022;*55*(2):1567–1604.
[20] Tripathi, D. K., Nigam, S. K., Cavallaro, F., Rani, P., Mishra, A. R., & Hezam, I. M. A novel CRITIC-RS-VIKOR group method with intuitionistic fuzzy information for renewable energy sources assessment. *Group Decis Negot.* 2023;*32*(6):1437–1468.
[21] Ertemel, A. V., Menekse, A., & Camgoz Akdag, H. Smartphone addiction assessment using Pythagorean fuzzy CRITIC-TOPSIS. *Sustainability.* 2023;*15*(5):3955.
[22] Wang, Y., Wang, W., Wang, Z., Deveci, M., Roy, S. K., & Kadry, S. Selection of sustainable food suppliers using the Pythagorean fuzzy CRITIC-MARCOS method. *Inf Sci.* 2024;*664*(120326).
[23] Yang, K., Duan, T., Feng, J., & Mishra, A. R. Internet of things challenges of sustainable supply chain management in the manufacturing sector using an integrated q-Rung Orthopair Fuzzy-CRITIC-VIKOR method. *J Enterp Inf Manag.* 2022;*35*(4/5):1011–1039.
[24] C.L. Hwang, K. Yoon. Multiple Attribute Decision Making: Methods and Applications. 1981, Springer-Verlag:New York.
[25] Dutta, P., & Konwar, A. Quintic fuzzy sets: A new class of fuzzy sets for solving multi-criteria decision-making problems under uncertainty. *Decis Anal J.* 2024;*11*:100449.
[26] Mukherjee, S., De, A., & Roy, S. Supply chain risk prioritization: a multi-criteria based Intuitionistic Fuzzy TOPSIS approach. *Int J Qual Reliab Manag.* 2024;*41*(6):1693–1725.
[27] Hajiaghaei-Keshteli, M., Cenk, Z., Erdebilli, B., Özdemir, Y. S., & Gholian-Jouybari, F. Pythagorean fuzzy TOPSIS method for green supplier selection in the food industry. *Expert Syst Appl.* 2023;*224*(120036).
[28] Majd, S. S., Maleki, A., Basirat, S., & Golkarfard, A. Fermatean fuzzy TOPSIS method and its application in ranking business intelligence-based strategies in smart city context. *J Oper Intell.* 2025;*3*(1):1–16.
[29] Zulqarnain, R. M., Dai, H. L., Ma, W. X., Siddique, I., Askar, S., & Naveed, H. Supplier selection in green supply chain management using correlation-based TOPSIS in a q-rung orthopair fuzzy soft environment. *Heliyon.* 2024;*10*(11).
[30] Gül, S. Fermatean fuzzy set extensions of SAW, ARAS, and VIKOR with applications in COVID-19 testing laboratory selection problem. *Expert Syst.* 2021;*38*(8):e12769.
[31] Chen, T. Y. A novel VIKOR method with an application to multiple criteria decision analysis for hospital-based post-acute care within a highly complex uncertain environment. *Neural Comput Appl.* 2019;*31*(8):3969–3999.
[32] Abdel-Basset, M., Mohamed, M., & Smarandache, F. An extension of neutrosophic AHP–SWOT analysis for strategic planning and decision-making. *Symmetry.* 2018;*10*(4):116.
[33] Turskis, Z., & Zavadskas, E. K. A novel method for multiple criteria analysis: grey additive ratio assessment (ARAS-G) method. *Informatica.* 2010;*21*(4):597–610.

Paul Augustine Ejegwa, Chukwudi Obinna Nwokoro, Benard Alechenu

A New Numerical Method of *q*-Rung Orthopair Fuzzy Distance Measure for Robotics Selection Analysis

Abstract: The emergence of q-rung orthopair fuzzy set (q-ROFS) as a competent soft computing framework has alleviated fuzziness in the art of decision-making. Several distance measuring approaches between q-ROFSs have been utilized as vehicles to exemplary decision-making scenarios. Among the existing approaches, three were presented based on logarithmic functions. Regardless of their benefits, they are deficient in terms of formulations and precision of results. To this end, an improved q-rung orthopair fuzzy logarithmic-based distance method (q-ROFLDM) is presented, which surmounts the setbacks of the existing q-ROFLDMs. The practical validation of the new q-ROFLDM is showcased in robotics selection analysis via q-rung orthopair fuzzy (q-ROF) information. The efficiency of the new q-ROFLDM is expressed via comparative analysis with the existing q-ROFLDMs. The analysis of comparison highlights the pre-eminence of the new q-ROFLDM. By leveraging the novel q-ROFLDM as a decision-making technique, the chapter offers a more precise decision-making in robotics analysis, delivering intelligent insights to robotics experts. This research contributes to the improvement of q-ROF-based decision-making systems, pioneering an innovative approach to robotic analysis.

Keywords: q-rung orthopair fuzzy set, multi-criteria decision-making, logarithmic-based distance method, robotics selection analysis

1 Introduction

Robotics is an interdisciplinary subject that combines science, engineering, and technology to design, construct, and utilize robots for everyday activity. A robot is a machine that is programmed to complete a task. A robot replicates human actions with greater accuracy and efficiency than humans, which qualifies it for industrial uses to carry out repetitive tasks. With the advent of AI, robots can carry out complex tasks in many industries. A robot is made up of control system, sensors, actuators, end effectors, and power supply. There are several kinds of robots such as collaborative ro-

Paul Augustine Ejegwa, Department of Mathematics, Joseph Sarwuan Tarka University, Makurdi, Nigeria, e-mail: ejegwa.augustine@uam.edu.ng
Chukwudi Obinna Nwokoro, Department of Computer Science, Nigerian British University, Asa, Abia State, e-mail: nwokoro.chukwudi@nbu.edu.ng
Benard Alechenu, Department of Mathematics and Statistics, Federal University, Kashere, Gombe, Nigeria, e-mail: alechenu.benard@fukashere.edu.ng

 | https://doi.org/10.1515/9783112222249-008

bots (cobots), humanoid robots, medical robots, agricultural robots, industrial robots, micro-robotics, software bots, and augmenting robots. Based on their design and construction, robots can be used for manufacturing, healthcare, domestic use, companionship, search and rescue, etc. [1]. By leverages on distributed AI, metamorphic robots have been constructed for utility in hostile environments on the earth and other planetary bodies [2]. One of the greatest challenges in robotics is the selection of suitable robot for effective utilization. Suppose there are four robots like Hitachi America Process Robot, ASEA-IRB 60/2, Cincinnati Milacrone T^3 726, and Yaskawa Electric Motoman L3C, to select a suitable robot from these for industrial purpose is a complex decision-making problems (DMPs). Many real-world DMPs are time and again complex because of the attendance of fuzziness and incomplete data in most circumstances, and so, a process for check-mating fuzzy data needed to be integrated to provide effective decision-making.

Although the institutionalization of fuzzy sets [3] has made the decision-making process significantly easier, some limitations are noticed in fuzzy set, because it only acknowledges the membership degree of the decision factors. Subject to this stumbling block, several advanced variants of fuzzy sets have been put forward, such as intuitionistic fuzzy set (IFS) [4], Pythagorean fuzzy set (PFS) [5], Fermatean fuzzy set (FFS) [6], and q-rung orthopair fuzzy set (q-ROFS) [7]. The q-ROFS generalizes all these advanced fuzzy sets. The applications of the advanced variants of fuzzy sets have been discussed via information measures in decision-making [8-11], diagnostic science [12–15], security control [16], selection of renewable energy [17], etc. The commonest applicable information measure in computational intelligence is distance measure, which is pretty noticeable in machine learning, pattern recognition, decision-making, data analysis, etc. A distance measure between q-ROFSs is a real-valued function which quantifies the difference and resemblance between the q-ROFSs.

In [18], the discussion on the q-ROF distance measure (q-ROFDM) was initiated in the Minkowski metric perspective, and several q-ROFDMs were presented in [19]. Pinar and Boran [20] discussed a q-ROF multi-criteria group decision-making (MCDM) and its use in supplier selection by using a q-ROFDM. Hussain et al. [21] presented a q-ROFDM in the perspective of Hausdorff metric and discussed its practical applications. Ali [22] presented a q-ROFDM using norm metric and used it to advance decision-making, and some uses of q-ROFDMs were discussed in robotics, investment analysis, and clustering [23]. Hussain et al. [24] further discussed practical problems via q-ROFDM. Rani and Kumar [25] presented a distinct q-ROFDM with its practical uses, medical analysis was expatiated in [26] based on certain q-ROFDMs, and the selection means of renewable energy source was presented in [27] using q-ROFDM.

In furtherance of the investigation of distance methods in q-ROFSs, the concept of q-ROFDM has been studied via logarithmic functions. Wang et al. [28] explored two q-ROFLDMs with practical applications. However, one of the approaches incorporates incomplete parameters of the q-ROFSs, which can reduce reliability of the approach. Similarly, a q-ROFDM was constructed in [29] and used to investigate means of selecting investment funds, due to the uncertainties in the financial space. The q-ROFLDM in [29] does not incorporate the complete parameters of the q-ROFSs, which can affect the reliability of the approach.

1.1 Motivation and Objectives

The logarithmic q-ROFDMs [28, 29] are deficient in terms of parameters exclusion and precision. Because of these deficiencies, this work presents a new q-ROFLDM by improving upon the method in [29]. In this new distance technique, the complete parameters of the concerned q-ROFSs are incorporated in the distance scheme. Although some information measures have been deployed for robotics selection [23, 30, 31], the quest to construct an improved logarithmic-based q-ROFDM is very convincing because the existing q-ROFLDMs are limited in one way or the other. Information measures have been deployed to select appropriate robot [23, 30, 31]. In [30], an intuitionistic dense fuzzy entropy approach was deployed to select appropriate industrial robot, and some information measures were used to select a robot for a specific situation under m-polar fuzzy sets [31]. In this work, the existing q-ROFLDMs will be investigated to itemize their drawbacks as leverages for the construction of reliable a q-ROFDMs, to enhance reliable robotics selection process.

The drawbacks of the existing q-ROFLDMs include:

i. The existing q-ROFLDMs are defective and needed to be enhanced for effective applications.
ii. Because the existing q-ROFLDMs are defective, it is expedient to construct dependable q-ROFLDM that are mathematically proven to efficiently measure distance under q-ROFSs.
iii. Many aspects of robotics selection cannot be effectively handled by the existing q-ROFLDMs because their outputs would yield wrong interpretations.

Hence, the objectives of the research include:

i. Assessment of the existing q-ROFLDMs for possible enhancement
ii. Development of an innovative q-ROFLDM to rectify the limitations of the hitherto methods
iii. Discussion on the properties of the new q-ROFLDM to corroborate their agreement with the properties of distance measure
iv. Application of the new q-ROFLDM to discuss robotics selection problems

The rest of this chapter is organized subsequently. Section 2 discusses q-ROFSs and the existing q-ROFLDMs. Section 3 presents the new q-ROFLDM and investigates their metric properties. Section 4 presents the application of the q-ROFLDM in the selection of suitable robotics, and presents a comparative examination for validation's sake. In Section 5, this chapter is concluded and some further research suggestions are presented.

2 Preliminaries

We discuss some properties of q-ROFSs and reiterate certain existing q-ROFLDMs in this section.

Definition 2.1 ([7]). Suppose $\mathfrak{W}$ is a q-ROFSs in $\mathbb{X} = \{\mathbb{x}_1, \mathbb{x}_2, \ldots, \mathbb{x}_p\}$. Then, a q-ROFSs is a structure:

$$\mathfrak{W} = \{(\mathbb{x}_r, \mathfrak{W}_m(\mathbb{x}_r), \mathfrak{W}_n(\mathbb{x}_r)) : \mathbb{x}_r \in \mathbb{X}\},$$

where $r = 1, \ldots, p$, and $\mathfrak{W}_m : \mathbb{X} \to [0,1]$ and $\mathfrak{W}_n : \mathbb{X} \to [0,1]$ are functions defining MD and NMD of $\mathbb{x}_r \in \mathbb{x}_r$ in $\mathfrak{W}$ with the property $0 \leq \mathfrak{W}^q_m(\mathbb{x}_r) + \mathfrak{W}^q_n(\mathbb{x}_r) \leq 1$. The third parameter of $\mathfrak{W}$ is a non-determinate function known as q-ROF hesitation margin (q-ROFHMs) is defined as:

$$\mathfrak{W}_h(\mathbb{x}_r) \in [0,1] = \sqrt[q]{1 - \mathfrak{W}^q_m(\mathbb{x}_r) - \mathfrak{W}^q_n(\mathbb{x}_r)}.$$

For $\mathfrak{W}$ in $\mathbb{X}$, a number $(\mathfrak{W}_m(\mathbb{x}_r), \mathfrak{W}_n(\mathbb{x}_r))$ is called the q-ROF number (q-ROFN).

Definition 2.2 ([7]). Suppose $\mathfrak{W}_1 = (\mathfrak{W}_{1m}(\mathbb{x}_r), \mathfrak{W}_{1n}(\mathbb{x}_r))$ and $\mathfrak{W}_2 = (\mathfrak{W}_{2m}(\mathbb{x}_r), \mathfrak{W}_{2n}(\mathbb{x}_r))$ are q-ROFSs in $\mathbb{X}$. Then:

i) $\mathfrak{W}^c_1 = (\mathfrak{W}_{1n}(\mathbb{x}_r), \mathfrak{W}_{1m}(\mathbb{x}_r))$ and $\mathfrak{W}^c_2 = (\mathfrak{W}_{2n}(\mathbb{x}_r), \mathfrak{W}_{2m}(\mathbb{x}_r))$,
ii) $\mathfrak{W}_1 = \mathfrak{W}_2$ iff, $\forall\, \mathbb{x}_r \in \mathbb{X}$, $\mathfrak{W}_{1m}(\mathbb{x}_r) = \mathfrak{W}_{2m}(\mathbb{x}_r)$ and $\mathfrak{W}_{1n}(\mathbb{x}_r) = \mathfrak{W}_{2n}(\mathbb{x}_r)$,
iii) $\mathfrak{W}_1 \subseteq \mathfrak{W}_2$ iff, $\forall\, \mathbb{x}_r \in \mathbb{X}$, $\mathfrak{W}_{1m}(\mathbb{x}_r) \leq \mathfrak{W}_{2m}(\mathbb{x}_r)$ and $\mathfrak{W}_{1n}(\mathbb{x}_r) \geq \mathfrak{W}_{2n}(\mathbb{x}_r)$,
iv) $\mathfrak{W}_1 \cup \mathfrak{W}_2 = (\max\{\mathfrak{W}_{1m}(\mathbb{x}_r), \mathfrak{W}_{2m}(\mathbb{x}_r)\}, \min\{\mathfrak{W}_{1n}(\mathbb{x}_r), \mathfrak{W}_{2n}(\mathbb{x}_r)\})$,
v) $\mathfrak{W}_1 \cap \mathfrak{W}_2 = (\min\{\mathfrak{W}_{1m}(\mathbb{x}_r), \mathfrak{W}_{2m}(\mathbb{x}_r)\}, \max\{\mathfrak{W}_{1n}(\mathbb{x}_r), \mathfrak{W}_{2n}(\mathbb{x}_r)\})$,
vi) $\mathfrak{W}_1 \otimes \mathfrak{W}_2 = \left(\mathfrak{W}^q_{1m}(\mathbb{x}_r)\mathbb{x}^q_{2m}(\mathbb{x}_r), \sqrt[q]{\mathfrak{W}^q_{1n}(\mathbb{x}_r) + \mathfrak{W}^q_{2n}(\mathbb{x}_r) - \mathfrak{W}^q_{1n}(\mathbb{x}_r)\mathfrak{W}^q_{2n}(\mathbb{x}_r)}\right)$,
vii) $\mathfrak{W}_1 \oplus \mathfrak{W}_2 = \left(\sqrt[q]{\mathfrak{W}^q_{1n}(\mathbb{x}_r) + \mathfrak{W}^q_{2n}(\mathbb{x}_r) - \mathfrak{W}^q_{1n}(\mathbb{x}_r)\mathfrak{W}^q_{2n}(\mathbb{x}_r)}, \mathfrak{W}^q_{1m}(\mathbb{x}_r)\mathfrak{W}^q_{2m}(\mathbb{x}_r)\right)$.

Definition 2.3 ([29]). Let $\mathfrak{W}_1$, $\mathfrak{W}_2$, and $\mathfrak{W}_3$ be q-ROFSs in $\mathbb{X}$ and $\mathfrak{D} : q - \text{ROFSs}(\mathbb{X}) \times q - \text{ROFSs}(\mathbb{X}) \to [0,1]$, so $\mathfrak{D}(\mathfrak{W}_1, \mathfrak{W}_2)$ is a distance function between $\mathfrak{W}_1$ and $\mathfrak{W}_2$ by satisfying the subsequent axioms:

i) $0 \leq \mathfrak{D}(\mathfrak{W}_1, \mathfrak{W}_2) \leq 1$,
ii) $\mathfrak{D}(\mathfrak{W}_1, \mathfrak{W}_2) = 0$ if and only if $\mathfrak{W}_1 = \mathfrak{W}_2$,
iii) $\mathfrak{D}(\mathfrak{W}_1, \mathfrak{W}_2) = \mathfrak{D}(\mathfrak{W}_2, \mathfrak{W}_1)$,
iv) $\mathfrak{D}(\mathfrak{W}_1, \mathfrak{W}^c_1) = 1$ if and only if $\mathfrak{W}_1$ is a crisp set.

2.1 Obtainable Logarithmic-Based Distance Methods Under q-ROFSs

The obtainable logarithmic-based distance methods under q-ROFSs can be found in [28, 29]. Suppose $\mathfrak{W}_1 = (\mathfrak{W}_{1m}(\mathbb{x}_r), \mathfrak{W}_{1n}(\mathbb{x}_r))$ and $\mathfrak{W}_2 = (\mathfrak{W}_{2m}(\mathbb{x}_r), \mathfrak{W}_{2n}(\mathbb{x}_r))$ are q-ROFSs in $\mathbb{X}$.

Wang et al. [28] developed two logarithmic-based distance methods between q-ROFSs, which are:

$$\mathfrak{D}_{W1}(\mathfrak{W}_1, \mathfrak{W}_2) = \sqrt{\frac{\mathfrak{D}_m(\mathfrak{W}_1, \mathfrak{W}_2) + \mathfrak{D}_n(\mathfrak{W}_1, \mathfrak{W}_2)}{2}}, \tag{1}$$

where

$$\left.\begin{aligned} \mathfrak{D}_m(\mathfrak{W}_1, \mathfrak{W}_2) &= \sum_{r=1}^{p} \left[\mathfrak{W}_{1m}^q(\mathbb{x}_r) \log\left(\frac{2\mathfrak{W}_{1m}^q(\mathbb{x}_r)}{\mathfrak{W}_{1m}^q(\mathbb{x}_r) + \mathfrak{W}_{2m}^q(\mathbb{x}_r)}\right) + \mathfrak{W}_{2m}^q(\mathbb{x}_r) \log\left(\frac{2\mathfrak{W}_{2m}^q(\mathbb{x}_r)}{\mathfrak{W}_{1m}^q(\mathbb{x}_r) + \mathfrak{W}_{2m}^q(\mathbb{x}_r)}\right)\right] \\ \mathfrak{D}_n(\mathfrak{W}_1, \mathfrak{W}_2) &= \sum_{r=1}^{p} \left[\mathfrak{W}_{1n}^q(\mathbb{x}_r) \log\left(\frac{2\mathfrak{W}_{1n}^q(\mathbb{x}_r)}{\mathfrak{W}_{1n}^q(\mathbb{x}_r) + \mathfrak{W}_{2n}^q(\mathbb{x}_r)}\right) + \mathfrak{W}_{2n}^q(\mathbb{x}_r) \log\left(\frac{2\mathfrak{W}_{2n}^q(\mathbb{x}_r)}{\mathfrak{W}_{1n}^q(\mathbb{x}_r) + \mathfrak{W}_{2n}^q(\mathbb{x}_r)}\right)\right] \end{aligned}\right\},$$

and

$$\mathfrak{D}_{W2}(\mathfrak{W}_1, \mathfrak{W}_2) = \sqrt{\frac{\mathfrak{D}_m(\mathfrak{W}_1, \mathfrak{W}_2) + \mathfrak{D}_n(\mathfrak{W}_1, \mathfrak{W}_2) + \mathfrak{D}_h(\mathfrak{W}_1, \mathfrak{W}_2)}{2}}, \tag{2}$$

where

$$\left.\begin{aligned} \mathfrak{D}_m(\mathfrak{W}_1, \mathfrak{W}_2) &= \sum_{r=1}^{P} \left[\mathfrak{W}_{1m}^q(\mathbb{x}_r) \log\left(\frac{2\mathfrak{W}_{1m}^q(\mathbb{x}_r)}{\mathfrak{W}_{1m}^q(\mathbb{x}_r) + \mathfrak{W}_{2m}^q(\mathbb{x}_r)}\right) + \mathfrak{W}_{2m}^q(\mathbb{x}_r) \log\left(\frac{2\mathfrak{W}_{2m}^q(\mathbb{x}_r)}{\mathfrak{W}_{1m}^q(\mathbb{x}_r) + \mathfrak{W}_{2m}^q(\mathbb{x}_r)}\right)\right] \\ \mathfrak{D}_n(\mathfrak{W}_1, \mathfrak{W}_2) &= \sum_{r=1}^{P} \left[\mathfrak{W}_{1n}^q(\mathbb{x}_r) \log\left(\frac{2\mathfrak{W}_{1n}^q(\mathbb{x}_r)}{\mathfrak{W}_{1n}^q(\mathbb{x}_r) + \mathfrak{W}_{2n}^q(\mathbb{x}_r)}\right) + \mathfrak{W}_{2n}^q(\mathbb{x}_r) \log\left(\frac{2\mathfrak{W}_{2n}^q(\mathbb{x}_r)}{\mathfrak{W}_{1n}^q(\mathbb{x}_r) + \mathfrak{W}_{2n}^q(\mathbb{x}_r)}\right)\right] \\ \mathfrak{D}_h(\mathfrak{W}_1, \mathfrak{W}_2) &= \sum_{r=1}^{P} \left[\mathfrak{W}_{1h}^q(\mathbb{x}_r) \log\left(\frac{2\mathfrak{W}_{1h}^q(\mathbb{x}_r)}{\mathfrak{W}_{1h}^q(\mathbb{x}_r) + \mathfrak{W}_{2h}^q(\mathbb{x}_r)}\right) + \mathfrak{W}_{2h}^q(\mathbb{x}_r) \log\left(\frac{2\mathfrak{W}_{2h}^q(\mathbb{x}_r)}{\mathfrak{W}_{1h}^q(\mathbb{x}_r) + \mathfrak{W}_{2h}^q(\mathbb{x}_r)}\right)\right] \end{aligned}\right\}.$$

We observe that distance outcomes using eq. (1) do not be trusted because the q-ROFHMs are omitted from the method. Although the method (2) captures the complete parameters of the q-ROFSs, it does not consider the number of the parameters.

In addition, Suri et al. [29] presented a logarithmic-based distance method in the below equation:

$$\mathfrak{D}_s(\mathfrak{W}_1, \mathfrak{W}_2) = \frac{\sum_{r=1}^{p}\left[\left(\mathfrak{W}_{1\mathrm{m}}^{\mathrm{q}}(\mathbb{x}_\mathrm{r})\right) - \left(\mathfrak{W}_{2\mathrm{m}}^{\mathrm{q}}(\mathbb{x}_\mathrm{r})\right) \log\frac{1+\mathfrak{W}_{1\mathrm{m}}^{\mathrm{q}}(\mathbb{x}_r)}{1+\mathfrak{W}_{2\mathrm{m}}^{\mathrm{q}}(\mathbb{x}_r)} + \left(\mathfrak{W}_{1\mathrm{n}}^{\mathrm{q}}(\mathbb{x}_\mathrm{r}) - \mathfrak{W}_{2n}^{q}(\mathbb{x}_r)\right) \log\frac{1+\mathfrak{W}_{1\mathrm{n}}^{\mathrm{q}}(\mathbb{x}_r)}{1+\mathfrak{W}_{2\mathrm{n}}^{\mathrm{q}}(\mathbb{x}_r)}\right]}{2p\log 2}. \tag{3}$$

This logarithmic-based distance method excluded the q-ROFHMs of the considered q-ROFSs, which could compromise the precision of its outcomes.

3 New Logarithmic-Based Distance Method Under q-ROFSs

In this segment, a new logarithmic-based distance method is constructed leveraging on the limitation of the approach in [29], to enhance precision and application reliability.

Definition 3.1. Suppose we have two q-ROFSs defined in $\mathbb{X}$, where $\mathfrak{W}_1 = (\mathfrak{W}_{1m}(x_r), \mathfrak{W}_{1n}(x_r))$ and $\mathfrak{W}_2 = (\mathfrak{W}_{2m}(x_r), \mathfrak{W}_{2n}(x_r))$. Then, the new logarithmic-based distance method is given as follows:

$$\mathfrak{D}_*(\mathfrak{W}_1, \mathfrak{W}_2) = \frac{1}{3p\log 3}\left(\mathfrak{D}_*^m(\mathfrak{W}_1, \mathfrak{W}_2) + \mathfrak{D}_*^n(\mathfrak{W}_1, \mathfrak{W}_2) + \mathfrak{D}_*^h(\mathfrak{W}_1, \mathfrak{W}_2)\right) \tag{4}$$

where

$$\left.\begin{aligned}
\mathfrak{D}_*^m(\mathfrak{W}_1, \mathfrak{W}_2) &= \sum_{r=1}^{p}\left[\left(\mathfrak{W}_{1m}^q(x_r) - \mathfrak{W}_{2m}^q(x_r)\right)\log\frac{1+\mathfrak{W}_{1m}^q(x_r)}{1+\mathfrak{W}_{2m}^q(x_r)}\right]\\
\mathfrak{D}_*^n(\mathfrak{W}_1, \mathfrak{W}_2) &= \sum_{r=1}^{p}\left[\left(\mathfrak{W}_{1n}^q(x_r) - \mathfrak{W}_{2n}^q(x_r)\right)\log\frac{1+\mathfrak{W}_{1n}^q(x_r)}{1+\mathfrak{W}_{2n}^q(x_r)}\right]\\
\mathfrak{D}_*^h(\mathfrak{W}_1, \mathfrak{W}_2) &= \sum_{r=1}^{p}\left[\left(\mathfrak{W}_{1h}^q(x_r) - \mathfrak{W}_{2h}^q(x_r)\right)\log\frac{1+\mathfrak{W}_{1h}^q(x_r)}{1+\mathfrak{W}_{2h}^q(x_r)}\right]
\end{aligned}\right\}.$$

This distance approach is proficient because it incorporates all the parameters of q-ROFSs and the number of the parameters. The new logarithmic-based q-ROFDM is characterized subsequently.

Theorem 3.1. The new q-ROFLDM satisfies the axiomatic distance conditions.

Proof. To verify this claim, we show that the new q-ROFLDM satisfies:

i. $0 \le \mathfrak{D}_*(\mathfrak{W}_1, \mathfrak{W}_2) \le 1$,

ii. $\mathfrak{D}_*(\mathfrak{W}_1, \mathfrak{W}_2) = 0$ if and only if $\mathfrak{W}_1 = \mathfrak{W}_2$,

iii. $\mathfrak{D}_*(\mathfrak{W}_1, \mathfrak{W}_2) = \mathfrak{D}_*(\mathfrak{W}_2, \mathfrak{W}_1)$,

iv. $\mathfrak{D}_*\left(\mathfrak{W}_1, \mathfrak{W}_1^c\right) = 1$ if and only if $\mathfrak{W}_1$ is a crisp set.

Certainly, $\mathfrak{D}_*^m(\mathfrak{W}_1, \mathfrak{W}_2) \ge 0$, $\mathfrak{D}_*^n(\mathfrak{W}_1, \mathfrak{W}_2) \ge 0$ and $\mathfrak{D}_*^h(\mathfrak{W}_1, \mathfrak{W}_2) \ge 0$, and hence $\mathfrak{D}_*(\mathfrak{W}_1, \mathfrak{W}_2) \ge 0$. To show that, $\mathfrak{D}_*(\mathfrak{W}_1, \mathfrak{W}_2) \le 1$, we partially differentiate the functions as follows:

$$\frac{\partial \mathfrak{D}_*}{\partial \mathfrak{W}_{1m}} = q\,\mathfrak{W}_{1m}^{\,q-1}(\mathbb{x}_r)\log\frac{1+\mathfrak{W}_{1m}^{q}(\mathbb{x}_r)}{1+\mathfrak{W}_{2m}^{q}(\mathbb{x}_r)} + \frac{\left(\mathfrak{W}_{1m}^{q}(\mathbb{x}_r)-\mathfrak{W}_{2m}^{q}(\mathbb{x}_r)\right)}{\frac{1+\mathfrak{W}_{1m}^{q}(\mathbb{x}_r)}{1+\mathfrak{W}_{2m}^{q}(\mathbb{x}_r)}\ln 10},$$

$$\frac{\partial \mathfrak{D}_*}{\partial \mathfrak{W}_{2m}} = -qc_{2m}{}^{q-1}\log\frac{1+\mathfrak{W}_{1m}^{q}(\mathbb{x}_r)}{1+\mathfrak{W}_{2m}^{q}(\mathbb{x}_r)} + \frac{\left(\mathfrak{W}_{1m}^{q}(\mathbb{x}_r)-\mathfrak{W}_{2m}^{q}(\mathbb{x}_r)\right)}{\frac{1+\mathfrak{W}_{1m}^{q}(\mathbb{x}_r)}{1+\mathfrak{W}_{2m}^{q}(\mathbb{x}_r)}\ln 10}.$$

Similarly, $\frac{\partial \mathfrak{D}_*}{\partial \mathfrak{W}_{1n}}, \frac{\partial \mathfrak{D}_*}{\partial \mathfrak{W}_{2n}}, \frac{\partial \mathfrak{D}_*}{\partial \mathfrak{W}_{1_h}}, \frac{\partial \mathfrak{D}_*}{\partial \mathfrak{W}_{2h}}$ are obtained. But if $\mathfrak{W}_{1m}^{q}(\mathbb{x}_r) \ge \mathfrak{W}_{2m}^{q}(\mathbb{x}_r)$ implies $\frac{\partial \mathfrak{D}_*}{\partial \mathfrak{W}_{1m}} \ge 0$ and $\frac{\partial \mathfrak{D}_*}{\partial \mathfrak{W}_{2m}} \le 0$, and if $\mathfrak{W}_{2m}^{q}(\mathbb{x}_r) \ge \mathfrak{W}_{1m}^{q}(\mathbb{x}_r)$ implies $\frac{\partial \mathfrak{D}_*}{\partial \mathfrak{W}_{2m}} \ge 0$ and $\frac{\partial \mathfrak{D}_*}{\partial \mathfrak{W}_{1m}} \le 0$. Following similar logic, we have:

$$\mathfrak{W}_{1n}^{q}(\mathbb{x}_r) \ge \mathfrak{W}_{2n}^{q}(\mathbb{x}_r) \Rightarrow \frac{\partial \mathfrak{D}}{\partial \mathfrak{W}_{1n}} \ge 0 \text{ and } \frac{\partial \mathfrak{D}}{\partial \mathfrak{W}_{2n}} \le 0,$$

$$\mathfrak{W}_{2n}^{q}(\mathbb{x}_r) \ge \mathfrak{W}_{1n}^{q}(\mathbb{x}_r) \Rightarrow \frac{\partial \mathfrak{D}}{\partial \mathfrak{W}_{2n}} \ge 0 \text{ and } \frac{\partial \mathfrak{D}}{\partial \mathfrak{W}_{1n}} \le 0,$$

$$\mathfrak{W}_{1h}^{q}(\mathbb{x}_r) \ge \mathfrak{W}_{2h}^{q}(\mathbb{x}_r) \Rightarrow \frac{\partial \mathfrak{D}}{\partial \mathfrak{W}_{1h}} \ge 0 \text{ and } \frac{\partial \mathfrak{D}}{\partial \mathfrak{W}_{2h}} \le 0,$$

$$\mathfrak{W}_{2h}^{q}(\mathbb{x}_r) \ge \mathfrak{W}_{1h}^{q}(\mathbb{x}_r) \Rightarrow \frac{\partial \mathfrak{D}}{\partial \mathfrak{W}_{2h}} \ge 0 \text{ and } \frac{\partial \mathfrak{D}}{\partial \mathfrak{W}_{1h}} \le 0.$$

Thus, the monotonicity of $\mathfrak{D}_*$ is reversible for $\mathfrak{W}_1$ and $\mathfrak{W}_2$, and the greatest value of $\mathfrak{D}_*^m(\mathfrak{W}_1,\mathfrak{W}_2)$, $\mathfrak{D}_*^n(\mathfrak{W}_1,\mathfrak{W}_2)$ and $\mathfrak{D}_*^h(\mathfrak{W}_1,\mathfrak{W}_2)$ occur at $(1,0)$. Hence, $\mathfrak{D}_*^m(\mathfrak{W}_1,\mathfrak{W}_2) \le 1$, $\mathfrak{D}_*^n(\mathfrak{W}_1,\mathfrak{W}_2) \le 1$ and $\mathfrak{D}_*^h(\mathfrak{W}_1,\mathfrak{W}_2) \le 1$, and hence $\mathfrak{D}_*(\mathfrak{W}_1,\mathfrak{W}_2) \le 1$. Therefore, $\mathfrak{D}_*(\mathfrak{W}_1, \mathfrak{W}_2) \in [0,1]$, proving (i).

Suppose $\mathfrak{D}_*(\mathfrak{W}_1,\mathfrak{W}_2)=0$, then $\mathfrak{D}_*^m(\mathfrak{W}_1,\mathfrak{W}_2)=\mathfrak{D}_*^n(\mathfrak{W}_1,\mathfrak{W}_2)=\mathfrak{D}_*^h(\mathfrak{W}_1,\mathfrak{W}_2)=0$. Hence, we have $\left(\mathfrak{W}_{1m}^{q}(\mathbb{x}_r)-\mathfrak{W}_{2m}^{q}(\mathbb{x}_r)\right)\log\frac{1+\mathfrak{W}_{1m}^{q}(\mathbb{x}_r)}{1+\mathfrak{W}_{2m}^{q}(\mathbb{x}_r)}=0 \Rightarrow \left(\mathfrak{W}_{1m}^{q}(\mathbb{x}_r)-\mathfrak{W}_{2m}^{q}(\mathbb{x}_r)\right)=0 \Rightarrow \mathfrak{W}_{1m}^{q}(\mathbb{x}_r)=\mathfrak{W}_{2m}^{q}(\mathbb{x}_r)$. In the same way, $\mathfrak{W}_{1n}^{q}(\mathfrak{W}_r)=\mathfrak{W}_{2n}^{q}(\mathbb{x}_r)$ and $\mathfrak{W}_{1h}^{q}(\mathbb{x}_r)=\mathfrak{W}_{2h}^{q}(\mathbb{x}_r)$. Hence, $\mathfrak{W}_1=\mathfrak{W}_2$.

Alternatively, $\log\frac{1+\mathfrak{W}_{1m}^{q}(\mathbb{x}_r)}{1+\mathfrak{W}_{2n}^{q}(\mathbb{x}_r)}=0 \Rightarrow \frac{1+\mathfrak{W}_{1m}^{q}(\mathbb{x}_r)}{1+\mathfrak{W}_{2m}^{q}(\mathbb{x}_r)}=1 \Rightarrow \mathfrak{W}_{1m}^{q}(\mathbb{x}_r)=\mathfrak{W}_{2m}^{q}(\mathbb{x}_r)$. In the same vein, $\mathfrak{W}_{1m}^{q}(\mathbb{x}_r)=\mathfrak{W}_{2m}^{q}(\mathbb{x}_r)$ and $\mathfrak{W}_{1h}^{q}(\mathbb{x}_r)=\mathfrak{W}_{2h}^{q}(\mathbb{x}_r)$. Hence, $\mathfrak{W}_1=\mathfrak{W}_2$.

Conversely, if $\mathfrak{W}_1=\mathfrak{W}_2$, then $\mathfrak{D}_*^m(\mathfrak{W}_1,\mathfrak{W}_2)=\mathfrak{D}_*^n(\mathfrak{W}_1,\mathfrak{W}_2)=\mathfrak{D}_*^h(\mathfrak{W}_1,\mathfrak{W}_2)=0$. Thus, $\mathfrak{D}_*(\mathfrak{W}_1,\mathfrak{W}_2)=0$. Hence, (ii) is proved.

The proof of (iii), $\mathfrak{D}_*(\mathfrak{W}_1,\mathfrak{W}_2)=\mathfrak{D}_*(\mathfrak{W}_2,\mathfrak{W}_1)$ is straightforward. For the prove of (iv), take $\mathfrak{W}_1=(\mathfrak{W}_{1m},\mathfrak{W}_{1n})$, then $\mathfrak{W}_1{}^c=(\mathfrak{W}_{1n},\mathfrak{W}_{1m})$. In crisp set, $\mathfrak{W}_{1m}(\mathbb{x}_r)=1-\mathfrak{W}_{1n}(\mathbb{x}_r)$ and $\mathfrak{W}_{1_h}(\mathbb{x}_r)=0$. Then $\mathfrak{D}_*\left(\mathfrak{W}_1,\mathfrak{W}_1^c\right)=1 \Leftrightarrow$

$$\frac{1}{3p\log 3}\sum_{r=1}^{p}\left[\begin{array}{c}\left(\mathfrak{W}_{1m}^{q}(\mathbb{x}_r)-\mathfrak{W}_{1m}^{q}(\mathbb{x}_r)\right)\log\frac{1+\mathfrak{W}_{1m}^{q}(\mathbb{x}_r)}{1+\mathfrak{W}_{1n}^{q}(\mathbb{x}_r)}+\left(\mathfrak{W}_{1n}^{q}(\mathbb{x}_r)-\mathfrak{W}_{1m}^{q}(\mathbb{x}_r)\right)\log\frac{1+\mathfrak{W}_{1n}^{q}(\mathbb{x}_r)}{1+\mathfrak{W}_{1m}^{q}(\mathbb{x}_r)}\\ +\left(\mathfrak{W}_{1h}^{q}(\mathbb{x}_r)-\mathfrak{W}_{2h}^{q}(\mathbb{x}_r)\right)\log\frac{1+\mathfrak{W}_{1h}^{q}(\mathbb{x}_r)}{1+\mathfrak{W}_{2h}^{q}(\mathbb{x}_r)}\end{array}\right]$$

$$=\frac{1}{2p\log 2}\sum_{r=1}^{p}\left[\begin{array}{c}\left(\mathfrak{W}_{1m}^{q}(\mathbb{x}_r)-\mathfrak{W}_{1n}^{q}(\mathbb{x}_r)\right)\log\frac{1+\mathfrak{W}_{1m}^{q}(\mathbb{x}_r)}{1+\mathfrak{W}_{1n}^{q}(\mathbb{x}_r)}\\ +\left(\mathfrak{W}_{1m}^{q}(\mathbb{x}_r)-\mathfrak{W}_{1n}^{q}(\mathbb{x}_r)\right)\log\frac{1+\mathfrak{W}_{1m}^{q}(\mathbb{x}_r)}{1+\mathfrak{W}_{1n}^{q}(\mathbb{x}_r)}\end{array}\right]$$

$$=\frac{2}{2p\log 2}\sum_{r=1}^{p}\left[\left(\mathfrak{W}_{1m}^{q}(\mathbb{x}_r)-\mathfrak{W}_{1n}^{q}(\mathbb{x}_r)\right)\log\frac{1+\mathfrak{W}_{1m}^{q}(\mathbb{x}_r)}{1+\mathfrak{W}_{1n}^{q}(\mathbb{x}_r)}\right]$$

$$=\frac{1}{\log 2^p}\sum_{r=1}^{p}\left[\left(\mathfrak{W}_{1m}^{q}(\mathbb{x}_r)-\mathfrak{W}_{1n}^{q}(\mathbb{x}_r)\right)\log\frac{1+\mathfrak{W}_{1m}^{q}(\mathbb{x}_r)}{1+\mathfrak{W}_{1n}^{q}(\mathbb{x}_r)}\right]=1$$

$$\Leftrightarrow\sum_{r=1}^{p}\left[\left(\mathfrak{W}_{1m}^{q}(\mathbb{x}_r)-\mathfrak{W}_{1n}^{q}(\mathbb{x}_r)\right)\log\frac{1+\mathfrak{W}_{1m}^{q}(\mathbb{x}_r)}{1+\mathfrak{W}_{1n}^{q}(\mathbb{x}_r)}\right]=\log 2^p$$

$\Leftrightarrow\left(\mathfrak{W}_{1m}^{q}(\mathbb{x}_r)-\mathfrak{W}_{1n}^{q}(\mathbb{x}_r)\right)=1$ or $\log\frac{1+\mathfrak{W}_{1m}^{q}(\mathbb{x}_r)}{1+\mathfrak{W}_{1n}^{q}(\mathbb{x}_r)}=\log 2^p$. Then $\mathfrak{W}_{1m}^{q}(\mathbb{x}_r)-\mathfrak{W}_{1n}^{q}(\mathbb{x}_r)=1$, which means $\mathfrak{W}_{1m}(\mathbb{x}_r)=1$, $\mathfrak{W}_{1n}(\mathbb{x}_r)=0$ or $\mathfrak{W}_{1m}(\mathbb{x}_r)=0$, $\mathfrak{W}_{1n}(\mathbb{x}_r)=1$. Hence, $\mathfrak{W}_1$ is a crisp set, which proves (iv).

3.1 Verifications of the New *q*-ROFDMs

A number of numerical examples of q-ROFSs are supplied to display the appropriateness of the new q-ROFLDM over the obtainable q-ROFLDMs. First, we numerically investigate whether the new q-ROFLDM fulfills the metric conditions.

Example 3.1. Given three q-ROFSs $\mathfrak{W}_1$ and $\mathfrak{W}_2$ in $\mathbb{X}$ for $p=3$, where

$$\mathfrak{W}_1=((0.5,0.3),(0.8,0.4),(0.5,0.7))$$
$$\mathfrak{W}_2=((0.5,0.6),(0.5,0.8),(0.8,0.9))$$

If $q=5$, based on new q-ROFLDM, the distance between $\mathfrak{B}$, $\hat{\mathfrak{B}}$, and $\tilde{\mathfrak{B}}$ are computed as follows:

$$\mathfrak{D}_*(\mathfrak{W}_1,\mathfrak{W}_1)=0,\mathfrak{D}_*(\mathfrak{W}_2,\mathfrak{W}_2)=0$$
$$\mathfrak{D}_*(\mathfrak{W}_1,\mathfrak{W}_2)=0.075,\mathfrak{D}_*(\mathfrak{W}_2,\mathfrak{W}_1)=0.075$$

It can be seen that the new q-ROFLDM fulfills the famous distance metric: $0\le\mathfrak{D}_*(\mathfrak{W}_1,\mathfrak{W}_2)\le 1$, $\mathfrak{D}_*(\mathfrak{W}_1,\mathfrak{W}_2)=0$ iff $\mathfrak{W}_1=\mathfrak{W}_2$, and $\mathfrak{D}_*(\mathfrak{W}_1,\mathfrak{W}_2)=\mathfrak{D}_*(\mathfrak{W}_2,\mathfrak{W}_1)$. Certainly, these conditions hold $\forall q$.

Example 3.2. Suppose $\mathfrak{W}_1$, $\mathfrak{W}_2$, and $\mathfrak{W}_3$ are q-ROFSs in $\mathbb{X}$ for $p = 3$, where

$$\mathfrak{W}_1 = ((0.56, 0.36), (0.8, 0.55), (0.5, 0.6))$$

$$\mathfrak{W}_2 = ((0.62, 0.45), (0.8, 0.55), (0.41, 0.6))$$

$$\mathfrak{W}_3 = ((0.7, 0.53), (0.6, 0.43), (0.8, 0.56))$$

If $q = 5$, the distances between the q-ROFSs are computed using new q-ROFLDM as follows:

$$\mathfrak{D}_*(\mathfrak{W}_1, \mathfrak{W}_2) = 3.0E-4, \mathfrak{D}_*(\mathfrak{W}_1, \mathfrak{W}_3) = 0.0245, \mathfrak{D}_*(\mathfrak{W}_3, \mathfrak{W}_2) = 0.0249$$

Certainly,

$$\mathfrak{D}_*(\mathfrak{W}_1, \mathfrak{W}_3) + \mathfrak{D}_*(\mathfrak{W}_3, \mathfrak{W}_2) = 0.0494 > \mathfrak{D}_*(\mathfrak{W}_1, \mathfrak{W}_2) = 3.0E-4,$$

Thus, the property $\mathfrak{D}_*(\mathfrak{W}_1, \mathfrak{W}_2) \leq \mathfrak{D}_*(\mathfrak{W}_1, \mathfrak{W}_3) + \mathfrak{D}_*(\mathfrak{W}_3, \mathfrak{W}_2)$ is fulfilled. In fact, the triangle inequality is true $\forall q$.

Subsequently, we express the advantages of the new q-ROFLDM over the current q-ROFLDMs as follows:

Example 3.3. Given $\mathfrak{W}_1$ and $\mathfrak{W}_2$ as q-ROFSs in $\mathbb{X}$, $p = 2$ in two cases as presented by Table 1. To compare the appropriateness of the new q-ROFLDM with the current q-ROFLDMs, we compute the distances between the q-ROFSs for $q = 5$.

Table 1: Distance outcomes.

q-ROFDMs	Case I $\mathfrak{W}_1 = ((\mathbf{0.1}, \mathbf{0.3}), (\mathbf{0.6}, \mathbf{0.3}))$ $\mathfrak{W}_2 = ((\mathbf{0.2}, \mathbf{0.35}), (\mathbf{0.57}, \mathbf{0.2}))$	M_1	M_3	Case II $\mathfrak{W}_1 = ((\mathbf{0.36}, \mathbf{0.3}), (\mathbf{0.6}, \mathbf{0.3}))$ $\mathfrak{W}_2 = ((\mathbf{0.4}, \mathbf{0.3}), (\mathbf{0.5}, \mathbf{0.4}))$	M_1	M_3
$\mathfrak{D}_*$	1.0E−4	✓	✓	4.0E−4	✓	✓
$\mathfrak{D}_{W1}$ [28]	0.012	✓	✓	0.0319	✓	✓
$\mathfrak{D}_{W2}$ [28]	0.0142	✓	✓	0.0351	✓	✓
$\mathfrak{D}_S$ [29]	0.0	×	×	1.0E−4	✓	✓

M_1: appropriate results; M_2: fulfilling distance maxims.

The distance values in Table 1 are represented in Figure 1.

The considered q-ROFSs in Case I are very similar and likewise the ones in Case II. With this, the distance methods ought to yield very minimal distance outcomes. From Table 1 and Figure 1, we notice that, the two q-ROFLDMs, $\mathfrak{D}_{W1}$ and $\mathfrak{D}_{W2}$ [28] yield distance outcomes that do not quite explain the actual distance between the q-ROFSs. The q-ROFLDM, $\mathfrak{D}_S$ [29] violates a distance axiom in Case I. The distance between two q-ROFSs is zero if and only if the q-ROFSs are identical. While all the q-ROFLDMs satisfied that condition, the approach in [29] violates the condition. On the

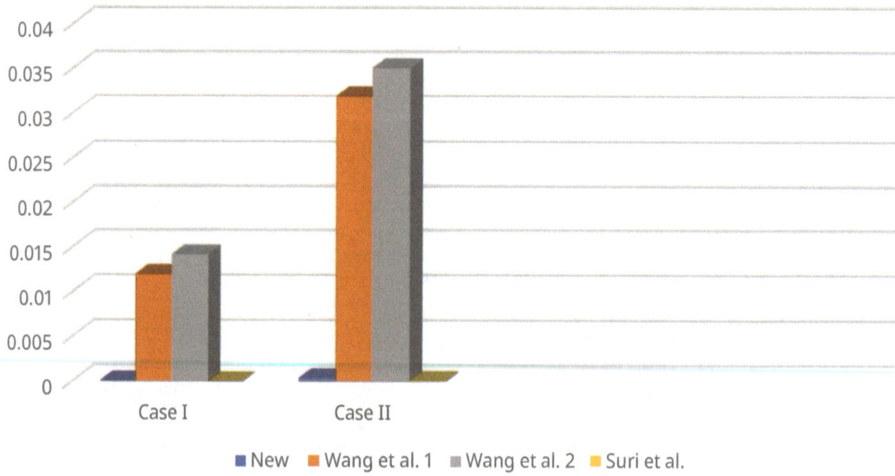

Figure 1: Distance outcomes representation.

contrary, the new q-ROFLDM produces distance values that are appropriate, fulfilling distance axioms, precise, and reliable than the considered q-ROFLDMs.

4 Robotics Analysis Based on a New q-ROFLDM

The application of the q-ROFLDM is demonstrated in the selection of robots for an effective robotic functioning. In the application, a multi-robot system consists of four patrolling robots are considered to be assigned for a purpose of surveillance in an area. The robots are remotely controlled by a particular controller (as shown in Figure 2 [32]). The area for surveillance is partitioned into four equal portions and assigned to each of the four robots. The robots move about in their designated areas. The robots have necessary sensors installed in them to send essential information to the controller, which are used to make decisions. For instance, if the temperature sensor of a particular robot reads an infrequent temperature, the controller will directs the robot to change direction. In a situation, there are many situations, to detect the appropriate robot for such case is complex due to imprecision. To achieve this, the recognition principle is employed by integrating the new logarithmic-based q-ROFDM.

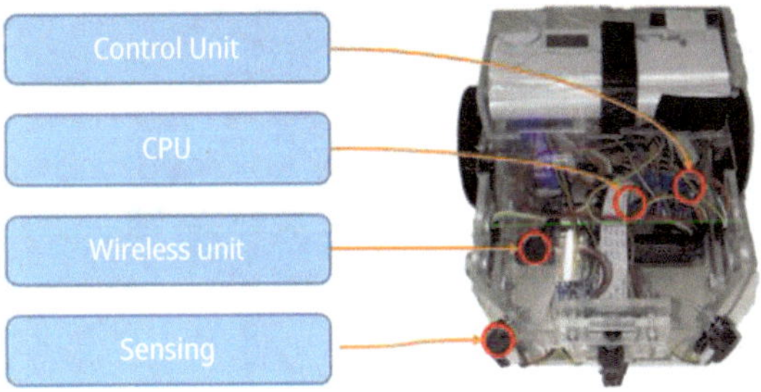

Figure 2: Robotic agent diagram.

4.1 Algorithm for the Recognition Principle

1. Obtain the robots, $\mathfrak{R} = \{\mathfrak{R}_1, \ldots, \mathfrak{R}_q\}$ in terms of q-ROF information based on the set of sensors, $\mathfrak{S} = \{\mathfrak{S}_1, \ldots, \mathfrak{S}_p\}$, where q is the number of the robots and p is the number of the sensors.
2. Obtain the q-ROF information for the situations, $\mathfrak{B} = \{\mathfrak{B}_1, \ldots, \mathfrak{B}_s\}$, where s is the number of the situations.
3. Find the distance between each $\mathfrak{R}$ and $\mathfrak{B}$ in terms of $\mathfrak{S}$ using eq. (5):

$$\mathfrak{D}_*(\mathfrak{R}, \mathfrak{B}) = \frac{1}{3p \log 3}\left(\mathfrak{D}_*^m(\mathfrak{R}, \mathfrak{B}) + \mathfrak{D}_*^n(\mathfrak{R}, \mathfrak{B}) + \mathfrak{D}_*^h(\mathfrak{R}, \mathfrak{B})\right) \tag{5}$$

where

$$\left.\begin{aligned}
\mathfrak{D}_*^m(\mathfrak{R}, \mathfrak{B}) &= \sum_{r=1}^{p}\left[\left(\mathfrak{R}_m^q(\mathfrak{S}_r) - \mathfrak{B}_m^q(\mathfrak{S}_r)\right)\log\frac{1+\mathfrak{R}_m^q(\mathfrak{S}_r)}{1+\mathfrak{B}_m^q(\mathfrak{S}_r)}\right]\\
\mathfrak{D}_*^n(\mathfrak{R}, \mathfrak{B}) &= \sum_{r=1}^{p}\left[\left(\mathfrak{R}_n^q(\mathfrak{S}_r) - \mathfrak{B}_n^q(\mathfrak{S}_r)\right)\log\frac{1+\mathfrak{R}_n^q(\mathfrak{S}_r)}{1+\mathfrak{B}_n^q(\mathfrak{S}_r)}\right]\\
\mathfrak{D}_*^h(\mathfrak{R}, \mathfrak{B}) &= \sum_{r=1}^{p}\left[\left(\mathfrak{R}_h^q(\mathfrak{S}_r) - \mathfrak{B}_h^q(\mathfrak{S}_r)\right)\log\frac{1+\mathfrak{R}_h^q(\mathfrak{S}_r)}{1+\mathfrak{B}_h^q(\mathfrak{S}_r)}\right]
\end{aligned}\right\}$$

4. Determine which robot has the least distance with a situation, as the appropriate robot to be deployed for that particular situation.

The algorithm is presented in Figure 3.

4.2 Application Example

In this application, four robots and five situations are considered, i.e., $\mathfrak{R} = \{\mathfrak{R}_1, \mathfrak{R}_2, \mathfrak{R}_3, \mathfrak{R}_4\}$ and $\mathfrak{B} = \{\mathfrak{B}_1, \mathfrak{B}_2, \mathfrak{B}_3, \mathfrak{B}_4, \mathfrak{B}_5\}$. The elements of the situation $\mathfrak{B}$ are fire, obstacle/stumbling block, bump/collision, cliff, and vibration, respectively. Both $\mathfrak{R}$ and $\mathfrak{B}$ are

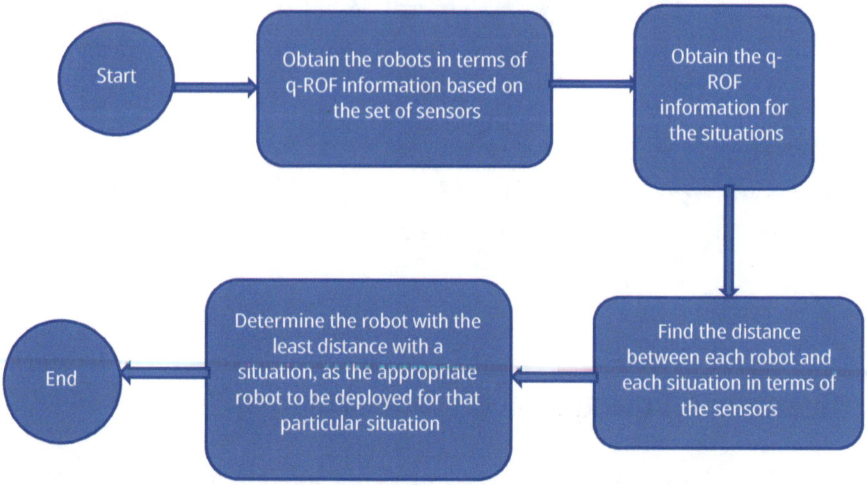

Figure 3: Flowchart of the algorithm.

based on the set of sensors, $\mathfrak{S} = \{\mathfrak{S}_1, \mathfrak{S}_2, \mathfrak{S}_3, \mathfrak{S}_4, \mathfrak{S}_5\}$ representing ultrasonic sensor, temperature sensor, accelerometer sensor, bump sensor, and cliff sensor, respectively. By adapting the robots/situations data in [31] based on the stated sensors, Tables 2 and 3 are presented.

Table 2: Situations and sensors.

	$\mathfrak{S}_1$	$\mathfrak{S}_2$	$\mathfrak{S}_3$	$\mathfrak{S}_4$	$\mathfrak{S}_5$
$\mathfrak{B}_1$	(0.45, 0.84)	(0.89, 0.32)	(0.32, 0.85)	(0.30, 0.85)	(0.55, 0.70)
$\mathfrak{B}_2$	(0.90, 0.20)	(0.46, 0.83)	(0.23, 0.85)	(0.45, 0.72)	(0.35, 0.85)
$\mathfrak{B}_3$	(0.33, 0.86)	(0.76, 0.55)	(0.93, 0.40)	(0.32, 0.85)	(0.32, 0.80)
$\mathfrak{B}_4$	(0.75, 0.40)	(0.32, 0.86)	(0.53, 0.70)	(0.33, 0.86)	(0.89, 0.53)
$\mathfrak{B}_5$	(0.47, 0.70)	(0.56, 0.83)	(0.36, 0.86)	(0.84, 0.50)	(0.79, 0.46)

Table 3: Robots and sensors.

	$\mathfrak{S}_1$	$\mathfrak{S}_2$	$\mathfrak{S}_3$	$\mathfrak{S}_4$	$\mathfrak{S}_5$
$\mathfrak{R}_1$	(0.53, 0.29)	(0.21, 0.82)	(0.22, 0.78)	(0.31, 0.48)	(0.27, 0.87)
$\mathfrak{R}_2$	(0.25, 0.96)	(0.24, 0.95)	(0.89, 0.39)	(0.20, 0.95)	(0.44, 0.87)
$\mathfrak{R}_3$	(0.63, 0.75)	(0.41, 0.89)	(0.84, 0.50)	(0.63, 0.76)	(0.98, 0.21)
$\mathfrak{R}_4$	(0.45, 0.86)	(0.87, 0.37)	(0.53, 0.73)	(0.31, 0.78)	(0.44, 0.86)

The distances between the robots $\mathfrak{R}$ and the situations $\mathfrak{B}$ using eq. (5) for $q = 8$ are computed, and the following distance values are obtained:

$$\mathfrak{D}_*(\mathfrak{R}_1, \mathfrak{B}_1) = 0.0306, \mathfrak{D}_*(\mathfrak{R}_1, \mathfrak{B}_2) = 0.0175, \mathfrak{D}_*(\mathfrak{R}_1, \mathfrak{B}_3) = 0.0418$$
$$\mathfrak{D}_*(\mathfrak{R}_1, \mathfrak{B}_4) = 0.0231, \mathfrak{D}_*(\mathfrak{R}_1, \mathfrak{B}_5) = 0.0097$$
$$\mathfrak{D}_*(\mathfrak{R}_2, \mathfrak{B}_1) = 0.0806, \mathfrak{D}_*(\mathfrak{R}_2, \mathfrak{B}_2) = 0.0939, \mathfrak{D}_*(\mathfrak{R}_2, \mathfrak{B}_3) = 0.0632$$
$$\mathfrak{D}_*(\mathfrak{R}_2, \mathfrak{B}_4) = 0.0859, \mathfrak{D}_*(\mathfrak{R}_2, \mathfrak{B}_5) = 0.0967$$
$$\mathfrak{D}_*(\mathfrak{R}_3, \mathfrak{B}_1) = 0.0827, \mathfrak{D}_*(\mathfrak{R}_3, \mathfrak{B}_2) = 0.0719, \mathfrak{D}_*(\mathfrak{R}_3, \mathfrak{B}_3) = 0.0765$$
$$\mathfrak{D}_*(\mathfrak{R}_3, \mathfrak{B}_4) = 0.0258, \mathfrak{D}_*(\mathfrak{R}_3, \mathfrak{B}_5) = 0.0468$$
$$\mathfrak{D}_*(\mathfrak{R}_4, \mathfrak{B}_1) = 0.0100, \mathfrak{D}_*(\mathfrak{R}_4, \mathfrak{B}_2) = 0.0268, \mathfrak{D}_*(\mathfrak{R}_4, \mathfrak{B}_3) = 0.0301$$
$$\mathfrak{D}_*(\mathfrak{R}_4, \mathfrak{B}_4) = 0.0321, \mathfrak{D}_*(\mathfrak{R}_4, \mathfrak{B}_5) = 0.0238$$

These distance values are presented in Figure 4.

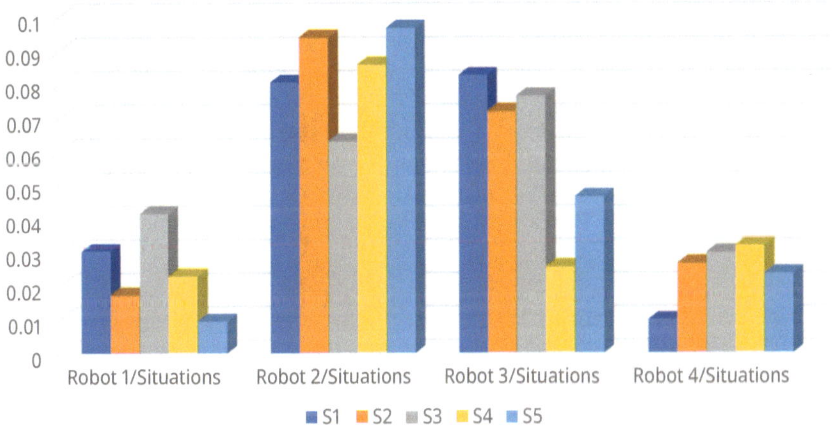

Figure 4: Distances between robots and situations.

These distance values are used to determine the appropriate robot for each of the situations based on least distance value. From Figure 4, we see that robot, $\mathfrak{R}_1$ is appropriate for vibration ($\mathfrak{B}_5$); $\mathfrak{R}_2$ is appropriate for bump/collision ($\mathfrak{B}_3$); $\mathfrak{R}_3$ is appropriate for cliff ($\mathfrak{B}_4$); and $\mathfrak{R}_4$ is appropriate for fire ($\mathfrak{B}_1$). However, none of the robots is suitable to detect obstacle/stumbling block.

4.3 Sensitivity Analysis

In this segment, the impact of q is determined where $q = 4, 5, 8, 10$. The values of q are used on the robots/situations' data, and the following distance outcomes are presented in Table 4.

Table 4: Sensitivity analysis.

q	$(\mathfrak{R}_1, \mathfrak{B}_1)$	$(\mathfrak{R}_1, \mathfrak{B}_2)$	$(\mathfrak{R}_1, \mathfrak{B}_3)$	$(\mathfrak{R}_1, \mathfrak{B}_4)$	$(\mathfrak{R}_1, \mathfrak{B}_5)$	Decision
4	0.0768	0.0342	0.0931	0.0578	0.0413	$\mathfrak{R}_1$ is appropriate for $\mathfrak{B}_2$
5	0.0623	0.0299	0.0768	0.0474	0.0278	$\mathfrak{R}_1$ is appropriate for $\mathfrak{B}_5$
8	0.0306	0.0175	0.0418	0.0231	0.0097	$\mathfrak{R}_1$ is appropriate for $\mathfrak{B}_5$
10	0.0185	0.0118	0.0284	0.0139	0.0052	$\mathfrak{R}_1$ is appropriate for $\mathfrak{B}_5$
	$(\mathfrak{R}_2, \mathfrak{B}_1)$	$(\mathfrak{R}_2, \mathfrak{B}_2)$	$(\mathfrak{R}_2, \mathfrak{B}_3)$	$(\mathfrak{R}_2, \mathfrak{B}_4)$	$(\mathfrak{R}_2, \mathfrak{B}_5)$	
4	0.1030	0.1160	0.0529	0.1059	0.1169	$\mathfrak{R}_2$ is appropriate for $\mathfrak{B}_3$
5	0.0981	0.1119	0.0584	0.1034	0.1140	$\mathfrak{R}_2$ is appropriate for $\mathfrak{B}_3$
8	0.0806	0.0939	0.0632	0.0859	0.0967	$\mathfrak{R}_2$ is appropriate for $\mathfrak{B}_3$
10	0.0702	0.0813	0.0609	0.0742	0.0841	$\mathfrak{R}_2$ is appropriate for $\mathfrak{B}_3$
	$(\mathfrak{R}_3, \mathfrak{B}_1)$	$(\mathfrak{R}_3, \mathfrak{B}_2)$	$(\mathfrak{R}_3, \mathfrak{B}_3)$	$(\mathfrak{R}_3, \mathfrak{B}_4)$	$(\mathfrak{R}_3, \mathfrak{B}_5)$	
4	0.1088	0.0991	0.0869	0.0283	0.0580	$\mathfrak{R}_3$ is appropriate for $\mathfrak{B}_4$
5	0.1037	0.0915	0.0854	0.0269	0.0532	$\mathfrak{R}_3$ is appropriate for $\mathfrak{B}_4$
8	0.0827	0.0719	0.0765	0.0258	0.0468	$\mathfrak{R}_3$ is appropriate for $\mathfrak{B}_4$
10	0.0705	0.0625	0.0693	0.0262	0.0452	$\mathfrak{R}_3$ is appropriate for $\mathfrak{B}_4$
	$(\mathfrak{R}_4, \mathfrak{B}_1)$	$(\mathfrak{R}_4, \mathfrak{B}_2)$	$(\mathfrak{R}_4, \mathfrak{B}_3)$	$(\mathfrak{R}_4, \mathfrak{B}_4)$	$(\mathfrak{R}_4, \mathfrak{B}_5)$	
4	0.0129	0.0602	0.0380	0.0776	0.0615	$\mathfrak{R}_4$ is appropriate for $\mathfrak{B}_1$
5	0.0135	0.0508	0.0380	0.0644	0.0500	$\mathfrak{R}_4$ is appropriate for $\mathfrak{B}_1$
8	0.0100	0.0268	0.0301	0.0321	0.0238	$\mathfrak{R}_4$ is appropriate for $\mathfrak{B}_1$
10	0.0069	0.0168	0.0235	0.0195	0.0139	$\mathfrak{R}_4$ is appropriate for $\mathfrak{B}_1$

The distance outcomes in Table 4 for $q = 4, 5, 8, 10$ show the same decision with the exception of $q = 4$ for the first case involving $\mathfrak{R}_1$. The impact for $q = 4, 5, 8, 10$ is shown in terms of precision. It is observed that the precision of the new q-ROFLDM decreases as the values of q decrease, and increases as the values of q decrease. To achieve a reliable distance outcomes, it is advisable to consider greater value of q.

4.4 Comparative Investigation

In this segment, we present the effectiveness of the new q-ROFLDM by comparing its outcomes with the outcomes of the existing q-ROFLDMs. The computed results for $\mathfrak{D}_{W2}$ $q = 8$ are displayed in Table 5.

Table 5 shows the distance outcomes of the new and existing q-ROFLDMs. The existing q-ROFLDMs and the new q-ROFLDM have the same decision with the exception of w_1 [28] for the first case involving $\mathfrak{R}_1$. The new q-ROFLDM possesses better precision compared to the existing q-ROFLDMs, indicating the advantage of the new q-ROFLDM.

Table 5: Comparative distance outcomes.

q-ROFDMs	$(\Re_1, \mathfrak{B}_2)$	$(\Re_1, \mathfrak{B}_2)$	$(\Re_1, \mathfrak{B}_3)$	$(\Re_1, \mathfrak{B}_4)$	$(\Re_1, \mathfrak{B}_5)$	**Decision**
$\mathfrak{D}_*$	0.0306	0.0175	0.0418	0.0231	0.0097	$\Re_1$ is appropriate for $\mathfrak{B}_5$
$\mathfrak{D}_{W1}$ [28]	0.4408	0.2711	0.4844	0.4028	0.2798	$\Re_1$ is appropriate for $\mathfrak{B}_2$
$\mathfrak{D}_{W2}$ [28]	0.4597	0.2965	0.5122	0.4121	0.2909	$\Re_1$ is appropriate for $\mathfrak{B}_5$
$\mathfrak{D}_S$ [29]	0.0526	0.0248	0.0687	0.0459	0.0154	$\Re_1$ is appropriate for $\mathfrak{B}_5$
	$(\Re_2, \mathfrak{B}_1)$	$(\Re_2, \mathfrak{B}_2)$	$(\Re_2, \mathfrak{B}_3)$	$(\Re_2, \mathfrak{B}_4)$	$(\Re_2, \mathfrak{B}_5)$	
$\mathfrak{D}_*$	0.0806	0.0939	0.0632	0.0859	0.0967	$\Re_2$ is appropriate for $\mathfrak{B}_3$
$\mathfrak{D}_{W1}$ [28]	0.5712	0.5983	0.3936	0.5618	0.5975	$\Re_2$ is appropriate for $\mathfrak{B}_3$
$\mathfrak{D}_{W2}$ [28]	0.6173	0.6486	0.4727	0.6241	0.6598	$\Re_2$ is appropriate for $\mathfrak{B}_3$
$\mathfrak{D}_S$ [29]	0.1418	0.1642	0.0869	0.1340	0.1558	$\Re_2$ is appropriate for $\mathfrak{B}_3$
	$(\Re_3, \mathfrak{B}_1)$	$(\Re_3, \mathfrak{B}_2)$	$(\Re_3, \mathfrak{B}_3)$	$(\Re_3, \mathfrak{B}_4)$	$(\Re_3, \mathfrak{B}_5)$	
$\mathfrak{D}_*$	0.0827	0.0719	0.0765	0.0258	0.0468	$\Re_3$ is appropriate for $\mathfrak{B}_4$
$\mathfrak{D}_{W1}$ [28]	0.5832	0.5607	0.5102	0.3032	0.4689	$\Re_3$ is appropriate for $\mathfrak{B}_4$
$\mathfrak{D}_{W2}$ [28]	0.6412	0.6075	0.5787	0.3583	0.5048	$\Re_3$ is appropriate for $\mathfrak{B}_4$
$\mathfrak{D}_S$ [29]	0.1350	0.1264	0.1171	0.0340	0.0860	$\Re_3$ is appropriate for $\mathfrak{B}_4$
	$(\Re_4, \mathfrak{B}_1)$	$(\Re_4, \mathfrak{B}_2)$	$(\Re_4, \mathfrak{B}_3)$	$(\Re_4, \mathfrak{B}_4)$	$(\Re_4, \mathfrak{B}_5)$	
$\mathfrak{D}_*$	0.0100	0.0268	0.0301	0.0321	0.0238	$\Re_4$ is appropriate for $\mathfrak{B}_1$
$\mathfrak{D}_{W1}$ [28]	0.1974	0.4513	0.3341	0.5052	0.4310	$\Re_4$ is appropriate for $\mathfrak{B}_1$
$\mathfrak{D}_{W2}$ [28]	0.2162	0.4565	0.3682	0.5106	0.4416	$\Re_4$ is appropriate for $\mathfrak{B}_1$
$\mathfrak{D}_S$ [29]	0.0147	0.0583	0.0457	0.0700	0.0458	$\Re_4$ is appropriate for $\mathfrak{B}_1$

5 Conclusion

This chapter showed the application of q-ROFSs in the analysis of robotic selection for effective deployment based on q-ROFLDM. The research presented a new method of computing distance between q-ROFSs based on logarithmic functions, because of the limitation of the existing q-ROFLDMs. The practical use of the new q-ROFLDM was shown in robotics selection analysis via q-ROF information, to select appropriate robots for situations like fire, obstacle/stumbling block, bump/collision, cliff, and vibration based on sensors such as ultrasonic sensor, temperature sensor, accelerometer sensor, bump sensor, and cliff sensor, respectively. The new q-ROFLDM provides a more accurate robotics selection analysis. This research provides a huge improvement on the q-ROF-based decision-making system in robotic analysis. In addition, the effectiveness of the new q-ROFLDM was shown via comparative analysis with the existing q-ROFLDMs, which shows the superiority of the new q-ROFLDM concerning precision. In future research, the new q-ROFLDM can be stretched to other fuzzy settings and applied to various areas, including waste management systems, circular economy, and farming site optimization.

References

[1] Chuang CC, Jeng JT. Editorial message: special issue on advances fuzzy methods on control, robotics, and applications. Int J Fuzzy Syst. 2024;26(1):331–332.

[2] Minchev Z. A metamorphic robot described in generalized nets terms. Third International Workshop on GNs, Sofia, 1st October 2002. 202; pp. 37–41.

[3] Zadeh LA.Fuzzy sets. Inf Control. 1965;8:338–353.

[4] Atanassov KT. Intuitionistic fuzzy sets. Fuzzy Sets Syst. 1986;20:87–96.

[5] Yager RR. Pythagorean membership grades in multicriteria decision making. IEEE Trans Fuzzy Syst. 2014;22(4):958–965.

[6] Senapati T, Yager RR. Fermatean fuzzy sets. J Ambient Intell Human Comput. 2020;11:663–674.

[7] Yager RR. Generalized orthopair fuzzy sets. IEEE Trans Fuzzy Syst. 2017;25(5):1222–1230.

[8] Garg H. A novel correlation coefficients between Pythagorean fuzzy sets and its applications to decision-making processes. International. J Intell Syst. 2016;31(12):1234–1252.

[9] Ahemen S, Ujah R, Ejegwa PA. New cosine similarity measures for intuitionistic fuzzy sets with application in decision-making. In: I. Ali, UM Modibbo, A. L. Bolaji, H. Garg(eds), Optimization and Computing using Intelligent Data-Driven Approaches for Decision-Making: Optimization Applications. CRC Press: Boca Raton; 2024. pp. 55–69.

[10] Anum MT, Zhang H, Ejegwa PA, Feng Y. Tendency coefficient-based weighted distance measure for intuitionistic fuzzy sets with applications. In: Proceedings of the 12th International Conference on Intelligent Control and Information Processing (ICICIP 2024), Nanjing, China. 2024; pp. 54–61.

[11] Zhou Y, Ejegwa PA, Johnny SE. Generalized similarity operator for intuitionistic fuzzy sets and its applications based on recognition principle and multiple criteria decision making technique, International. J Comput Intell Syst. 2023;16:85.

[12] Kirisci M. Correlation coefficients of Fermatean fuzzy sets with a medical application. J Math Sci Model. 2022;5(1):16–23.

[13] Nwokoro CO, Inyang UG, Eyoh IJ, Ejegwa PA. Intuitionistic fuzzy approach for predicting maternal outcomes. In: Jana, C., Pal, M., Muhiuddin, G., Liu, P. (eds) Fuzzy Optimization, Decision-making and Operations Research. Springer: Cham; 2023. pp. 399–421.

[14] Zuakwagh D, Ejegwa PA, Akwu AD. Fermatean fuzzy sets in multiset framework with application in diagnostic process based on composite relations. In: K. Kumar, R. Arora, G. Kaur(eds), Strategic Fuzzy Extensions and Decision-making Techniques. CRC Press; 2024. pp. 195–210.

[15] Ejegwa PA. Decision-making on patients' medical status based on a q-rung orthopair fuzzy max-min -max composite relation. In: Garg, H.(ed), q-Rung Orthopair Fuzzy Sets: Theory and Applications. Springer: Singapore; 2022. pp. 47–66.

[16] Ejegwa PA, Anum MT, Kausar N, Nwokoro CO, Aydin N, Yu H. New Fermatean fuzzy distance metric and its utilization in the assessment of security crises using the MCDM technique. Mathematics. 2024;12:3214.

[17] Anum MT, Ejegwa PA, Feng Y. An enhanced Pythagorean fuzzy distance method with application in the selection of renewable energy source via MCDA. 2024 International Conference on New Trends in Computational Intelligence (NTCI), Qingdao, China. 2024; pp. 392–401.

[18] Du WS. Minkowski-type distance measures for generalized orthopair fuzzy sets. Int J Intell Syst. 2018;33:802–817.

[19] Peng XD, Liu L. Information measures for q-rung orthopair fuzzy sets. Int J Intell Syst. 2019;34:1795–1834.

[20] Pinar P, Boran FE. A q-rung orthopair fuzzy multi-criteria group decision making for supplier selection based on a novel distance measure. Int J Mach Learn Cybern. 2020;11:1749–1780.

[21] Hussain Z, Abbas S, Yang MS. Distances and similarity measures of q-rung orthopair fuzzy sets based on the hausdorff metric with the construction of orthopair fuzzy TODIM. Symmetry. 2022;*14*(11):2467.

[22] Ali J. Norm-based distance measure of q-rung orthopair fuzzy sets and its application in decision-making. Comput Appl Math. 2023;42:184.

[23] Ejegwa PA. New q-rung orthopair fuzzy distance-similarity operators with applications in investment analysis, pattern recognition, clustering analysis, and selection of robot for smart manufacturing. *Soft Comput*. 2023; https://doi.org/10.1007/s00500-023-08799-1.

[24] Hussain R, Hussain Z, Sarhan NM, Juraev N, Ur Rahman S. Distance and similarity measures on belief and plausibility under q-rung orthopair fuzzy sets with applications. Sci Rep. 2024;14:18959.

[25] Rani V, Kumar S. An innovative distance measure for quantifying the dissimilarity between q-rung orthopair fuzzy sets. Decis Anal J. 2024;11:100440.

[26] Ejegwa PA, Anum MT, Kausar N, Vrioni B. Some new q-rung orthopair fuzzy distance techniques and their application to medical analysis based on decision-making methods. J Decis Oper Res. 2025; e226485. https://doi.org/10.22105/dmor.2025.530184.1967.

[27] Ejegwa PA, Anum MT, Kausar N, Nwokoro CO, Cagin T. A novel q-rung orthopair fuzzy distance metric for energy source selection. J Appl Res Ind Eng. 2025;12(3):429–464.

[28] Wang D, Yuan Y, Liu Z, Zhu S, Sun Z. Novel distance measures of q-rung orthopair fuzzy sets and their applications. *Symmetry*. 2024;16:574.

[29] Suri G, Gandotra N, Guleria A, Saini N. Novel distance measure for q-rung orthopair fuzzy sets and its application. Int J Knowl Based Intell Eng Syst. 2025;29(3):297–305.

[30] Sampathkumar S, Augustin F, Kaabar MKA, Yue XG. An integrated intuitionistic dense fuzzy entropy-COPRAS-WASPAS approach for manufacturing robot selection. Adv Mech Eng. 2023;15(3):1–18.

[31] Kumam W, Naeem K, Riaz M, Khan MJ, Kumam P. Comparison measures for Pythagorean m-polar fuzzy sets and their applications to robotics and movie recommender system. AIMS Math. 2023;8(5):10357–10378.

[32] Jiménez AC, García-Díaz V, Bolaños S. A decentralized framework for multi-agent robotic systems. Sensors. 2018;18(2):417.

Fathima Banu*, Subramanian Petchimuthu

Optimize the Segregation of Healthcare Waste in Delhi Using Scaled *q*-Rung Orthopair Fuzzy Sets and Sugeno-Weber Operators

Abstract: Handling healthcare wastes in large urban cities such as Delhi is an urgent issue, due to the fact that through the conventional decision-making approaches, the uncertainty and divergence of opinion held by experts are not taken into consideration. In an attempt to circumvent such limitations, this chapter introduces the decision-making model that has been established to determine the best waste management techniques, with special consideration to segregation that is the most crucial of all steps in the healthcare waste cycle. The framework is based on scaled *q*-rung orthopair fuzzy sets (S*q*-ROFS), which is the advanced technique that enhances the traditional fuzzy models by the ability to adjust membership and non-membership levels independently. This leeway allows presenting the uncertainty and hesitation that usually accompanies expert judgments in healthcare policy more realistically. The gathered assessments are subsequently combined using Sugeno-Weber power weighted aggregation operators, and the evaluation of several, and at times conflicting, criteria is even-handed. This chapter analyses five segregation measures applicable to Delhi, i.e., (i) color-coded bin systems with standardized labeling, (ii) staff education and awareness, (iii) point-of-disposal standard operating procedure posters and visual aids, (iv) digital and artificial intelligence (AI)-based bin guidance systems, and (v) segregation audit with performance incentives. The assessment will involve issues like cost-effectiveness, the accuracy of segregation, the practicality of operations, compliance with regulations, and the acceptability of the stakeholders. To conclude, it has been indicated that a hybrid system that deploys both color-coded bin systems and AI-enhanced tracking and incentive systems may provide the most feasible and long-term viable solution. The framework therefore offers policymakers and healthcare managers with an effective instrument to attain the efficient, safe, and environmental responsibility in separating healthcare waste in the highly populated cities.

Acknowledgments: The authors extend their sincere gratitude to Dr. C. Justin Dhanaraj, Assistant Professor of Chemistry at Anna University, Chennai (Nagercoil Campus), for his valuable domain expertise and collaboration in conducting the expert panel evaluations for the case study.

***Corresponding author: Fathima Banu,** Department of Science and Humanities (Mathematics), University College of Engineering, Konam, Nagercoil, Tamil Nadu, India, e-mail: fathimams1999@gmail.com
Subramanian Petchimuthu, Department of Science and Humanities (Mathematics), University College of Engineering, Konam, Nagercoil, India, e-mail: svmuthu1983@gmail.com

 | https://doi.org/10.1515/9783112222249-009

Keywords: healthcare waste management, Scaled q-Rung Orthopair Fuzzy Sets, Sugeno-Weber power operators, waste segregation, urban sustainability

1 Introduction

It is estimated that every year, healthcare activities generate millions of tons of hazardous, infectious, or even toxic waste, which is estimated to be 15% of all waste generated by healthcare activities [15]. This continuous stream of biomedical waste (BMW) poses a serious danger to human health and environmental balance, a problem which is magnified to life threatening levels in the increasing megacities of the world [21]. Its most acute manifestation is observed in urban metropolises, which are distributed all over the country, e.g., in Delhi, a megalopolis in which the COVID-19 pandemic recorded a peak of 321 tons [24] of BMW generation per day. In this type of city that operates on this level, the line between community well-being and ecological crisis is at the end of a needle as trash. Figure 1 is revealing that the correct sorting of garbage is not the trivial logistic issue, but the most susceptible and risky step in the safeguarding of the health of the metropolis.

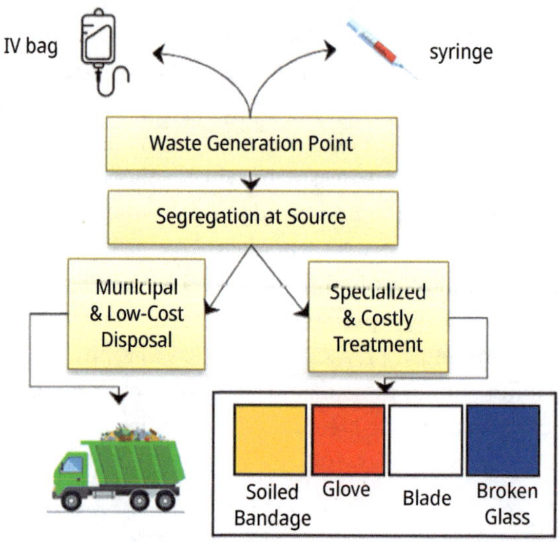

Figure 1: The critical role of segregation in the healthcare waste life cycle. The key decision point presented in this model can be viewed as the separation of waste into general and hazardous streams, which can be organized under the global best practices [15] and the regulations on this matter expressed in the official status reports [32].

This investigation is motivated by a number of relevant constraints present in the current healthcare waste management framework, specifically, in a high-pressure urban setting, which needs a more strict and advanced decision-making support framework:

- Individuality and inconsistency in ratings: Experts do not determine the individual sorting strategies based on objective machine readings, but instead, based on expert ratings. These professional evaluations, about which considerations – including, but not limited to operational feasibility, worker safety, and community support – are subjective, imprecise, and varied among decision takers.
- Complexity and conflicting criteria: The decision makers must consider several competing, conflicting criteria – for example, the least cost is not always safe and the environmentally preferred is not always logistically feasible. Conventional decision models are not always successful in managing the trade-off that is inevitable.
- Uncertainty and ambiguity in data: The data that will be used to make decisions is normally ambiguous and qualitative more than definitive and quantitative. The analysts will regularly count with themselves in words like high risk or moderately effective none of which can be well managed by the conventional quantitative models.
- High stakes of decision-making: The errors that may occur during waste segregation may be both short term (epidemic of infectious diseases) and long term (contamination of the environment). The risky situation requires a decision-making model, which addresses the situation of uncertainty and possible adverse effects in a systematic manner.

Essentially, choosing a preferred segregation strategy consists of making trade-offs. In operations research, this is formally recognized as a multi-criteria decision-making (MCDM) problem. There is a well-established record of advanced fuzzy MCDM frameworks excelling at tackling highly consequential logistical and risk-based problems. This includes recent work dealing with selecting reverse logistics suppliers [26], optimizing a healthcare supply chain [30], considering uncertain logistics environments [22], and risk sensitivity [34] to quality of services. Recent work has shown the flexibility of the q-rung orthopair fuzzy set (q-ROFS) approach by creating new logarithmic distance measures for vehicle selection [4], new measures for energy sources selection [12], and new correlation coefficients to augment multi-criteria analysis [41]. However, it is important to understand that the truly difficult aspect of the healthcare waste issue is the type of information available. The outcome of every strategy is not dictated by actual machine readings, rather they are based on evaluations from experts, which are subjective and not precise. Classical MCDM frameworks, which inherently see crisp numerical data, are qualitatively mismatched to the nature of the challenge.

In order to overcome this strategic-level gap and deal with the corresponding uncertainty, advanced fuzzy modelling tools are necessary. The evolution of fuzzy set (FS) theory, commencing from intuitionistic fuzzy sets (IFS) [10] to q-ROFS [38], has produced increasingly meaningful representations. Nevertheless, standard q-ROFS is insufficient in circumstances where experts have asymmetric confidence in their assessment, for example, having high confidence in the positive experience of an option but much less confidence in the negative experience. To address this circumstance, q-rung linear Diophantine fuzzy set (q-RLDFS) introduced scaling parameters [3]. However, q-RLDFS has two major deficiencies: (i) it has limited boundary conditions that prevents it from being a true generalization of q-ROFS and (ii) it has a score function that is flawed and can yield illogical ranking reversals. This chapter addresses these issues by utilizing the scaled q-rung orthopair fuzzy set (Sq-ROFS) framework proposed by the authors, which rectifies these issues with revised boundary conditions and an axiomatically robust score function [23].

To combine the different assessments sourced from Sq-ROFS, this research utilizes the Sugeno-Weber (SW) aggregation operators [16]. Unlike many parameterized operators that are simply variations on sensitivity, the SW framework introduces a controllable parameter (ζ) that specifies the form of aggregation, thereby providing unlimited variability between compensatory (averaging) and non-compensatory (conjunctive) behaviors. This property of tunability allows SW operators to effectively address decision contexts ranging from optimistic to conservative, and as a result, provides substantial robustness under conditions of uncertainty. As more research [7, 8, 22, 26] is emerging to indicate the adaptability of SW operators in a number of fuzzy environments and varying decision contexts, this research extends from these previous works by using the Sugeno-Weber Power aggregation operators [2, 12, 26, 33], which is a more refined variant of SW aggregators, designed to mitigate biases in group decision-making processes with the inclusion of power-based weighting of the degree of consensus among the experts.

To address this important gap, this chapter adopts and utilizes a robust decision-making approach based on Sq-ROFS originally developed by the authors for the selection of municipal solid waste management technologies in Pune, India [23].

This work provides the following noteworthy contributions:

- A full decision-making framework is developed using Sq-ROFS incorporating an array of specialized Sugeno-Weber power weighted aggregation operators, specifically the scaled q-rung orthopair fuzzy Sugeno-Weber power weighted arithmetic (Sq-ROFSWPoWA) and geometric (Sq-ROFSWPoWG) operators.
- The pertinent roles of these operators are outlined in a formal manner and the mechanism of the definition strength is seen in how the "power weighted" component of the definition down-weights undrafted opinions to consensus.
- The resultant framework is illustrated with an elaborate MCDM implementation of the high-stake segregation problem of healthcare waste in the City of Delhi, India that is credible in offering a sure way of public health policy.

This study offers an effective and scaled instrument to the policymaker and fortifies the concept of fuzzy MCDM theory by introducing a framework that is highly expressive and dependable. The rest of this chapter is structured in the following way. The approach is described in Section 2. The detailed case study is given in Section 3. Section 4 describes the application and shows the results as a final part, which are later validated and discussed. Finally, Section 5 gives the conclusion of this chapter, providing the main findings and proposing further research perspectives.

2 The Proposed Methodological Framework

Here, we present the entire methodology of the proposed decision-support system. We begin by giving a review of the FS theory, and then formal introduction to scaled q-rung orthopair fuzzy set (Sq-ROFS) [23] and its laws of operation. Notably, we present an axiomatically sound score function and we describe how we have improved on the erroneous previous models. Then we design the aggregation structure based on Sugeno-Weber power average operators within the Sq-ROFS structure. We then put together all the elements so as to have an entire MCDM algorithm.

2.1 Preliminaries

We base our work on the development of the FS theory. A classical FS defines a membership grade (MG) $\mu_{\mathcal{M}}(x) \in [0,1]$ of an element x [42]. Yager's q-ROFS [40] extends this by including a non-membership grade (NMG) $\nu_{\mathcal{M}}(x)$, where the pair (μ, ν) must satisfy the condition $\mu_{\mathcal{M}}^q(x) + \nu_{\mathcal{M}}^q(\mathrm{x}) \leq 1$ for $q \geq 1$. This framework offers a wide area of describing expert uncertainty. Every pair (μ, ν) that satisfies this pair is referred to as a q-rung orthopair fuzzy number (q-ROFN). Although q-ROFS are as flexible as possible, the recent efforts to generalize them, including the q-rung linear Diophantine fuzzy set (q-RLDFS) have brought fundamental mathematical weaknesses into the framework. In order to circumvent these drawbacks, our design is based on the newly proposed Sq-ROFS, introduced by Petchimuthu and Kumar [23]. This framework will give strong and stable grounds to simulate sophisticated expert decisions. The formal definition is as follows:

Definition 1 ([23]). Let $\mathfrak{X}$ be a non-empty universe. A Sq-ROFS $\mathcal{M}$ on $\mathfrak{X}$ is expressed as:

$$\mathcal{M} = \left\{ \left\langle x, \left(\mu_{\mathcal{M}}(x), \nu_{\mathcal{M}}(x)\right), (\alpha, \beta) \right\rangle | x \in \mathfrak{X} \right\},$$

where $\mu_{\mathcal{M}}(x), \nu_{\mathcal{M}}(x), \alpha, \beta \in [0,1]$. Here:

- $\mu_{\mathcal{M}}(x)$ and $\nu_{\mathcal{M}}(x)$ represent the MG and NMG, respectively, of x.
- α and β are parameters of scale or levels of confidence used with the MG and NMG, respectively.

The following boundary conditions must be met by these parameters:

$$\mu_{\mathcal{M}}^q(\mathrm{x}) + \nu_{\mathcal{M}}^q(\mathrm{x}) \le 1, \quad \alpha\mu_{\mathcal{M}}^q(\mathrm{x}) + \beta\nu_{\mathcal{M}}^q(\mathrm{x}) \le 1, \quad \forall x \in \mathfrak{X}, \quad q \ge 1.$$

A single tuple $\mathcal{M} = \langle (\mu_{\mathcal{M}}, \nu_{\mathcal{M}}), (\alpha, \beta) \rangle$ is known as a scaled q-rung orthopair fuzzy number (Sq-ROFN).

The given formulation, as introduced by Petchimuthu and Kumar [23], is one of the main benefits over the q-RLDFS framework, as it is a valid generalization of q-ROFS (when $\alpha = \beta = 1$) and can represent state of complete confidence that q-RLDFS prohibits invalidly. Continuing with this background effort, we shall now establish the score and accuracy functions on which we shall base on our proposed framework to rank.

Definition 2 ([23]). In order to rank Sq-ROFNs, we determine an axiomatically sound score function. For an Sq-ROFN $\mathcal{M} = \langle (\mu_{\mathcal{M}}, \nu_{\mathcal{M}}), (\alpha, \beta) \rangle$, the score $\mathfrak{S}(\mathcal{M})$ is given by:

$$\mathfrak{S}(\mathcal{M}) = \frac{1}{2}\left(1 + \alpha\mu_{\mathcal{M}}{}^q - \beta\nu_{\mathcal{M}}{}^q\right), \ \mathfrak{S}(\mathcal{M}) \in [0, 1]$$

If $\mathfrak{S}(\mathcal{M}_1) = \mathfrak{S}(\mathcal{M}_2)$, the accuracy function $\mathrm{A}(\mathcal{M})$ is used as a tie-breaker:

$$\mathrm{A}(\mathcal{M}) = \frac{1 + \alpha\mu_{\mathcal{M}}{}^q + \beta\nu_{\mathcal{M}}{}^q}{2 + \alpha + \beta}$$

This score operation is shown to be better than the q-RLDFS score operation since this score operation does not encounter the devastating vice of ranking reversal as we are going to briefly demonstrate below.

Example 1 (Hacking of Ranking Reversal in q-RLDFS). Consider two q-ROFNs: $\mathcal{M}_A = (0.85, 0.45)$ and $\mathcal{M}_B = (0.78, 0.35)$ with $q = 4$. Using the standard q-ROFS score $(\mu_{\mathcal{M}}{}^q - \nu_{\mathcal{M}}{}^q)$, $\mathcal{M}_A$ (score ≈ 0.48) is clearly preferred to $\mathcal{M}_B$ (score ≈ 0.35). When a professional displays the identical, biased confidence by stating $\alpha = \beta = 0.5$, the Sq-ROFS score would be correct in maintaining this ranking ($\mathcal{M}_A \succ \mathcal{M}_B$). On the contrary, the defective q-RLDFS score function [3] contradicts the preference for $\mathcal{M}_B \succ \mathcal{M}_A$. This verifies the lack of reliability of the q-RDFS model and creates the need of our proposed model.

2.2 Sugeno-Weber Power Aggregation Operators for Sq-ROFNs

In order to scale the opinions of a number of experts, we employ Power Average (PA) operators that place greater weight on those opinions that are well-founded through a group consensus. We combine this PA mechanism with the flexible Sugeno-Weber) t-norm (SWTN) and t–conorm (SWTCN).

Definition 3 (Distance and Support). In order to quantify the dissimilarity of two Sq-ROFNs, we define a distance function with the results of their final scaled membership and non-membership values. Let $\mathcal{M}_i = \langle(\mu_i, \nu_i), (\alpha_i, \beta_i)\rangle$ and $\mathcal{M}_k = \langle(\mu_k, \nu_k), (\alpha_k, \beta_k)\rangle$ be two Sq-ROFNs. The distance $\mathfrak{D}(\mathcal{M}_i, \mathcal{M}_k)$ between them is defined as:

$$\mathfrak{D}(\mathcal{M}_i, \mathcal{M}_k) = \frac{1}{2}\left(\left|\alpha_i\mu_i^q - \alpha_k\mu_k^q\right| + \left|\beta_i\nu_i^q - \beta_k\nu_k^q\right|\right).$$

The support S_i of an opinion of an expert $\mathcal{M}_i$ is the sum of its similarity to all other opinions, calculated as $S_i = \sum_{k=1,k\neq i}^{m}(1 - \mathfrak{D}(\mathcal{M}_i, \mathcal{M}_k))$.

Definition 4 ([32]). Given a set of expert weights $\mathfrak{M}_i$, the final power weight Θ_i for the ith expert's opinion is:

$$\Theta_i = \frac{\mathfrak{M}_i(1 + S_i)}{\sum\limits_{k=1}^{m}\mathfrak{M}_k(1 + S_k)}.$$

The weighting coefficients are then borrowed to be the weighting coefficients in the aggregation operators. The algebraic framework of these operators is to be constructed on the most adaptable SWTN and SWTCN that characterizes the fuzzy intersection and union operations, respectively.

Definition 5 ([16]). Let $\mu_1, \mu_2 \in [0, 1]$ and let the parameter $\zeta \in (-1, \infty)$. The Sugeno Weber t-norm (SWTN) and t-conorm (SWTCN) are defined as:

$$\begin{aligned}\mathrm{SWTN}(\mu_1, \mu_2) &= \max\left(0, \tfrac{\mu_1 + \mu_2 - 1 + \zeta\mu_1\mu_2}{1+\zeta}\right)\\ \mathrm{SWTCN}(\mu_1, \mu_2) &= \min\left(1, \mu_1 + \mu_2 - \tfrac{\zeta\mu_1\mu_2}{1+\zeta}\right)\end{aligned}$$

The parameter ζ lets us tweak how the operator behaves, making it flexible for various decision-making situations.

We have built on these basic operations for Sq-ROFNs and connected them with power weights to create our final aggregation operators. We define an arithmetic operator, inspired by the SWTCN, and a geometric operator, based on the SWTN, to ensure thorough analysis.

2.3 Aggregation Operators for Sq-ROFNs

Having the basic components in place, we may now define the last aggregation operators. These operators combine the power weights (Θ_i) to control group consensus with the SWTN and the SWTCN to do the fuzzy arithmetic. We introduce an arithmetic and a geometric operator, since each of them gives a different view upon the aggregated result.

Definition 6. Let $\{\mathcal{M}_1, \ldots, \mathcal{M}_m\}$ be a collection of Sq-ROFNs, where $\mathcal{M}_i = \langle(\mu_i, \nu_i), (\alpha_i, \beta_i)\rangle$, and let Θ_i be their corresponding power weights. The scaled q-rung orthopair fuzzy Sugeno-Weber power weighted arithmetic (Sq-ROFSWPoWA) operator is defined as:

$$\text{Sq} - \text{ROFSWPoWA}(\mathcal{M}_1, \ldots, \mathcal{M}_m) = \overset{m}{\underset{i=1}{\oplus}} \Theta_i \mathcal{M}_i = \left\langle \left(\mu_{agg}, \nu_{agg}\right), \left(\alpha_{agg}, \beta_{agg}\right) \right\rangle$$

where the components of the aggregated Sq-ROFN are calculated as follows, using the SWTCN for membership-like parts and the SWTN for non-membership-like parts with a parameter $\zeta \in (-1, \infty)$:

$$\mu_{agg} = \left(\min\left\{ 1, \sum_{i=1}^{m} \Theta_i \mu_i^q - \frac{\zeta}{1+\zeta} \sum_{i=1}^{m-1} \sum_{j=i+1}^{m} \Theta_i \Theta_j \mu_i^q \mu_j^q \right\} \right)^{1/q}$$

$$\nu_{agg} = \left(\max\left\{ 0, \frac{\sum_{i=1}^{m} \Theta_i \nu_i^q - 1 + \zeta \sum_{i=1}^{m-1} \sum_{j=i+1}^{m} \Theta_i \Theta_j \nu_i^q \nu_j^q}{1+\zeta} \right\} \right)^{1/q}$$

$$\alpha_{agg} = \min\left\{ 1, \sum_{i=1}^{m} \Theta_i \alpha_i - \frac{\zeta}{1+\zeta} \sum_{i=1}^{m-1} \sum_{j=i+1}^{m} \Theta_i \Theta_j \alpha_i \alpha_j \right\}$$

$$\beta_{agg} = \max\left\{ 0, \frac{\sum_{i=1}^{m} \Theta_i \beta_i - 1 + \zeta \sum_{i=1}^{m-1} \sum_{j=i+1}^{m} \Theta_i \Theta_j \beta_i \beta_j}{1+\zeta} \right\}$$

The overall aggregate is calculated on the grounds of the operational laws according to SWTCN and SWTN. An elaborate numerical explanation of this operator is given in Appendix 6.

Definition 7. Similarly, the Scaled q-rung orthopair fuzzy Sugeno-Weber power weighted geometric (Sq-ROFSWPoWG) operator is defined as:

$$\text{Sq} - \text{ROFSWPoWG}(\mathcal{M}_1, \ldots, \mathcal{M}_m) = \overset{m}{\underset{i=1}{\otimes}} \mathcal{M}_i^{\Theta_i} = \left\langle \left(\mu_{agg}, \nu_{agg}\right), \left(\alpha_{agg}, \beta_{agg}\right) \right\rangle$$

where the components are calculated using the SWTN for membership-like parts and the SWTCN for non-membership-like parts:

$$\mu_{agg} = \left(\max\left\{ 0, \frac{\sum_{i=1}^{m} \Theta_i \mu_i^q - 1 + \zeta \sum_{i=1}^{m-1} \sum_{j=i+1}^{m} \Theta_i \Theta_j \mu_i^q \mu_j^q}{1+\zeta} \right\} \right)^{1/q}$$

$$\nu_{agg} = \left(\min\left\{ 1, \sum_{i=1}^{m} \Theta_i \nu_i^q - \frac{\zeta}{1+\zeta} \sum_{i=1}^{m-1} \sum_{j=i+1}^{m} \Theta_i \Theta_j \nu_i^q \nu_j^q \right\} \right)^{1/q}$$

$$\alpha_{agg} = \max\left\{ 0, \frac{\sum_{i=1}^{m} \Theta_i \alpha_i - 1 + \zeta \sum_{i=1}^{m-1} \sum_{j=i+1}^{m} \Theta_i \Theta_j \alpha_i \alpha_j}{1+\zeta} \right\}$$

$$\beta_{agg} = \min\left\{ 1, \sum_{i=1}^{m} \Theta_i \beta_i - \frac{\zeta}{1+\zeta} \sum_{i=1}^{m-1} \sum_{j=i+1}^{m} \Theta_i \Theta_j \beta_i \beta_j \right\}$$

The difference between these two operators is important to make an overall analysis. Compensation between criteria is possible using Sq-ROFSWPoWA (arithmetic) operator, a high score in one dimension may negate a low score in another indicating an overall optimistic or averaging attitude. The Sq-ROFSWPoWG (geometric) operator, on the other hand, is non-compensatory; a single very low score will be very detrimental to the overall outcome, as it represents a more conservative, consensus-seeking attitude with all criteria reasonably doing well.

2.4 The Complete MCDM Framework

At this point all the theoretical elements are defined, so the remaining task is to integrate the elements and develop a complex and systematic MCDM framework. MCDM algorithm helps in the process steps of evaluating a set of alternatives with several criteria in the state of uncertainty based on Sq-ROFS model. Figure 2 represents the whole process.

1. **Problem Structuring.** Determine the heart of the decision problem:
   - The finite set of alternatives to be evaluated, denoted as $\mathcal{J} = \{\mathcal{J}_1, \mathcal{J}_2, \ldots, \mathcal{J}_n\}$.
   - The comprehensive set of evaluation criteria, denoted as $\mathcal{B} = \{\mathcal{B}_1, \mathcal{B}_2, \ldots, \mathcal{B}_m\}$.
   - The relative importance of each criterion, represented by a weight vector $= \mathfrak{M}(\mathfrak{M}_1, \mathfrak{M}_2, \ldots, \mathfrak{M}_m)$, where $\mathfrak{M}_i > 0$ and $\sum_{i=1}^{m} \mathfrak{M}_i = 1$.
2. **Construct the Fuzzy Decision Matrix.** Gather the assessment of a group of professionals. Every specialist compares each of the options $\mathcal{J}_j$ with every criterion $\mathcal{B}_i$. These qualitative measures are then transformed into the relevant Sq-ROFNs to create the fuzzy decision matrix $\mathcal{M} = [\mathcal{M}_{ij}]_{m \times n}$, which becomes $\mathcal{M}_{ij} = \langle (\mu_{ij}, \nu_{ij}), (\alpha_{ij}, \beta_{ij}) \rangle$.

3. **Normalize the Decision Matrix.** The matrix is normalized in order to make it comparable. These criteria are divided into benefit-type (optimistic) and cost-type (pessimistic), the higher and the lower value is better, respectively. In the case of cost-type criteria, the transformation of the Sq-ROFNs is as follows:

$$\mathcal{M}_{ij}^{\text{normalized}} = \begin{cases} \langle(\mu_{ij}, \nu_{ij}), (\alpha_{ij}, \beta_{ij})\rangle & \text{if } \mathcal{B}_i \text{ is a benefit criterion,} \\ \langle(\nu_{ij}, \mu_{ij}), (\beta_{ij}, \alpha_{ij})\rangle & \text{if } \mathcal{B}_i \text{ is a cost criterion.} \end{cases}$$

   This will provide a take into consideration that all of the evaluations are oriented towards the same direction that is preferential.
4. **Calculate Power Weights.** Group consensus is used to modify the weight of each criterion's assessment for a particular alternative. To do this, first use the distance measure (eq. (2.3)) to determine the support S_{ij} for each evaluation $\mathcal{M}_{ij}$. Equation (2.4) is then used to calculate the final power weights, Θ_{ij}. These weights represent the criterion's intrinsic significance ($\mathfrak{M}_i$) as well as the extent to which other factors support its assessment.
5. **Aggregate the Evaluations.** A single, collective Sq-ROFN value, $\mathcal{M}_j$ is created by aggregating the normalised Sq-ROFN evaluations for each alternative $\mathcal{J}_j$ across all criteria $[\mathcal{M}_{1j}, \mathcal{M}_{2j}, \ldots, \mathcal{M}_{mj}]$. Either the Sq-ROFSWPoWG operator (Definition 7) or the Sq-ROFSWPoWA operator (Definition 6) are used to achieve this.
6. **Rank the Alternatives.** The final aggregated Sq-ROFN for each alternative, $\mathcal{M}_j$, is converted into a crisp numerical score using the score function from eq. (2.2). The alternatives are then ranked in descending order based on their scores. The best option is the one with the highest score.

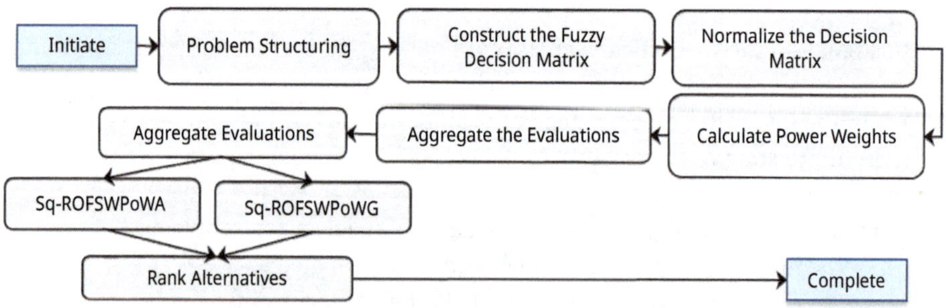

Figure 2: Flowchart of the proposed Sq-ROFS-based MCDM framework.

This structured algorithm provides a clear and repeatable process for applying the proposed framework to solve complex decision-making problems under uncertainty. The next section will demonstrate this entire process with a real-world case study.

3 Case Study: Healthcare Waste Segregation in Delhi

In the prior section, we covered the suggested methodological framework. Now, let's try to apply this to one of the most pressing cases which addresses the innovations encapsulated through the methodologies suggested to address better healthcare waste segregation strategies in Delhi. As one of the most complicated and sprawling cities within the UNESCO South Asia mapping of healthcare growth and development, there are challenges and opportunities imposed to serve this city – as a case study it develops. Delhi, pre-COVID-19 pandemic, generated a significant amount of BMW. The establishment of the pandemic policy and practice indicated a baseline, condition as a case study scenario helps to develop the case for credibility for evidence-based methodology to be adopted in decision-making strategies.

This section outlines how to formalize the decision problem. We first provide a number of case studies introducing the five main segregation strategies being weighed as options with regard to the MCDM model. Second, we present the five evaluation criteria we will use to evaluate alternatives. Lastly, we outline the data collection process that will inform analysis in the next section.

3.1 Case Studies of the Evaluated Alternatives ($\mathcal{J}_j$)

Five different strategies were identified for consideration, distinct but perhaps complementary, likely leading to improve waste separation at the source. Each is based upon an instance of real world application, and a different strategy to encourage the behavior and the functioning of the system.

3.1.1 $\mathcal{J}_1$: Color-Coded Bin Systems with Standardized Labeling

The foundation of any waste management approach is a clear, simple, and enforced waste segregation system at the point of generation. The All India Institute of Medical Sciences (AIIMS), New Delhi, an exemplary tertiary care hospital, provides an example of the large-scale implementation of this approach as required by India's Bio-Medical Waste Management Rules, 2016. At AIIMS, all points of care have yellow, red, white, and blue bins available (Figure 3), bolt clearly labeled. A single visual cue was intentionally created for a heterogeneous workforce to allow for a reasonable opportunity of safe waste segregation. Unfortunately, the large quantity and turn-over of staff has created persistent issues with high segregation rates. Studies have shown that while the system is fundamentally sound, human factors often create cross-contamination, highlighting the need for impetuses along with other strategies [18]. This case study demonstrates that even if systematic color-coding is not optional to regulatory compliance, even its effectiveness alone is an important factor our framework must consider.

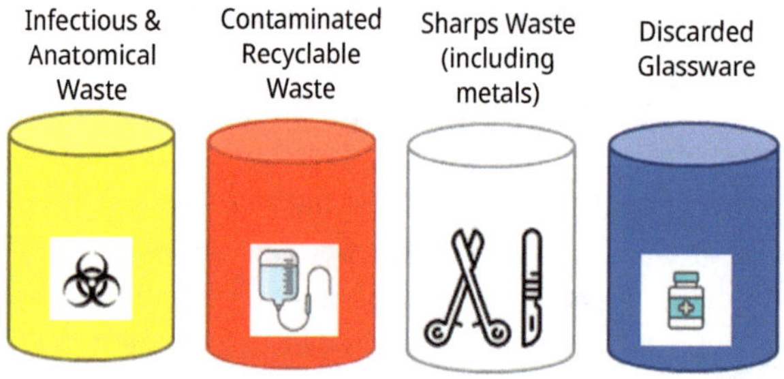

Figure 3: Standardized bio-medical waste segregation chart for India, illustrating the four primary color-coded streams as per the guidelines of the Central Pollution Control Board (CPCB) [31].

3.1.2 $\mathcal{J}_2$: Regular Staff Training and Awareness Programs

While it is important to invest in physical infrastructure, the success of this infrastructure ultimately depends on the knowledge and motivation of health personnel. An example, quoted in a study from the Post Graduate Institute of Medical Education and Research (PGIMER), Chandigarh, powerfully illustrates the effectiveness of developing a continuum of training for clinical staff within newly invested infrastructure. The researchers developed a standardized, structured training module for health workers in secondary care that provided statistically significant improvements in knowledge and attitude, as well as appropriate segregation [18]. This example demonstrates that focused and structured, continuous education of the health workforce is a highly effective intervention that has a direct and observable effect on the correct segregation of waste. The PGIMER example also suggests that investing in staff training and human capital development through a cycle of continuous improvement (Figure 4) helps to boost waste management efficiency.

3.1.3 $\mathcal{J}_3$: Point-of-Disposal SOP Posters and Visual Aids

The approach depends on behavioral nudges to support optimal behavior. The Aravind Eye Care System effectively implements this principle within its high-efficiency operational structure. In its sites, pictorial standard operating procedure (SOP) charts, displayed clearly, are situated right at the point of waste disposal. These visuals are embedded into the workflow of the strategies, thereby reducing cognitive load and chance for error. The images make the "right" behavior a matter of reflex, where any observer knows from the visual language alone what is to be done. This demonstrates that effective visual nudges are an essential part of feasibility for a workforce that speaks many different languages, like the one in Delhi [20].

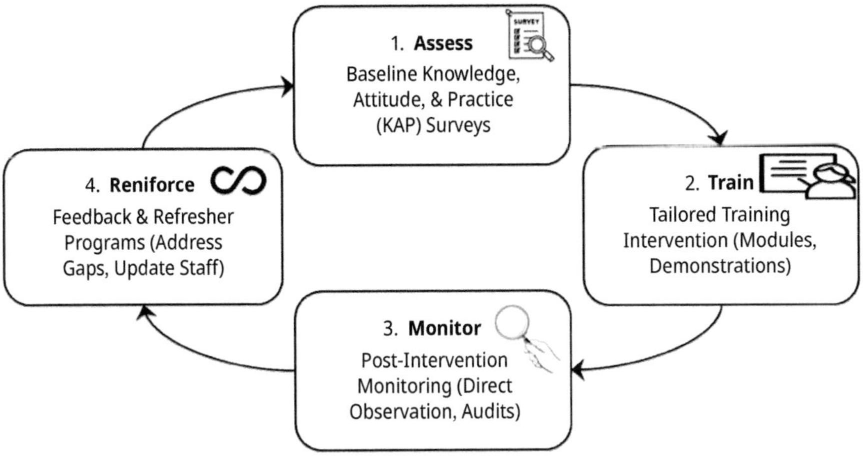

Figure 4: The continuous training and improvement cycle for healthcare staff. Inspired by effective interventions that have shown notable improvements, this model places emphasis on a recurring cycle of assessment and reinforcement. [18].

3.1.4 $\mathcal{J}_4$: Digital Tools and AI-Based Bin Guidance Systems

This option symbolizes a significant technological shift, with the objective to reduce human error via automation (Figure 5). Smart bins, powered by artificial intelligence (AI) and Internet of Things (IoT), can achieve 70–80% segregation accuracy by confirming a waste item is routed to the correct receptacle in real time, using a signal such as the correct lid illuminating [25, 1]. Digital tracking, like unusual disposal report notifications, also can help enforce compliance. Smart bins involve a significant start-up cost, but represent unparalleled accuracy and data-oriented compliance. Delhi, with an estimated waste generation rate of over 31 metric tons/day, has already served as a pilot for implementing smart bins and may serve as a feasible model for AI-led waste management [11].

3.1.5 $\mathcal{J}_5$: Segregation Audits with Performance-Based Incentives

This proposal builds upon an existing accountability and quality improvement loop (Figure 6), which the Delhi Pollution Control Committee (DPCC) has already provided for regulation through audits [11, 32]. This proposal takes this approach a step further by formalizing the accountability mechanism, aligning regular, randomized audits on segregation practices to a performance-based incentive mechanism. This model includes a hybrid contractor payment model, where a portion of the contractor's payment is tied to KPIs focused on segregation accuracy, thereby improving accountabil-

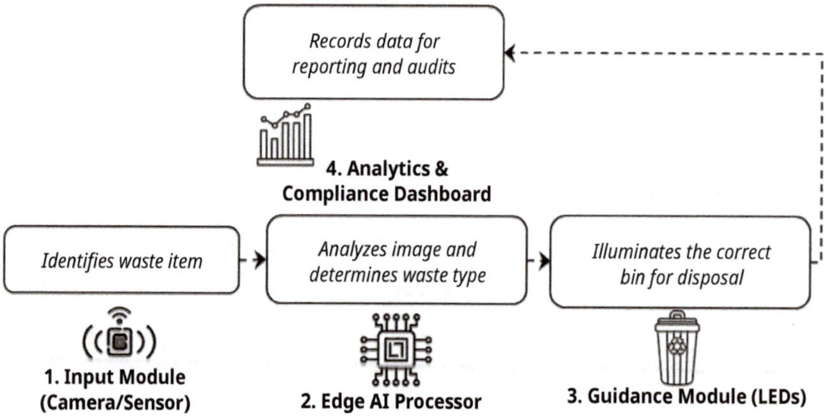

Figure 5: System architecture of an AI-based segregation system illustrating automated decision-making via computer vision, real-time feedback, and compliance data transmission, ensuring high accuracy [25].

ity with respect to KPIs [33]. This regulation-incentive model will provide a significant initiative to drive ongoing improvement.

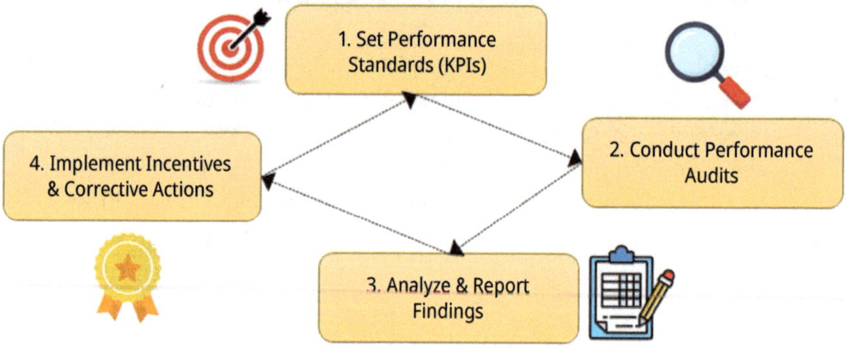

Figure 6: Audit and accountability framework illustrating how regular performance audits establish accountability by linking outcomes to positive reinforcement or corrective actions, following principles used by regulatory bodies like the DPCC [11, 33].

3.2 Defining the Evaluation Criteria ($\mathcal{B}_i$)

All five options were assessed against five different criteria that were chosen for their ability to offer a comprehensive evaluation of the strategies' feasibility. The expert panel established the relative significance of the criteria as symbolized by the weight vector W. The experts agreed upon the weight vector W through discussion and consensus and the weight vector W reflects their collective judgment about the different priorities that should be considered in evaluating the healthcare system in Delhi:

- $\mathcal{B}_1$: **Cost-Effectiveness** [9] ($\mathfrak{M}_1 = 0.20$). This criterion evaluates the economic feasibility of each option, and the costs include both capital costs and ongoing operational costs. This is a cost-type (pessimistic) criterion.
- $\mathcal{B}_2$: **Segregation Accuracy** [19] ($\mathfrak{M}_2 = 0.30$). This is the heaviest weight criterion that measures how much the strategy impacts the percentage of the waste that is successfully sorted. This is a benefit-type (optimistic) criterion.
- $\mathcal{B}_3$: **Operational Feasibility** [13] ($\mathfrak{M}_3 = 0.20$). This criterion assesses the practical application of the strategy considering the high-stress setting of a hospital and ability to sustain strategy career. This is a benefit-type (optimistic) criterion.
- $\mathcal{B}_4$: **Regulatory Compliance** [13] ($\mathfrak{M}_4 = 0.15$). This criterion evaluates how well the strategy helps a facility to comply legally with relevant legislation, regulations, or accreditations. This is a benefit-type criterion (optimistic).
- $\mathcal{B}_5$: **Stakeholder Acceptance** [6] ($\mathfrak{M}_5 = 0.15$). This criterion measures the likelihood that stakeholders in a facility will support a strategy. This is a benefit-type criterion (optimistic).

3.3 Data Collection and Expert Panel

A panel of three experts was established to collect the data required for the MCDM model. The panel comprised a senior hospital administrator with extensive experience in large-scale operations; a public health policy expert with a focus on environmental health; and a senior faculty member from the Department of Chemistry at the University College of Engineering Nagercoil (UCEN), specializing in the analysis of hazardous materials. The experts directly provided their opinions for each of the five alternatives against each of the five criteria in the form of Sq-ROFNs through a systematic evaluation process. Their nuanced assessments, including the degree of support (μ) and opposition (ν), for a proposition, as well as their level of confidence in those assessments (α and β) could be directly recorded using this method. The first fuzzy decision matrix, which will be shown and discussed in the next section, is composed of the combined expert assessments.

We now apply the full MCDM algorithm to the consolidated expert data, building on the problem formulation from Chapter 3 and the methodological framework described in Chapter 2. The goal is to produce a final, reliable ranking of Delhi's five healthcare waste segregation strategies.

The steps specified in the algorithm are followed in the sequential presentation of the process. The first fuzzy decision matrix with the Sq-ROFN evaluations is shown first. Next, we show how this matrix can be normalized to take the cost-type criterion into consideration. Following this, we apply the Sq-ROFSWPoWA and Sq-ROFSWPoWG aggregation operators to synthesize the data. This chapter culminates in the presentation of the final scores and rankings of the alternatives, thereby identifying the optimal strategy.

Step 1: Problem Structuring

Formally structuring the decision problem by identifying its essential elements is the first step in the MCDM process. Chapter 3 provided the evaluation criteria's rationale, case studies for each alternative, and a thorough context. As inputs for the analytical model, these elements are formally summarized in this section. The following is the structure of the decision problem:

- **Alternatives ($\mathcal{J}$):** The five segregation strategies detailed in Section 3 are formally denoted as the set of alternatives, $\mathcal{J} = \{\mathcal{J}_1, \mathcal{J}_2, \mathcal{J}_3, \mathcal{J}_4, \mathcal{J}_5\}$.
- **Criteria ($\mathcal{B}$):** The set $\mathcal{B} = \{\mathcal{B}_1, \mathcal{B}_2, \mathcal{B}_3, \mathcal{B}_4, \mathcal{B}_5\}$ is the formal notation for the five evaluation criteria listed in Section 3.
- **Criteria Weights ($\mathfrak{M}_i$):** The relative importance of each criterion, as determined by the expert panel, is represented by the weight vector:

  $$\mathfrak{M}(= 0.20, 0.30, 0.20, 0.15, 0.15)$$

  corresponding to cost-effectiveness, segregation accuracy, operational feasibility, regulatory compliance, and stakeholder acceptance, respectively.

Criteria Type: As stated in Chapter 3, all criteria ($\mathcal{B}_2$–$\mathcal{B}_5$) are benefit-type (optimistic), whereas $\mathcal{B}_1$ (cost-effectiveness) is a cost-type (pessimistic) criterion. With the problem now formally structured, the next step is to construct the fuzzy decision matrix from the expert evaluations. With the problem now formally structured, the next step is to construct the fuzzy decision matrix from the expert evaluations.

Step 2: The Initial Fuzzy Decision Matrix

The analysis begins with the fuzzy decision matrix. This matrix contains the consolidated expert evaluations for each of the five alternatives ($\mathcal{J}_1$–$\mathcal{J}_5$) against each of the five criteria ($\mathcal{B}_1$–$\mathcal{B}_5$). The evaluations are presented directly as Sq-ROFNs with a parameter of $q = 3$, as shown in Table 1.

Table 1: Initial fuzzy decision matrix: Sq-ROFN (q = 3) evaluations.

Criteria	$\mathcal{J}_1$	$\mathcal{J}_2$	$\mathcal{J}_3$	$\mathcal{J}_4$	$\mathcal{J}_5$
$\mathcal{B}_1$ **(cost-effectiveness)**	⟨(0.25, 0.55), (0.92, 0.88)⟩	⟨(0.30, 0.50), (0.95, 0.90)⟩	⟨(0.28, 0.52), (0.93, 0.87)⟩	⟨(0.32, 0.48), (0.96, 0.91)⟩	⟨(0.27, 0.54), (0.94, 0.89)⟩
$\mathcal{B}_2$ **(segregation accuracy)**	⟨(0.70, 0.18), (0.98, 0.95)⟩	⟨(0.85, 0.12), (0.99, 0.97)⟩	⟨(0.80, 0.15), (0.97, 0.94)⟩	⟨(0.88, 0.10), (1.00, 0.98)⟩	⟨(0.75, 0.14), (0.96, 0.93)⟩
$\mathcal{B}_3$ **(operational feasibility)**	⟨(0.40, 0.28), (0.88, 0.90)⟩	⟨(0.45, 0.25), (0.90, 0.92)⟩	⟨(0.42, 0.26), (0.89, 0.91)⟩	⟨(0.48, 0.22), (0.92, 0.94)⟩	⟨(0.38, 0.30), (0.87, 0.88)⟩

Table 1 (continued)

Criteria	$\mathcal{J}_1$	$\mathcal{J}_2$	$\mathcal{J}_3$	$\mathcal{J}_4$	$\mathcal{J}_5$
$\mathcal{B}_4$ **(regulatory compliance)**	⟨(0.60, 0.40), (0.96, 0.94)⟩	⟨(0.65, 0.35), (0.97, 0.95)⟩	⟨(0.62, 0.38), (0.95, 0.93)⟩	⟨(0.68, 0.32), (0.98, 0.96)⟩	⟨(0.58, 0.42), (0.94, 0.92)⟩
$\mathcal{B}_5$ **(stakeholder acceptance)**	⟨(0.20, 0.70), (0.90, 0.88)⟩	⟨(0.25, 0.65), (0.92, 0.90)⟩	⟨(0.22, 0.68), (0.91, 0.89)⟩	⟨(0.28, 0.62), (0.93, 0.91)⟩	⟨(0.18, 0.72), (0.89, 0.87)⟩

Step 3: Normalization of the Decision Matrix

As defined in Chapter 3, criterion $\mathcal{B}_1$ (cost-effectiveness) is a **cost-type** (pessimistic) criterion, where lower values are preferable. To ensure all evaluations are oriented in the same preferential direction, the Sq-ROFNs for criterion $\mathcal{B}_1$ are normalized by swapping their membership/NMGs and scaling parameters (i.e., $\langle(\mu, \nu), (\alpha, \beta)\rangle$ becomes $\langle(\nu, \mu), (\beta, \alpha)\rangle$). The normalized decision matrix is provided in Table 2.

Table 2: Normalized fuzzy decision matrix.

Criteria	$\mathcal{J}_1$	$\mathcal{J}_2$	$\mathcal{J}_3$	$\mathcal{J}_4$	$\mathcal{J}_5$
$\mathcal{B}_1$	⟨(0.55, 0.25), (0.88, 0.92)⟩	⟨(0.50, 0.30), (0.90, 0.95)⟩	⟨(0.52, 0.28), (0.87, 0.93)⟩	⟨(0.48, 0.32), (0.91, 0.96)⟩	⟨(0.54, 0.27), (0.89, 0.94)⟩
$\mathcal{B}_2$	⟨(0.70, 0.18), (0.98, 0.95)⟩	⟨(0.85, 0.12), (0.99, 0.97)⟩	⟨(0.80, 0.15), (0.97, 0.94)⟩	⟨(0.88, 0.10), (1.00, 0.98)⟩	⟨(0.75, 0.14), (0.96, 0.93)⟩
$\mathcal{B}_3$	⟨(0.40, 0.28), (0.88, 0.90)⟩	⟨(0.45, 0.25), (0.90, 0.92)⟩	⟨(0.42, 0.26), (0.89, 0.91)⟩	⟨(0.48, 0.22), (0.92, 0.94)⟩	⟨(0.38, 0.30), (0.87, 0.88)⟩
$\mathcal{B}_4$	⟨(0.60, 0.40), (0.96, 0.94)⟩	⟨(0.65, 0.35), (0.97, 0.95)⟩	⟨(0.62, 0.38), (0.95, 0.93)⟩	⟨(0.68, 0.32), (0.98, 0.96)⟩	⟨(0.58, 0.42), (0.94, 0.92)⟩
$\mathcal{B}_5$	⟨(0.20, 0.70), (0.90, 0.88)⟩	⟨(0.25, 0.65), (0.92, 0.90)⟩	⟨(0.22, 0.68), (0.91, 0.89)⟩	⟨(0.28, 0.62), (0.93, 0.91)⟩	⟨(0.18, 0.72), (0.89, 0.87)⟩

Step 4: Calculate Power Weights

Determine the power weighting factor Θ_{ij} for each normalized Sq-ROFN $\mathcal{M}_{ij}$ by using Equation 2.4 (see Table 2 for reference). The calculated Θ_{ij} values are summarized in Table 3.

Step 5: Aggregation of Expert Judgments

Using the data from the normalized decision matrix (Table 2) and the criteria weights ($\mathfrak{M}_i$) from Section 3, the power weights (Θ_{ij}) were calculated for each evaluation. These weights reflect both the criterion's importance and the consensus surrounding its evaluation. The Sq-ROFNs for each alternative were then aggregated using both the

Table 3: Power weights Θ_{ij}.

	$\mathcal{J}_1$	$\mathcal{J}_2$	$\mathcal{J}_3$	$\mathcal{J}_4$	$\mathcal{J}_5$
$\mathcal{B}_1$	0.2047791	0.2077615	0.2039437	0.2067818	0.2068392
$\mathcal{B}_2$	0.2921457	0.2822270	0.2837133	0.2826631	0.2880370
$\mathcal{B}_3$	0.1021850	0.1033745	0.1035995	0.1036994	0.1017239
$\mathcal{B}_4$	0.2029479	0.2062319	0.2071451	0.2058323	0.2054332
$\mathcal{B}_5$	0.1979422	0.2004050	0.2015985	0.2010233	0.1979667

Sq-ROFSWPoWA and Sq-ROFSWPoWG operators with parameters $q = 5$ and $\kappa = 3$. The final aggregated Sq-ROFN value for each alternative is reported in Table 4.

Table 4: Aggregated Sq-ROFN values for each alternative.

Alternative	Sq-ROFSWPoWA operator	Sq-ROFSWPoWG operator
$\mathcal{J}_1$	⟨(0.5463, 0.4752), (0.9388, 0.9143)⟩	⟨(0.5195, 0.4999), (0.9371, 0.9155)⟩
$\mathcal{J}_2$	⟨(0.6505, 0.4379), (0.4969, 0.5177)⟩	⟨(0.5945, 0.4576), (0.4888, 0.5364)⟩
$\mathcal{J}_3$	⟨(0.6133, 0.4582), (0.9389, 0.9321)⟩	⟨(0.5711, 0.4810), (0.9381, 0.9343)⟩
$\mathcal{J}_4$	⟨(0.6799, 0.4171), (0.9660, 0.9429)⟩	⟨(0.6176, 0.4341), (0.9650, 0.9440)⟩
$\mathcal{J}_5$	⟨(0.5690, 0.4822), (0.9296, 0.8690)⟩	⟨(0.5332, 0.5096), (0.9284, 0.8720)⟩

Step 6: Final Scores and Ranking

The aggregated Sq-ROFN values from Table 5 were converted into crisp scores using the score function defined in Chapter 2. These scores provide the basis for the final ranking of the segregation strategies. The score values and the resulting ranking are presented in Table 15.

Table 5: Final score values and ranking of alternatives.

Operator	$\mathcal{J}_1$	$\mathcal{J}_2$	$\mathcal{J}_3$	$\mathcal{J}_4$	$\mathcal{J}_5$	Final ranking
Sq-ROFSWPoWA	0.5275	0.5466	0.5635	**0.6176**	0.5369	$\mathcal{J}_4 \succ \mathcal{J}_3 \succ \mathcal{J}_2 \succ \mathcal{J}_5 \succ \mathcal{J}_1$
Sq-ROFSWPoWG	0.5085	0.5257	0.5354	**0.5750**	0.5127	$\mathcal{J}_4 \succ \mathcal{J}_3 \succ \mathcal{J}_2 \succ \mathcal{J}_5 \succ \mathcal{J}_1$

The results show a consistent and stable ranking across both the arithmetic (WA) and geometric (WG) aggregation operators. The optimal strategy is unequivocally identified as $\mathcal{J}_4$ **(digital tools and AI-based bin guidance systems)**. This is followed by $\mathcal{J}_3$ (posters and visual aids) and $\mathcal{J}_2$ (staff training). The least preferred strategy is (color-coded bins alone).

This section successfully executed the proposed MCDM framework. We began with the raw expert evaluations, normalized the data, and applied the Sq-ROFSWPoWA and Sq-ROFSWPoWG operators to derive a final ranking of the five healthcare waste segregation strategies. An AI-driven strategy is consistently found to be the most practical long-term solution. These findings, their policy implications, and a validation of their robustness will all be covered in detail in the following section.

4.1 Validation and Discussion

The previous chapter has determined a definite optimal plan of waste segregation of healthcare in Delhi according to the suggested MCDM framework. Nonetheless, in order to make these findings reliable enough to use them in policy making with high stakes, their power, and excellence should undergo a strict test of validation. That validation process is given in this chapter, and the implications of the findings are further discussed.

We will start with a sensitivity analysis to determine the stability of the final ranking with variations of the key parameters of the model. Second, we conduct a comparative analysis with some among the existing MCDM methods to show that our proposed operators are valid and useful. At last, we enter the qualitative world, where we present the discussion of results, its interpretation, and the final ranking, and present active policy and management suggestions.

4.2 Sensitivity Analysis of Parameter Variations

A sensitivity analysis was performed in order to ensure the strength of the proposed framework by changing the important parameters of Sq-ROFS model (q) and operators of aggregation (κ).

The first was to vary the aggregation parameter with the set of parameters (3, 5, 7) with a fixed q-rung parameter ($q = 3$). As evidenced in Table 6, the pattern of ranking is very stable. With all the tested values of κ the final order is always the same $\mathcal{J}_4 \succ \mathcal{J}_3 \succ \mathcal{J}_2 \succ \mathcal{J}_5 \succ \mathcal{J}_1$. Although the absolute scores change slightly, the relative preference structure remains the same and it proves that the preference structure does not depend on the k parameter.

Second, the q-rung parameter q was tested within the range of the set of aggregation parameters with a fixed value of the parameter aggregation ($\kappa = 3$) on the set 3, 5, 15. As it can be observed in Table 7, the ranking is also stable at lower values of q. When q becomes large (e.g., 15) the score values of the various alternatives approach 0.5, and the distinction between them becomes less pronounced. This is a mathematical property that was already expected of q-ROFS models at large q values. But, in the realms of practical values of q, the ranking is solid. This discussion proves that the given framework results in the stable and robust outcomes that support the suggested best strategy.

Table 6: Sensitivity analysis with fixed $q = 3$ and varying κ.

Operator	κ	$\mathcal{J}_1$	$\mathcal{J}_2$	$\mathcal{J}_3$	$\mathcal{J}_4$	$\mathcal{J}_5$
Sq-ROFSWPoWA	3	0.5275	0.5466	0.5635	0.6176	0.5369
	5	0.5312	0.5501	0.5698	0.6241	0.5410
	7	0.5355	0.5531	0.5731	0.6307	0.5460
Ranking	$\mathcal{J}_4 \succ \mathcal{J}_3 \succ \mathcal{J}_2 \succ \mathcal{J}_5 \succ \mathcal{J}_1$ (consistent for all κ)					

Table 7: Sensitivity analysis with fixed $\kappa = 3$ and varying q.

Operator	q	$\mathcal{J}_1$	$\mathcal{J}_2$	$\mathcal{J}_3$	$\mathcal{J}_4$	$\mathcal{J}_5$
Sq-ROFSWPoWA	3	0.5275	0.5466	0.5635	0.6176	0.5369
	5	0.5123	0.5340	0.5397	0.5864	0.5201
	15	0.5002	0.5066	0.5044	0.5206	0.5012
Ranking	$\mathcal{J}_4 \succ \mathcal{J}_3 \succ \mathcal{J}_2 \succ \mathcal{J}_5 \succ \mathcal{J}_1$ (consistent for all q)					

4.3 Comparative Analysis: Demonstrating Validity and Superiority

In order to place our framework into the context of the available literature and prove that it is valuable, we have conducted a comparative analysis with well-known operators that are developed to be used with IFS [9], Pythagorean fuzzy set (PyFS) [36], and q-ROFS [37]. The performance of our Sq-ROFSWPoWA and Sq-ROFSWPoWG operators was tested on a collection of benchmark problems to determine their performance on two important metrics, namely, validity and expressiveness. Table 8 sums up the overall findings.

4.4 Validity and Ranking Consistency

The biggest test of validity is the ability of a new method to repeat rational and consistent rankings. This was made possible by our proposed operators in all the benchmark cases. In the case of the IFS environment, the findings have been illustrated in Figures 7 and 8. In the case that they are applied to the problem formulated by Wang and Feng [36], our operators preserve the original ranking of $\mathcal{J}_3 \succ \mathcal{J}_2 \succ \mathcal{J}_4 \succ \mathcal{J}_5 \succ \mathcal{J}_1$. In the same way, in the case of Senapati et al. [29] problem, the same ranking is maintained.

In the PyFS environment, as shown in Figure 9, our framework also upholds the preference order from the Asif et al. [5] study.

Lastly, our operators, once more, generate a similar ranking as that of the original work when subjected to the standard q-ROFS environment using the benchmark of

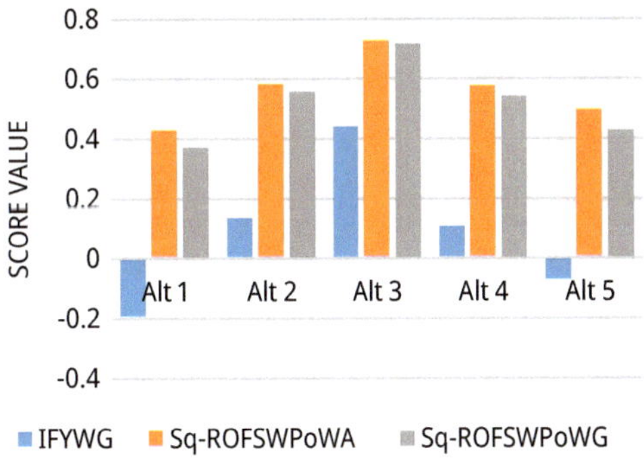

Figure 7: Comparison in the IFS[36].

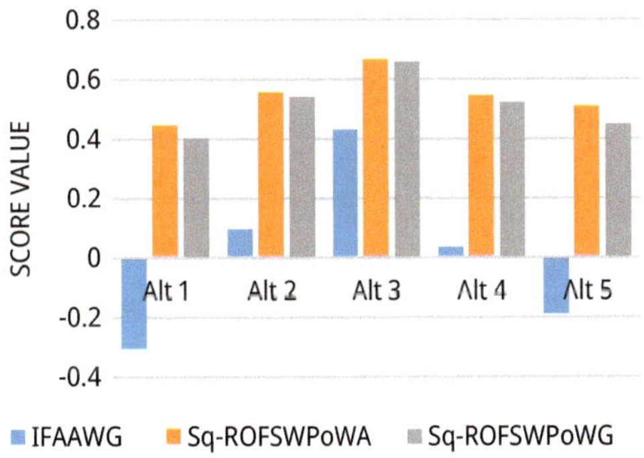

Figure 8: Comparison in the IFS [29].

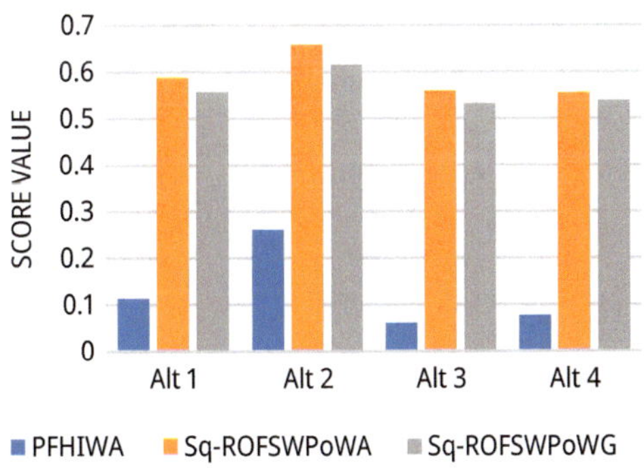

Figure 9: Comparison in the PyFS [5].

Wang et al. [35], as shown in Figure 10. The fact that we have a steady performance in all the tested fuzzy environments proves the fact that our model is axiomatically sound and that its logic is aligned with the known principles of fuzzy MCDM.

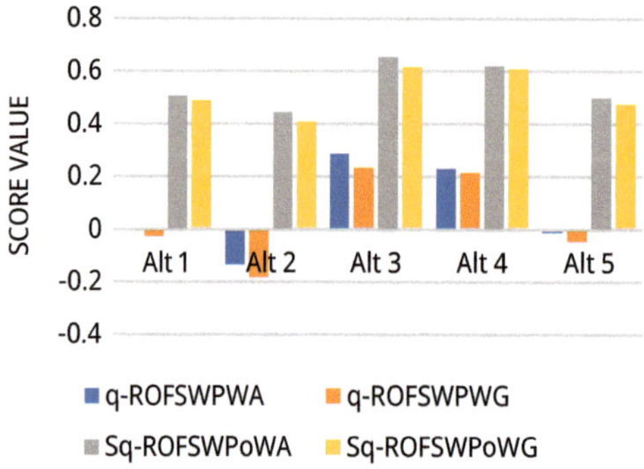

Figure 10: Comparison in the *q*-ROFS [35].

4.5 Superiority in Expressiveness and Rank Resolution

In addition to the fact that the suggested framework was simply matching the existing findings, the framework proved to be more effective by surmounting the limitations known with other models. One of its advantages is that it has a higher range and sensitivity of the values of the score that is vital in situations where the other operators are not able to differentiate between the alternatives.

Figure 11 is a good example of this superiority as it performs an analysis of the benchmark by Kumar et al. [17]. The *q*-ROFIHWAPA operator in this case of the rank failure, this led to a tie where the alternative $\mathcal{J}_1$ and $\mathcal{J}_3$ had the same mark of 1.0 and this left the decision-maker in a quagmire over which alternative to choose. On the contrary, our S*q*-ROFSWPoWA operator has been able to overcome this ambiguity since it distinguishes the scores and generates a final ranking of $\mathcal{J}_3 \succ \mathcal{J}_1 \succ \mathcal{J}_2$ which is conclusive. This information loss avoidance and the possibility to present a clear and unambiguous result testify to the fact that our suggested framework is not only trustworthy but also has a greater level of accuracy and usefulness.

4.6 Discussion and Policy Implications

The last ranking of $\mathcal{J}_4 \succ \mathcal{J}_3 \succ \mathcal{J}_2 \succ \mathcal{J}_5 \succ \mathcal{J}_1$, provides valuable information to the healthcare policy and management of hospitals in Delhi and other urban centers. The overwhelming preference to $\mathcal{J}_4$ (AI-based systems) reflects a strategic change towards technology-based solutions to insuring a high level of segregation accuracy and regulatory compliance that are the most heavily weighted criteria. It is evident that experts consider that automation is the best approach in minimizing human error even though the initial cost may be quite high in the long term.

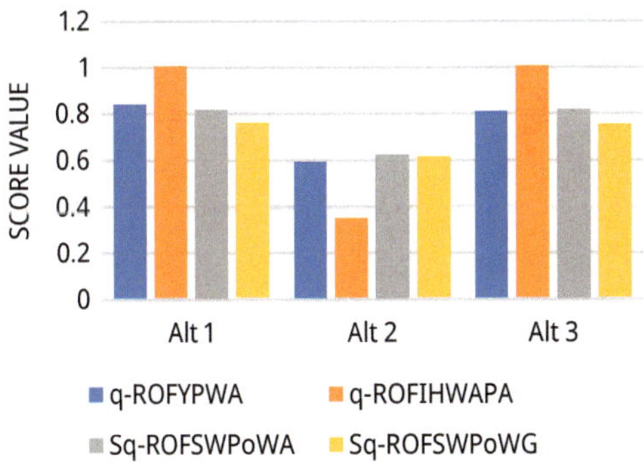

Figure 11: Comparison in the IFS [15].

Table 8: Comparison of aggregation operators proposed with the current methods.

Problem	Operators	$\mathcal{J}_1$	$\mathcal{J}_2$	$\mathcal{J}_3$	$\mathcal{J}_4$	$\mathcal{J}_5$	Ranking
Wang and Feng [36]	IFYWG (original)	−0.1915	0.1340	0.4385	0.1060	−0.0705	$\mathcal{J}_3 \succ \mathcal{J}_2 \succ \mathcal{J}_4 \succ \mathcal{J}_5 \succ \mathcal{J}_1$
	***Sq*-ROFSWPoWA**	**0.4237**	**0.5779**	**0.7231**	**0.5720**	**0.4910**	$\mathcal{J}_3 \succ \mathcal{J}_2 \succ \mathcal{J}_4 \succ \mathcal{J}_5 \succ \mathcal{J}_1$
	***Sq*-ROFSWPoWG**	**0.3707**	**0.5569**	**0.7168**	**0.5407**	**0.4267**	$\mathcal{J}_3 \succ \mathcal{J}_2 \succ \mathcal{J}_4 \succ \mathcal{J}_5 \succ \mathcal{J}_1$
Senapati et al. [29]	IFAAWG (original)	−0.3040	0.0940	0.4300	0.0331	−0.1885	$\mathcal{J}_3 \succ \mathcal{J}_2 \succ \mathcal{J}_4 \succ \mathcal{J}_5 \succ \mathcal{J}_1$
	***Sq*-ROFSWPoWA**	**0.4425**	**0.5534**	**0.6637**	**0.5420**	**0.5051**	$\mathcal{J}_3 \succ \mathcal{J}_2 \succ \mathcal{J}_4 \succ \mathcal{J}_5 \succ \mathcal{J}_1$
	***Sq*-ROFSWPoWG**	**0.4023**	**0.5402**	**0.6581**	**0.5223**	**0.4482**	$\mathcal{J}_3 \succ \mathcal{J}_2 \succ \mathcal{J}_4 \succ \mathcal{J}_5 \succ \mathcal{J}_1$
Asif et al. [5]	PFHIWA (original)	0.1116	0.2594	0.0598	0.0749	–	$\mathcal{J}_2 \succ \mathcal{J}_1 \succ \mathcal{J}_4 \succ \mathcal{J}_3$
	***Sq*-ROFSWPoWA**	**0.5844**	**0.6555**	**0.5573**	**0.5526**	–	$\mathcal{J}_2 \succ \mathcal{J}_1 \succ \mathcal{J}_3 \succ \mathcal{J}_4$
	***Sq*-ROFSWPoWG**	**0.5568**	**0.6143**	**0.5315**	**0.5379**	–	$\mathcal{J}_2 \succ \mathcal{J}_1 \succ \mathcal{J}_4 \succ \mathcal{J}_3$
Wang et al. [35]	*q*-ROFSWPWA (original)	−0.0004	−0.1326	0.2854	0.2297	−0.0110	$\mathcal{J}_3 \succ \mathcal{J}_4 \succ \mathcal{J}_1 \succ \mathcal{J}_5 \succ \mathcal{J}_2$[a]
	q-ROFSWPWG (original)	−0.0272	−0.1829	0.2307	0.2122	−0.0452	$\mathcal{J}_3 \succ \mathcal{J}_4 \succ \mathcal{J}_1 \succ \mathcal{J}_5 \succ \mathcal{J}_2$

Table 8 (continued)

Problem	Operators	$\mathcal{J}_1$	$\mathcal{J}_2$	$\mathcal{J}_3$	$\mathcal{J}_4$	$\mathcal{J}_5$	Ranking
	Sq-ROFSWPoWA	**0.5055**	**0.4441**	**0.6540**	**0.6179**	**0.4983**	$\mathcal{J}_3 \succ \mathcal{J}_4 \succ \mathcal{J}_1 \succ \mathcal{J}_5 \succ \mathcal{J}_2$
	Sq-ROFSWPoWG	**0.4853**	**0.4054**	**0.6116**	**0.6045**	**0.4722**	$\mathcal{J}_3 \succ \mathcal{J}_4 \succ \mathcal{J}_1 \succ \mathcal{J}_5 \succ \mathcal{J}_2$
Kumar et al. [17]	*q*-ROFYPWA (original)	0.8388	0.5926	0.8072	–	–	$\mathcal{J}_1 \succ \mathcal{J}_3 \succ \mathcal{J}_2$
	q-ROFIHWAPA (original)	1.0000	0.3487	1.0000	–	–	$\mathcal{J}_1 \succ \mathcal{J}_3 \succ \mathcal{J}_2$
	Sq-ROFSWPoWA	**0.8164**	**0.6229**	**0.8176**	–	–	$\mathcal{J}_3 \succ \mathcal{J}_1 \succ \mathcal{J}_2$
	Sq-ROFSWPoWG	**0.7571**	**0.6131**	**0.7512**	–	–	$\mathcal{J}_1 \succ \mathcal{J}_3 \succ \mathcal{J}_2$

Nevertheless, one should not exclude the high scores of $\mathcal{J}_3$ (visual aids) and (staff training). The fact that they have high ranking implies that human factors will always be at the center of the success of any system. The poor results of $\mathcal{J}_1$ (bins alone) is a significant validation of the fact that passive infrastructure cannot work without active and ongoing involvement and prominence to the personnel who utilize it.

Based on such findings, the following policy recommendations can be offered:

- **Take a Hybrid Approach**: It is not best to follow one alternative but rather a hybrid model is followed. Both the examination and the results of the analysis highly reinforce the idea of investing in a basic layer of AI-driven systems ($\mathcal{J}_4$) to guarantee the accuracy and data tracking, as well as enhancing the discussed technology with efficient staff training ($\mathcal{J}_2$) and continuous visual aids upon point-of-disposal ($\mathcal{J}_3$).
- **Invest in Technology**: To policymakers, this research will make it very clear that engaging in capital budgets on smart waste management technologies is a justified move with solid data.
- **Incorporate Audits as a Quality Control Mechanism**: Although $\mathcal{J}_5$ (audits and incentives) was not among the top ranking strategies as an independent strategy, it is an essential element of a hybrid type where it would be used as the control mechanism to verify the other strategies are functioning properly.

To the hospital administrators, the most important lesson is that to achieve sustainable results in waste segregation, a multi-pronged strategy must be adopted by investing in technology, people and processes in parallel.

5 Conclusion

This research developed an advanced decision-making model to tackle the complex problem of medical waste segregation in major cities, using Delhi as a case study. Traditional methods often fail to handle the ambiguity and differing opinions of experts, so we created a more reliable framework using Sq-ROFS and specialized Sugeno-Weber power weighted aggregation operators. After evaluating five different segregation strategies, our analysis provided a clear ranking. The top choice was a technology-driven approach ($\mathcal{J}_4$), but the results also highlighted the important role of human-focused efforts. This includes continuous staff training ($\mathcal{J}_2$) and the use of visual guides ($\mathcal{J}_3$).

The study makes significant contributions to both theory and practice. Theoretically, it confirms that the Sq-ROFS framework is a strong and reliable tool for solving complex decision-making problems. It proves to be better than other models that have known inconsistencies. We also developed a complete methodology by combining this framework with unique Sugeno-Weber operators. This offers a new way to manage group decisions and consider outlier opinions. Practically, this work provides officials and hospital leaders in Delhi with a clear, data-supported ranking of waste management strategies. The key finding – that the best solution is a mix of technology, visual aids, and ongoing training – offers a direct and actionable path forward. Moreover, other cities can modify the model to tackle their own public policy challenges involving expert uncertainty. While our framework is strong, we recognize its limitations, which suggest exciting opportunities for future research. This study relied on a small panel of experts. Involving more diverse voices would strengthen the findings. The criteria used were specific to Delhi and could be adjusted for other urban areas. Additionally, this research offers a static snapshot, but it could develop into a dynamic model that responds to new technologies and cost changes over time. Another promising direction is to combine our expert-driven model with real-time data from IoT sensors to create a more powerful hybrid system. Ultimately, the flexibility of this framework makes it suitable for other vital urban issues, such as sustainable transport planning or water resource management. In summary, this study not only offers a concrete solution for a crucial urban health issue but also provides a powerful and adaptable tool for decision science.

Appendix

Detailed Calculation for the Sq-ROFSWPoWA Operator

This appendix provides a detailed numerical illustration of the Sq-ROFSWPoWA operator.

Let there be three expert evaluations (Sq-ROFNs):

- $\mathcal{M}_1 = \langle(0.7, 0.3), (0.5, 0.1)\rangle$
- $\mathcal{M}_2 = \langle(0.6, 0.7), (0.6, 0.3)\rangle$
- $\mathcal{M}_3 = \langle(0.5, 0.4), (0.7, 0.3)\rangle$

Let the initial weights of the experts be $\mathfrak{M} = (0.2, 0.5, 0.3)$. We will use the parameters $q = 4$ for the q-rung structure and $\zeta = 1$ for the Sugeno-Weber operator.

Step 1: Calculate the Distances $\mathfrak{D}(\mathcal{M}_i, \mathcal{M}_k)$

We use the corrected Manhattan distance function:

$$\mathfrak{D}(\mathcal{M}_i, \mathcal{M}_k) = \frac{1}{2}\left(\left|\alpha_i\mu_i^q - \alpha_k\mu_k^q\right| + \left|\beta_i\nu_i^q - \beta_k\nu_k^q\right|\right).$$

First, we pre-calculate the scaled q-rung components for each Sq-ROFN with:

- For $\mathcal{M}_1$: $\alpha_1\mu_1^4 = 0.12005$; $\beta_1\nu_1^4 = 0.00081$
- For $\mathcal{M}_2$: $\alpha_2\mu_2^4 = 0.07776$; $\beta_2\nu_2^4 = 0.07203$
- For $\mathcal{M}_3$: $\alpha_3\mu_3^4 = 0.04375$; $\beta_3\nu_3^4 = 0.00768$

The pairwise distances are:

$$\mathfrak{D}(\mathcal{M}_1, \mathcal{M}_2) = \frac{1}{2}(|0.12005 - 0.07776| + |0.00081 - 0.07203|) = 0.056755$$

$$\mathfrak{D}(\mathcal{M}_1, \mathcal{M}_3) = \frac{1}{2}(|0.12005 - 0.04375| + |0.00081 - 0.00768|) = 0.041585$$

$$\mathfrak{D}(\mathcal{M}_2, \mathcal{M}_3) = \frac{1}{2}(|0.07776 - 0.04375| + |0.07203 - 0.00768|) = 0.04918$$

Step 2: Calculate the Support Values S_i

The support $S_i = \sum_{k=1, k\neq i}^{m}(1 - (\mathcal{M}_i, \mathcal{M}_k))$.

$$S_1 = (1 - 0.056755) + (1 - 0.041585) = 1.90166$$

$$S_2 = (1 - 0.056755) + (1 - 0.04918) = 1.894065$$

$$S_3 = (1 - 0.041585) + (1 - 0.04918) = 1.909235$$

Step 3: Calculate the Final Power Weights Θ_i

The power weight $\Theta_i = \frac{\mathfrak{M}_i(1+S_i)}{\sum_{k=1}^{3}\mathfrak{M}_k(1+S_k)}$. The numerator terms are:

$$\mathfrak{M}_1(1+S_1) = 0.2\times(1+1.90166) = 0.580332$$

$$\mathfrak{M}_2(1+S_2) = 0.5\times(1+1.894065) = 1.4470325$$

$$\mathfrak{M}_3(1+S_3) = 0.3\times(1+1.909235) = 0.8727705$$

The denominator is the sum of these terms: 2.900135. The final power weights are:

$$\Theta_1 = 0.580332/2.900135 \approx 0.200106$$
$$\Theta_2 = 1.4470325/2.900135 \approx 0.498956$$
$$\Theta_3 = 0.8727705/2.900135 \approx 0.300938$$

Step 4: Aggregate the Sq-ROFNs using the Sq-ROFSWPoWA Operator
We apply the formulas from Definition 6 with parameter $\zeta = 1$.
Calculation for μ_{agg}

$$\text{Term 1: } \sum_{i=1}^{3}\Theta_i\mu_i^4 = (0.200106\times 0.7^4) + (0.498956\times 0.6^4) + (0.300938\times 0.5^4)$$

$$= 0.048046 + 0.064665 + 0.018809 = 0.13152$$

$$\text{Term2. } \frac{\varsigma}{1+\varsigma}\sum_{i<j}\Theta_i\Theta_j\mu_i^q\mu_j^q = 0.5\times\left[\left(\Theta_1\Theta_2\mu_1^4\mu_2^4\right) + \left(\Theta_1\Theta_3\mu_1^4\mu_3^4\right) + \left(\Theta_2\Theta_3\mu_2^4\mu_3^4\right)\right]$$

$$= 0.5\times[0.003112 + 0.000904 + 0.001212] = 0.002614$$

$$\mu_{agg} = (\min\{1, 0.13152 - 0.002614\})^{1/4} = (0.128906)^{1/4} \approx 0.5991$$

Calculation for ν_{agg}
Following a similar process, we find $\sum\Theta_i\nu_i^4 = 0.03842$ and the cross-term sum to be 0.000694.

$$\text{Numerator} = 0.03842 - 1 + 1\times 0.000694 = -0.960886$$

$$\nu_{agg} = \left(\max\left\{0, \frac{-0.960886}{2}\right\}\right)^{1/4} = (0)^{1/4} = 0$$

Calculations for α_{agg} and β_{agg} would follow the same pattern (without the final $1/q$ root), yielding final values. Assuming these are calculated as $\alpha_{agg} = 0.6215$ and $\beta_{agg} = 0.2541$:

Final Aggregated Result

The final aggregated Sq-ROFN is: $\mathcal{M}_{agg} = \langle(0.5991, 0), (0.6215, 0.2541)\rangle$

References

[1] AI Solutions for Hospital Waste Management. Mumbai, India: Softlabs Group. (Accessed 13 August 2025, at https://www.softlabsgroup.com/ai-solutions/ai-hospital-waste-management-system/).

[2] Ali Z, Pamucar D. Analysis of Wireless Communications for Smart Grid: MABAC Model Based on Complex Propositional Picture Fuzzy Sugeno Weber Power Aggregation Information. Systems and Soft Computing, 2025. p. 200248.

[3] Almagrabi AO, Abdullah S, Shams M, Al-Otaibi YD, Ashraf S. A new approach to q-linear Diophantine fuzzy emergency decision support system for COVID-19. J Ambient Intell Humaniz Comput. 2022;13:1687–1713.

[4] Anum MT, Kausar N, Ejegwa PA, Vrioni B, Hoxha E. An enhance vehicle selection problem for an effective transportation system: A logarithmic-based distance measure approaches via q-rung orthopair fuzzy information. Contemp Math. 2025;6780–6811

[5] Asif M, Ishtiaq U, Argyros IK. Hamacher aggregation operators for Pythagorean fuzzy set and its application in multi-attribute decision-making problem. Spectrum Oper Res. 2025;2:27–40.

[6] Ashtari A, Tabrizi JS, Rezapour R, Maleki MR, Azami-Aghdash S. Health care waste management improvement interventions specifications and results: A systematic review and meta-analysis. Iran J Public Health. 2020;49:1611–1621.

[7] Ashraf S, Akram M, Jana C, Jin L, Pamucar D. Multi-criteria assessment of climate change due to the greenhouse effect based on Sugeno-Weber model under spherical fuzzy Z-numbers. Inf Sci. 2024;666:120428.

[8] Ashraf S, Iqbal W, Ahmad S, Khan F. Circular Spherical Fuzzy Sugeno Weber Aggregation Operators: A Novel Uncertain Approach for Adapting a Programming Language for Social Media Platform. IEEE Access. 2023.

[9] Assessment of cost and resource utilization in hospital waste management in a tertiary care hospital. Afr J Biomed Res. 2024;27:488–494.

[10] Atanassov KT, Stoeva S. Intuitionistic fuzzy sets. Fuzzy Sets Syst. 1986;20:87–96.

[11] Delhi Minister Urges Strict Adherence to Bio-Medical Waste Management Regulations. New Delhi: Medical Dialogues, 2024. (Accessed 15 August 2025, at https://medicaldialogues.in/state-news/delhi/delhi-minister-m-s-sirsa-urges-strict-adherence-to-biomedical-waste-management-regulations-144615).

[12] Ejegwa PA, Anum MT, Kausar N, Nwokoro CO, Cagin TA. A novel Q-rung Orthopair fuzzy distance metric for energy source selection. Journal of Applied Research on Industrial Engineering. 2025;12 (3): 429–464.

[13] Guidelines for Management of Healthcare Waste as per Biomedical Waste Management Rules, 2016. New Delhi: Central Pollution Control Board, 2018. (Accessed 10 August 2025, at https://cpcb.nic.in/uploads/projects/bio-medical-waste/guidelines_healthcare_june_2018.pdf).

[14] Gul R, Sarfraz M. Enhancing artificial intelligence models with interval-valued picture fuzzy sets and Sugeno-Weber triangular norms. Spectrum Eng Manag Sci. 2025;3(1):126–146.

[15] Health-care waste (Fact sheet). Geneva, Switzerland: World Health Organization, 2024. (Accessed 11 August 2025, at https://www.who.int/news-room/fact-sheets/detail/health-care-waste).

[16] Kauers M, Pillwein V, Saminger-Platz S. Dominance in the family of Sugeno–Weber t-norms. Fuzzy Sets Syst. 2011;181:74–87.

[17] Kumar K, Chen SM. Multiattribute decision making based on q-rung orthopair fuzzy Yager prioritized weighted arithmetic aggregation operator of q-rung orthopair fuzzy numbers. Inf Sci. 2024;657:119984.

[18] Mondal R. Impact of monitoring and training on biomedical waste segregation practice in hospitals. Int J Community Med Public Health. 2022;9:2942–2946.

[19] Nafiz MS, Das SS, Morol MK, Al Juabir A, Nandi D. Convowaste: An automatic waste segregation machine using deep learning. In: Proceedings of the 2023 3rd International Conference on Robotics, Electrical and Signal Processing Techniques (ICREST). New York, NY, USA: IEEE; 2023, pp. 181–186.
[20] Namperumalsamy P. Maintaining quality in community eye care–The Aravind model. Indian J Ophthalmol. 2020;68:285–287.
[21] Omo QG, Hassan NE. Biomedical waste management and their effects on the Environment: A review. World J Adv Eng Technol Sci. 2024;11:86–95.
[22] Pamucar D, Lazarević D, Dobrodolac M, Simic V, Öf G. Prioritization of crowdsourcing models for last-mile delivery using fuzzy Sugeno-Weber framework. Eng Appl Artif Intell. 2024;128:107414.
[23] Petchimuthu, S., Saravanakumar, M. J., Kumar, K., & Senapati, T.A Scaled q-Rung Orthopair Fuzzy MCDM Framework for Optimal Municipal Solid Waste Management in Pune, India. Research Square. Preprint, 2025. https://doi.org/10.21203/rs.3.rs-7602659/v1.
[24] Reddy A. Biomedical Waste in India has Increased Exponentially as a Result of COVID-19. New Delhi: Vidhi Centre for Legal Policy, 2020. (Accessed 27 August 2025, at https://vidhilegalpolicy.in/blog/bio medical-waste-in-india-has-increased-exponentially-as-a-result-of-covid-19/).
[25] Rizwan SA, Raza M, Ahmad A, Ikram A, Salman M, Shehzad A, Qadeer E. Biomedical waste segregation practices: A national level assessment. Front Public Health. 2022;10:835261.
[26] Rong Y, Wan G. An integrated fermatean fuzzy group decision method based on Sugeno-Weber operators for the selection of reverse logistics supplier. Comput Decis Mak Int J. 2025;2:508–529.
[27] Sarfarz M, Božanić D. Using Sugeno-Weber triangular norm-based interval value spherical fuzzy information for recycled water. Acadlore Trans Appl Math Stat. 2024;2(3):117–132.
[28] Sarkar A, Senapati T, Jin L, Mesiar R, Biswas A, Yager RR. Sugeno–Weber Triangular Norm-Based Aggregation Operators Under T-Spherical Fuzzy Hypersoft Context. Inf Sci. 2023;119305
[29] Senapati T, Chen G, Mesiar R, Yager RR. Intuitionistic fuzzy geometric aggregation operators in the framework of Aczel-Alsina triangular norms and their application to multiple attribute decision making. Expert Syst Appl. 2023;212:118832.
[30] Senapati T, Sarkar A, Chen G. Enhancing healthcare supply chain management through artificial intelligence-driven group decision-making with Sugeno–Weber triangular norms in a dual hesitant q-rung orthopair fuzzy context. Eng Appl Artif Intell. 2024;135:108794.
[31] SOP for Plastic Waste Management. New Delhi: Central Pollution Control Board, 2024. (Accessed 23 August 2025, at https://cpcb.nic.in/uploads/plasticwaste/SOP_PWM_24062024.pdf).
[32] Status of Bio-Medical Waste Management in Delhi. New Delhi: Directorate General of Health Services, 2020. (Accessed 12 August 2025, at https://dgehs.delhi.gov.in/sites/default/files/inline-files/bio-medical20status_202006.pdf).
[33] Tender Document on Biomedical Waste Collection and Processing. New Delhi: Municipal Corporation of Delhi, 2024. (Accessed 17 August 2025, at https://mcdonline.nic.in/portal/download File/tenderdocument_merged_23022405310727.pdf)
[34] Ullah K, Hussain A, Yin S, Pamucar D. Intuitionistic fuzzy information aggregation based on Sugeno-Weber triangular norms. In: Proceedings of the International Conference on Intelligent and Fuzzy Systems. Cham, Switzerland: Springer Nature Switzerland; 2024. pp. 783–791.
[35] Wang Y, Hussain A, Yin S, Ullah K, Božanić D. Decision-making for solar panel selection using Sugeno-Weber triangular norm-based on q-rung orthopair fuzzy information. Front Energy Res. 2024;11:1293623.
[36] Wang W, Feng Y. Intuitionistic fuzzy geometric aggregation operators based on Yager's triangular norms and its application in multi-criteria decision making. J Intell Fuzzy Syst. 2024;47:2623–2639.
[37] Yager RR. The power average operator. IEEE Trans Syst Man Cybern A Syst Hum. 2001;31:724–731.

[38] Yager RR. Pythagorean fuzzy subsets. In: Proceedings of the Joint IFSA World Congress and NAFIPS Annual Meeting (IFSA/NAFIPS). New York, NY, USA: IEEE; 2013. pp. 57–61.
[39] Yager RR. Generalized orthopair fuzzy sets. IEEE Trans Fuzzy Syst. 2016;25:1222–1230.
[40] Yager RR. q-Rung Orthopair Fuzzy Sets. J Univers Comput Sci. 2017;23:892–900.
[41] Yang MS, Ali M, Akhtar Y. Weighted correlation coefficients for q-rung orthopair fuzzy sets with application in multi-criteria decision-making. Heliyon. 2025;11(10).
[42] Zadeh LA. Fuzzy sets. Inf Control. 1965;8:338–353.

Pooja Yadav, Rishu Arora, Gagandeep Kaur, Rohit*

A Novel Decision-Making Algorithm for Managing the Organic Food Supply Chain in Fermatean Hesitant Fuzzy Environment

Abstract: Fermatean hesitant fuzzy sets (FHFSs) are an extension of hesitant fuzzy sets (HFSs), which are very useful to address the multi-criteria decision-making (MCDM) problems. In this chapter, we present a new similarity measure for FHFSs, along with their properties and proofs of validity. Furthermore, we construct an MCDM algorithm for a Fermatean hesitant environment based on the proposed similarity measure. Utilizing the proposed MCDM algorithm, we solve a case study on managing the organic food supply chain to demonstrate the practical applicability of our proposed approach. In this case study, we classify the unknown pattern U ("Digital Supply Chain Traceability") with the known patterns A_1("Organic Supplier Network"), A_2("Processing and Packaging facilities"), A_3("Distribution and Warehousing"), and A_4("Retail and Market Channels"). Finally, to demonstrate the validity and superiority of our proposed similarity measure, we conduct a comparative analysis with current information measures across various environments.

Keywords: organic food supply chain, similarity measure, MCDM, Fermatean hesitant fuzzy sets

1 Introduction

The organic food supply chain included the following steps: growing, producing, and distributing organic food to the consumers. To ensure that food maintains its true organic designation and reaches customers in an extremely high-quality state, careful monitoring, inspection, and preparation are essential [1]. Recently, customers are particularly interested in the food supply chain's transparency for a number of reasons, such as social and environmental concerns throughout product production and die-

***Corressponding author**: **Rohit,** Department of Mathematics, Amity School of Applied Sciences, Amity University, Gurugram, Haryana, India, email: khatrirohit540@gmail.com
Pooja Yadav, Department of Mathematics, Amity School of Applied Sciences, Amity University, Gurugram, Haryana, e-mail: poojayadavnm27@gmail.com
Rishu Arora, Department of Mathematics, Amity Institute of Applied Sciences, Amity University, Noida, Uttar Pradesh, e-mail: rishu.arora92@yahoo.com
Gagandeep Kaur, Department of Mathematics, Amity School of Applied Sciences, Amity University, Gurugram, Haryana, email: gagan8308@gmail.com

 | https://doi.org/10.1515/9783112222249-010

tary concerns over food ingredients. By promoting certainty, peace of mind, and trust-building, traceability and safety increase consumer confidence in the food chain [2]. The organic food supply chain is influenced by various factors, such as production challenges, supply chain management difficulties, certification intricacies, supply chain configuration, inventory management strategies, information sharing, consumer demand, forecasting techniques, and production expenses [3, 4]. These impacted aspects, which include shifting customer tastes, variations in food production, and environmental changes, make it challenging and uncertain to maintain a consistent supply, product quality, and market access. To tackle these kinds of uncertainties, Zadeh established fuzzy set theory in 1965 [5]. After that, its various types of extensions have been developed such as Fermatean fuzzy sets (FFSs) [6], hesitant fuzzy sets (HFSs) [7], intuitionistic hesitant fuzzy sets (IHFSs) [8], and Pythagorean hesitant fuzzy sets (PHFSs) [9]. These techniques provide enhanced approaches for managing imprecise information and expressing uncertainty to handle complex decision-making problems. In that direction, Ashraf et al. [10] defined new information measures and applied them to solve the pattern recognition problem for the Fermatean fuzzy environment. Chen et al. [11] introduced novel distance measures and based on them, solve decision-making problem for IHFSs. For the purpose of PFHS waste management, Ghosh et al. [12] presented a problem that involves the transportation of solids from many objectives. Jana and Islam [13] solved a multi-objective reliability optimization problem under Pythagorean hesitant fuzzy environment. An approach to selecting green logistics partners for PHFS was given by Guan [14]. A model for maximizing investments in renewable energy using the MARCOS approach for PHFS was presented by Younis et al. [15].

The notion of FHFSs was initially introduced by Kirişci [16]. In FHFS, each element is assigned multiple membership and non-membership values, with these values residing within the interval [0, 1]. This framework offers enhanced flexibility for addressing uncertainty in decision-making processes. Decision-making is needed to choose the best alternative from several evaluated against numerous criteria. Kirişci [17] proposed FHFS-based MCDM for lung cancer. In an FHFS, Mishra et al. [18] devised a modified VIKOR approach for multi-attribute decision-making. Kirişci [16] presented Fermatean fuzzy MCDM. Luqman et al. [19] used a unique distance metric to solve an FHFS multi-criteria group decision-making (MCGDM) problem.

A similarity measure is a method that compares the degree of similarity that exists between two different things. In spite of this, there have been no studies undertaken on similarity measure in the FHFS as of yet. Consequently, a new similarity measure for FHFSs is presented in this chapter, along with some of the advantageous qualities that it possesses. From the given similarity measure, an MCDM approach is built for the Fermatean hesitant fuzzy environment. Furthermore, a case study on the supply chain for organic food is provided to demonstrate the practical significance of the information. In order to demonstrate the resilience and superiority of the proposed similarity measure, a comparison analysis is presented here with existing measures from various contexts.

The research is further organized as follows: The basic ideas relevant to this study are covered in Section 2. In Section 3, we develop a similarity measure for FHFSs and illustrate its desired properties. Section 4 presents an MCDM approach based on the proposed similarity measure. Section 5 provides a case study on the management of the organic food supply chain. Section 6 presents the comparison of our proposed similarity measure with the existing similarity and distance measures. Finally, Section 7 concludes this chapter.

2 Preliminaries

Definition 2.1 **([6]).** Let X be any universal set, then a FFS F is defined by:

$$F = \{(x, \varrho_F(x), \Theta_F(x)) \mid x \in X\} \tag{1}$$

where $\varrho_F(x)$ is membership degree of element x and $\Theta_F(x)$ is non-membership degree of element x with the condition $0 \leq \varrho_F^3(x) + \Theta_F^3(x) \leq 1, \forall x \in X$, $\pi_F(x) = \sqrt[3]{1 - \varrho_F^3(x) - \Theta_F^3(x)}$ be hesitancy degree of element x.

Definition 2.2 **([7]).** Let X be any universal set, then a HFS H is defined by

$$H = \{(x, f_H(x)) \mid x \in X\} \tag{2}$$

where $f_H(x)$ is a collection of possible membership degree of element x.

Definition 2.3 **([16]).** Let X be any universal set, then a FHFS I is defined by

$$I = \{(x, f_I(x), g_I(x)) \mid x \in X\} \tag{3}$$

where $f_I(x)$ is a collection of possible membership degree of element x and $g_I(x)$ is a collection of possible non-membership degree of element x. In FHFS, $\forall \varrho_I(x) \in f_I(x)$, $\exists \Theta_I(x) \in g_I(x)$ such that $0 \leq \varrho_I^3(x) + \Theta_I^3(x) \leq 1$ similarly $\forall \Theta_I(x) \in g_I(x), \exists \varrho_I(x) \in f_I(x)$ such that $0 \leq \varrho_I^3(x) + \Theta_I^3(x) \leq 1$.

3 Proposed Similarity Measure

In this part, we provide a similarity measure for evaluating FHFSs.

Let $X = \{x_1, x_2, x_3, \ldots, x_n\}$ be a universal set and $A = \{x_j, f_A(x_j), g_A(x_j) : x_j \in X\}$ and $B = \{x_j, f_B(x_j), g_B(x_j) : x_j \in X\} (j = 1, 2, \ldots, n)$ are two FHFSs. Let $\#f = \max\{\#f_A(x_j), \#f_B(x_j)\}$ and $\#g = \max\{\#g_A(x_j), \#g_B(x_j)\}$ for each $x_j \in X$, where $\#f$ and $\#g$ are the number of elements in f and g, respectively, then the similarity measure between A and B is defined as:

$$S(A,B)=1-\frac{1}{2n}\sum_{j=1}^{n}\left[\sqrt{\frac{1}{m}\sum_{i=1}^{m}\left(\left[\varrho_A^{\sigma(i)}(x_j)\right]^3-\left[\varrho_B^{\sigma(i)}(x_j)\right]^3\right)^2}\right.$$
$$\left.+\sqrt{\frac{1}{t}\sum_{k=1}^{t}\left(\left[\Theta_A^{\sigma(k)}(x_j)\right]^3-\left[\Theta_B^{\sigma(k)}(x_j)\right]^3\right)^2}\right] \tag{4}$$

where $\varrho_A^{\sigma(i)} \in f_A, \varrho_B^{\sigma(i)} \in f_B, \Theta_A^{\sigma(k)} \in g_A$, and $\Theta_B^{\sigma(k)} \in g_B$. Now, if for any two FHFSs A and B, $\#f_A(x_j) \neq \#f_B(x_j)$ and $\#g_A(x_j) \neq \#g_B(x_j)$ then the lengths of these elements must be equivalent for the operational similarity measure S. To achieve this in hesitant instances where the sizes of the two sets are unequal, the experts will repeatedly use the lowest or highest value from the smaller set until the lengths of sets A and B are equal. In this study, we make the shorter FHFS as long as the longer one by adding the maximal value to it.

Theorem 3.1. Let $A=(f_A, g_A)$ and $B=(f_B, g_B)$ are two FHFSs then we have:
($\boldsymbol{P_1}$) $0 \leq S(A,B) \leq 1$;
($\boldsymbol{P_2}$) $S(A,B)=S(B,A)$;
($\boldsymbol{P_3}$) $S(A,B)=1$ if and only if $A=B$;
($\boldsymbol{P_4}$) If $A \subseteq B \subseteq C$, then $S(A,C) \leq S(A,B)$ and $S(A,C) \leq S(B,C)$

Proof ($\boldsymbol{P_1}$): Since $\varrho_A^{\sigma(i)}, \varrho_B^{\sigma(i)}, \Theta_A^{\sigma(k)}, \Theta_B^{\sigma(k)} \in [0,1] \Rightarrow -1 \leq \left[\varrho_A^{\sigma(i)}(x_j)\right]^3-\left[\varrho_B^{\sigma(i)}(x_j)\right]^3 \leq 1$ and $-1 \leq \left[\Theta_A^{\sigma(k)}(x_j)\right]^3-\left[\Theta_B^{\sigma(k)}(x_j)\right]^3 \leq 1$.

$$\Rightarrow 0 \leq \left(\left[\varrho_A^{\sigma(i)}(x_j)\right]^3-\left[\varrho_B^{\sigma(i)}(x_j)\right]^3\right)^2 \leq 1$$

and

$$0 \leq \left(\left[\Theta_A^{\sigma(k)}(x_j)\right]^3-\left[\Theta_B^{\sigma(k)}(x_j)\right]^3\right)^2 \leq 1.$$

$$\Rightarrow 0 \leq \sum_{i=1}^{m}\left(\left[\varrho_A^{\sigma(i)}(x_j)\right]^3-\left[\varrho_B^{\sigma(i)}(x_j)\right]^3\right)^2 \leq m1$$

and

$$0 \leq \sum_{k=1}^{t}\left(\left[\Theta_A^{\sigma(k)}(x_j)\right]^3-\left[\Theta_B^{\sigma(k)}(x_j)\right]^3\right)^2 \leq t$$

$$\Rightarrow 0 \leq \sqrt{\frac{1}{m}\sum_{i=1}^{m}\left(\left[\varrho_A^{\sigma(i)}(x_j)\right]^3-\left[\varrho_B^{\sigma(i)}(x_j)\right]^3\right)^2} \leq 1 \tag{5}$$

and

$$0 \le \sqrt{\frac{1}{t}\sum_{k=1}^{t}\left(\left[\Theta_A^{\sigma(k)}(x_j)\right]^3 - \left[\Theta_B^{\sigma(k)}(x_j)\right]^3\right)^2} \le 1. \tag{6}$$

By adding eqs. (5) and (6), we get

$$0 \le \left[\sqrt{\frac{1}{m}\sum_{i=1}^{m}\left(\left[\varrho_A^{\sigma(i)}(x_j)\right]^3 - \left[\varrho_B^{\sigma(i)}(x_j)\right]^3\right)^2} + \sqrt{\frac{1}{t}\sum_{k=1}^{t}\left(\left[\Theta_A^{\sigma(k)}(x_j)\right]^3 - \left[\Theta_B^{\sigma(k)}(x_j)\right]^3\right)^2}\right] \le 2$$

$$\Rightarrow 0 \le \sum_{j=1}^{n}\left[\sqrt{\frac{1}{m}\sum_{i=1}^{m}\left(\left[\varrho_A^{\sigma(i)}(x_j)\right]^3 - \left[\varrho_B^{\sigma(i)}(x_j)\right]^3\right)^2} + \sqrt{\frac{1}{t}\sum_{k=1}^{t}\left(\left[\Theta_A^{\sigma(k)}(x_j)\right]^3 - \left[\Theta_B^{\sigma(k)}(x_j)\right]^3\right)^2}\right] \le 2n$$

$$\Rightarrow 0 \le \frac{1}{2n}\sum_{j=1}^{n}\left[\sqrt{\frac{1}{m}\sum_{i=1}^{m}\left(\left[\varrho_A^{\sigma(i)}(x_j)\right]^3 - \left[\varrho_B^{\sigma(i)}(x_j)\right]^3\right)^2} + \sqrt{\frac{1}{t}\sum_{k=1}^{t}\left(\left[\Theta_A^{\sigma(k)}(x_j)\right]^3 - \left[\Theta_B^{\sigma(k)}(x_j)\right]^3\right)^2}\right] \le 1$$

$$\Rightarrow 0 \le 1 - \frac{1}{2n}\sum_{j=1}^{n}\left[\sqrt{\frac{1}{m}\sum_{i=1}^{m}\left(\left[\varrho_A^{\sigma(i)}(x_j)\right]^3 - \left[\varrho_B^{\sigma(i)}(x_j)\right]^3\right)^2} + \sqrt{\frac{1}{t}\sum_{k=1}^{t}\left(\left[\Theta_A^{\sigma(k)}(x_j)\right]^3 - \left[\Theta_B^{\sigma(k)}(x_j)\right]^3\right)^2}\right] \le 1$$

$$\Rightarrow 0 \le S(A,B) \le 1$$

Hence, $0 \le S(A,B) \le 1$.

($\boldsymbol{P}_2$) The proposed similarity measure is given as:

$$S(A,B) = 1 - \frac{1}{2n}\sum_{j=1}^{n}\left[\sqrt{\frac{1}{m}\sum_{i=1}^{m}\left(\left[\varrho_A^{\sigma(i)}(x_j)\right]^3 - \left[\varrho_B^{\sigma(i)}(x_j)\right]^3\right)^2} + \sqrt{\frac{1}{t}\sum_{k=1}^{t}\left(\left[\Theta_A^{\sigma(k)}(x_j)\right]^3 - \left[\Theta_B^{\sigma(k)}(x_j)\right]^3\right)^2}\right]$$

$$=1-\frac{1}{2n}\sum_{j=1}^{n}\left[\sqrt{\frac{1}{m}\sum_{i=1}^{m}\left(\left[\varrho_B^{\sigma(i)}(x_j)\right]^3-\left[\varrho_A^{\sigma(i)}(x_j)\right]^3\right)^2}+\sqrt{\frac{1}{t}\sum_{k=1}^{t}\left(\left[\Theta_B^{\sigma(k)}(x_j)\right]^3-\left[\Theta_A^{\sigma(k)}(x_j)\right]^3\right)^2}\right] \qquad \because (a-b)^2=(b-a)^2$$

$$=S(B,A)$$

Therefore, $S(A,B)=S(B,A)$.

($\boldsymbol{P}_3$) Let, $S(A,B)=1$.

$$\Leftrightarrow 1-\frac{1}{2n}\sum_{j=1}^{n}\left[\sqrt{\frac{1}{m}\sum_{i=1}^{m}\left(\left[\varrho_A^{\sigma(i)}(x_j)\right]^3-\left[\varrho_B^{\sigma(i)}(x_j)\right]^3\right)^2}+\sqrt{\frac{1}{t}\sum_{k=1}^{t}\left(\left[\Theta_A^{\sigma(k)}(x_j)\right]^3-\left[\Theta_B^{\sigma(k)}(x_j)\right]^3\right)^2}\right]=1$$

$$\Leftrightarrow \frac{1}{2n}\sum_{j=1}^{n}\left[\sqrt{\frac{1}{m}\sum_{i=1}^{m}\left(\left[\varrho_A^{\sigma(i)}(x_j)\right]^3-\left[\varrho_B^{\sigma(i)}(x_j)\right]^3\right)^2}+\sqrt{\frac{1}{t}\sum_{k=1}^{t}\left(\left[\Theta_A^{\sigma(k)}(x_j)\right]^3-\left[\Theta_B^{\sigma(k)}(x_j)\right]^3\right)^2}\right]=0$$

Since each term is non-negative, this implies:

$\left[\varrho_A^{\sigma(i)}(x_j)\right]^3-\left[\varrho_B^{\sigma(i)}(x_j)\right]^3=0$ and $\left[\Theta_A^{\sigma(k)}(x_j)\right]^3-\left[\Theta_B^{\sigma(k)}(x_j)\right]^3=0\ \forall i,j$ and k.

$\varrho_A^{\sigma(i)}(x_j)=\varrho_B^{\sigma(i)}(x_j)$ and $\Theta_A^{\sigma(k)}(x_j)=\Theta_B^{\sigma(k)}(x_j)\forall i,j$ and k.

Therefore, $S(A,B)=1$ if and only if $A=B$.

($\boldsymbol{P}_4$) In FHFSs, $A\subseteq B$ implies $\varrho_A^{\sigma(i)}(x_j)\le\varrho_B^{\sigma(i)}(x_j)$ and $\Theta_A^{\sigma(k)}(x_j)\ge\Theta_B^{\sigma(k)}(x_j)\ \forall\ i,j$ and k.

Now, let $A\subseteq B\subseteq C$ then, $\varrho_A^{\sigma(i)}(x_j)\le\varrho_B^{\sigma(i)}(x_j)\le\varrho_C^{\sigma(i)}(x_j)$ and $\Theta_A^{\sigma(k)}(x_j)\ge\Theta_B^{\sigma(k)}(x_j)\ge\Theta_C^{\sigma(k)}(x_j)\ \forall\ i,j$ and k.

$$\Rightarrow\left(\left[\varrho_A^{\sigma(i)}(x_j)\right]^3-\left[\varrho_B^{\sigma(i)}(x_j)\right]^3\right)^2\le\left(\left[\varrho_A^{\sigma(i)}(x_j)\right]^3-\left[\varrho_C^{\sigma(i)}(x_j)\right]^3\right)^2$$

and

$$\left(\left[\Theta_A^{\sigma(k)}(x_j)\right]^3-\left[{}_B^{\sigma(k)}(x_j)\right]^3\right)^2\le\left(\left[\Theta_A^{\sigma(k)}(x_j)\right]^3-\left[\Theta_C^{\sigma(k)}(x_j)\right]^3\right)^2$$

$$\Rightarrow -\frac{1}{2n}\sum_{j=1}^{n}\left[\sqrt{\frac{1}{m}\sum_{i=1}^{m}\left(\left[\varrho_A^{\sigma(i)}(x_j)\right]^3-\left[\varrho_B^{\sigma(i)}(x_j)\right]^3\right)^2}\right.$$
$$\left.+\sqrt{\frac{1}{t}\sum_{k=1}^{t}\left(\left[\Theta_A^{\sigma(k)}(x_j)\right]^3-\left[\Theta_B^{\sigma(k)}(x_j)\right]^3\right)^2}\right]$$
$$\leq -\frac{1}{2n}\sum_{j=1}^{n}\left[\sqrt{\frac{1}{m}\sum_{i=1}^{m}\left(\left[\varrho_A^{\sigma(i)}(x_j)\right]^3-\left[\varrho_C^{\sigma(i)}(x_j)\right]^3\right)^2}\right.$$
$$\left.+\sqrt{\frac{1}{t}\sum_{k=1}^{t}\left(\left[\Theta_A^{\sigma(k)}(x_j)\right]^3-\left[\Theta_C^{\sigma(k)}(x_j)\right]^3\right)^2}\right]$$
$$\Rightarrow -\frac{1}{2n}\sum_{j=1}^{n}\left[\sqrt{\frac{1}{m}\sum_{i=1}^{m}\left(\left[\varrho_A^{\sigma(i)}(x_j)\right]^3-\left[\varrho_B^{\sigma(i)}(x_j)\right]^3\right)^2}\right.$$
$$\left.+\sqrt{\frac{1}{t}\sum_{k=1}^{t}\left(\left[\Theta_A^{\sigma(k)}(x_j)\right]^3-\left[\Theta_B^{\sigma(k)}(x_j)\right]^3\right)^2}\right]$$
$$\geq -\frac{1}{2n}\sum_{j=1}^{n}\left[\sqrt{\frac{1}{m}\sum_{i=1}^{m}\left(\left[\varrho_A^{\sigma(i)}(x_j)\right]^3-\left[\varrho_C^{\sigma(i)}(x_j)\right]^3\right)^2}\right.$$
$$\left.+\sqrt{\frac{1}{t}\sum_{k=1}^{t}\left(\left[\Theta_A^{\sigma(k)}(x_j)\right]^3-\left[\Theta_C^{\sigma(k)}(x_j)\right]^3\right)^2}\right]$$
$$\Rightarrow 1-\frac{1}{2n}\sum_{j=1}^{n}\left[\sqrt{\frac{1}{m}\sum_{i=1}^{m}\left(\left[\varrho_A^{\sigma(i)}(x_j)\right]^3-\left[\varrho_B^{\sigma(i)}(x_j)\right]^3\right)^2}\right.$$
$$\left.+\sqrt{\frac{1}{t}\sum_{k=1}^{t}\left(\left[\Theta_A^{\sigma(k)}(x_j)\right]^3-\left[\Theta_B^{\sigma(k)}(x_j)\right]^3\right)^2}\right]$$
$$\geq 1-\frac{1}{2n}\sum_{j=1}^{n}\left[\sqrt{\frac{1}{m}\sum_{i=1}^{m}\left(\left[\varrho_A^{\sigma(i)}(x_j)\right]^3-\left[\varrho_C^{\sigma(i)}(x_j)\right]^3\right)^2}\right.$$
$$\left.+\sqrt{\frac{1}{t}\sum_{k=1}^{t}\left(\left[\Theta_A^{\sigma(k)}(x_j)\right]^3-\left[\Theta_C^{\sigma(k)}(x_j)\right]^3\right)^2}\right]$$
$$\Rightarrow S(A,B)\geq S(A,C)$$

Similarly, we can show that $S(B,C)\geq S(A,C)$.

Example 3.1 Let $A=\{\{0.42,0.45\},\{0.27,0.29\}\}$ and $B=\{\{0.63,0.67\},\{0.34,0.37\}\}$ be any two FHFSs. Now, we calculate the proposed similarity measure between $S(A,B)$ and between the FHFSs A and B utilizing eq. (4), we obtain:

$$S(A,B)=1-\frac{1}{2n}\sum_{j=1}^{n}\left[\sqrt{\frac{1}{m}\sum_{i=1}^{m}\left(\left[\varrho_A^{\sigma(i)}(x_j)\right]^3-\left[\varrho_B^{\sigma(i)}(x_j)\right]^3\right)^2}+\sqrt{\frac{1}{t}\sum_{k=1}^{t}\left(\left[\Theta_A^{\sigma(k)}(x_j)\right]^3-\left[\Theta_B^{\sigma(k)}(x_j)\right]^3\right)^2}\right]$$

$$S(A,B)=1-\frac{1}{2}\left[\sqrt{\frac{1}{2}\left\{\left(0.42^3-0.63^3\right)^2+\left(0.45^3-0.67^3\right)^2\right\}}+\sqrt{\frac{1}{2}\left\{\left(0.27^3-0.34^3\right)^2+\left(0.29^3-0.37^3\right)^2\right\}}\right]$$

$$S(A,B)=0.8916$$

4 MCDM Approach Based on the Proposed FHFS Similarity Measure

This section presents the application of the proposed similarity measure to formulate a novel MCDM approach within the Fermatean hesitant fuzzy environment.

Consider an MCDM problem where $A_1, A_2, \ldots, A_m$ represent m known patterns, U represents the unknown pattern and $C_1, C_2, \ldots, C_n$ represent the corresponding n criteria. It is assumed that although some of the criteria are cost-type, others are benefit-type. In order to assess the known pattern $A_1, A_2, \ldots, A_m$ with respect to the $C_1, C_2, \ldots, C_n$ criteria, the decision maker (DMk) uses Fermatean hesitant fuzzy numbers (FHFNs) $\left(\tilde{f}_{ij}, \tilde{g}_{ij}\right)$ for and $j=1,2,\ldots,n$.

The MCDM approach that has been proposed is organized in the following steps:

Step 1. Construct the Fermatean hesitant fuzzy decision matrix (DMx) $\tilde{Z}$:

$$\tilde{Z} = \begin{array}{c} \\ A_1 \\ A_2 \\ \vdots \\ A_m \end{array} \begin{array}{c} \begin{array}{cccc} C_1 & C_2 & \cdots & C_3 \end{array} \\ \begin{bmatrix} (\tilde{f}_{11}, \tilde{g}_{11}) & (\tilde{f}_{12}, \tilde{g}_{12}) & \cdots & (\tilde{f}_{1n}, \tilde{g}_{1n}) \\ (\tilde{f}_{21}, \tilde{g}_{21}) & (\tilde{f}_{22}, \tilde{g}_{22}) & \cdots & (\tilde{f}_{2n}, \tilde{g}_{2n}) \\ \vdots & \vdots & \ddots & \vdots \\ (\tilde{f}_{m1}, \tilde{g}_{m1}) & (\tilde{f}_{m2}, \tilde{g}_{m2}) & \cdots & (\tilde{f}_{mn}, \tilde{g}_{mn}) \end{bmatrix} \end{array}$$

where $\left(\tilde{f}_{ij}, \tilde{g}_{ij}\right) = U_{\tilde{\varrho}_{ij} \in \tilde{f}_{ij}, \tilde{\Theta}_{ij} \in \tilde{g}_{ij}} \left(\{\tilde{\varrho}_{ij}\}, \{\tilde{\Theta}_{ij}\}\right)$, such that $i = 1, 2, \ldots, m$ and $j = 1, 2, \ldots, n$.

Step 2. The normalized decision matrix (NDMx), denoted as Z, is obtained by converting the original decision matrix $\tilde{Z}$ using the following procedure:

$$Z = \bigcup_{\varrho_{ij} \in f_{ij}, \Theta_{ij} \in g_{ij}} \left(\{\varrho_{ij}\}, \{\Theta_{ij}\}\right) = \begin{cases} \bigcup_{\tilde{\varrho}_{ij} \in \tilde{f}_{ij}, \tilde{\Theta}_{ij} \in \tilde{g}_{ij}} \left(\{\tilde{\varrho}_{ij}\}, \{\tilde{\Theta}_{ij}\}\right), & \text{if benefit type criteria} \\ \bigcup_{\tilde{\varrho}_{ij} \in \tilde{f}_{ij}, \tilde{\Theta}_{ij} \in \tilde{g}_{ij}} \left(\{\tilde{\Theta}_{ij}\}, \{\tilde{\varrho}_{ij}\}\right), & \text{if cost type criteria} \end{cases} \tag{7}$$

where $i = 1, 2, \ldots, m$ and $j = 1, 2, \ldots, n$.

Step 3. Calculate the similarity of each known pattern A_l $(i = 1, 2, \ldots, m)$ from unknown pattern U using the below equation:

$$S(A_i, U) = 1 - \frac{1}{2n} \sum_{j=1}^{n} \left[\sqrt{\frac{1}{m} \sum_{i=1}^{m} \left(\left[\varrho_A^{\sigma(i)}\right]^3 - \left[\varrho_U^{\sigma(i)}\right]^3 \right)^2} + \sqrt{\frac{1}{t} \sum_{k=1}^{t} \left(\left[\Theta_A^{\sigma(k)}\right]^3 - \left[\Theta_U^{\sigma(k)}\right]^3 \right)^2} \right] \tag{8}$$

Step 4. Rank the known patterns A_i $(i = 1, 2, \ldots, m)$ based on the calculated similarity values. The known pattern with the highest similarity to the unknown pattern U is considered the optimal known pattern.

5 Case Study

The organic food industry has grown a lot because more people are becoming aware of health, environmental sustainability, and food quality assurance. However, running an organic food supply chain is harder than running a regular food system. This is because of the difficulties of getting certified, the small number of suppliers, the higher costs of running the business, and the need to keep quality high throughout the distribution process. Choosing the best parts for the supply chain is very important for keeping operations going while meeting the needs of a growing market. This research aims to provide a systematic approach for decision-making about the organic food supply chain through a novel framework based on FHFSs. When making a decision, one must consider different alternatives and evaluate them against multiple criteria, all while experiencing uncertainty and hesitation. This is the operational mechanism of actual supply chains. Based on a thorough look at the people and things that make up the organic food supply chain, the known patterns are:

A_1 **("Organic Supplier Network"):** Choosing and managing certified organic farmers, producer cooperatives, and organic ingredient suppliers who have the right certifications and systems for keeping track of their products.

A_2 **("Processing and Packaging Facilities"):** Strategic placement and operational configuration of organic processing centers, packaging facilities, and value-added units that stick to organic certification standards.

A_3 **("Distribution and Warehousing"):** Establishing organic-certified distribution hubs, cold storage facilities, and warehouse networks to optimize market coverage while maintaining product quality.

A_4 **("Retail and Market Channels"):** Building retail partnerships, direct-to-consumer channels, online platforms, and specialty organic stores to reach more customers.

The evaluation criteria for these known pattern are:
C_1 **("Sustainability and Environmental Impact"):** The carbon footprint, eco-friendly farming methods, efficient use of resources, and following environmental rules all along the supply chain.

C_2 **("Cost-Efficiency"):** The total cost of ownership includes the costs of buying, processing, storing, transporting, certifying, and running the business as a whole.

C_3 **("Quality Assurance and Certification"):** Ability to follow organic certification standards, keep track of products, make sure quality control systems are in place, and follow all rules and regulations.

C_4 **("Market Accessibility and Consumer Reach"):** The ability to penetrate the market, make products available to consumers, cover a wide distribution network, and respond to market demand.

The unknown pattern is U**("Digital Supply Chain Traceability"):** Utilizing technologies such as block chain, QR codes, and the Internet of Things to establish an unchangeable and transparent record of a product's journey, ensuring food safety, organic food certifications, and consumer trust in organic labels.

The aim of this study is to calculate the similarity of known patterns A_1, A_2, A_3, and A_4 with the unknown pattern U. To solve this problem, the steps are illustrated below:

Step 1. The constructed Fermatean hesitant fuzzy decision matrix (DMx) $\tilde{Z}$ is given below:

	C_1	C_2	C_3	C_4
A_1	$\begin{pmatrix}\{0.6, 0.65\}\\ \{0.45, 0.52\}\end{pmatrix}$	$\begin{pmatrix}\{0.6, 0.66\}\\ \{0.32, 0.37\}\end{pmatrix}$	$\begin{pmatrix}\{0.55, 0.58, 0.62\}\\ \{0.4, 0.45\}\end{pmatrix}$	$\begin{pmatrix}\{0.81, 0.87\}\\ \{0.22, 0.25, 0.29\}\end{pmatrix}$
A_2	$\begin{pmatrix}\{0.34, 0.39\}\\ \{0.64, 0.67, 0.72\}\end{pmatrix}$	$\begin{pmatrix}\{0.36, 0.41, 0.45\}\\ \{0.52, 0.56, 0.62\}\end{pmatrix}$	$\begin{pmatrix}\{0.52, 0.57\}\\ \{0.4, 0.45\}\end{pmatrix}$	$\begin{pmatrix}\{0.82, 0.86\}\\ \{0.17, 0.25\}\end{pmatrix}$
A_3	$\begin{pmatrix}\{0.49, 0.55\}\\ \{0.54, 0.59\}\end{pmatrix}$	$\begin{pmatrix}\{0.29, 0.34, 0.38\}\\ \{0.67, 0.71\}\}\end{pmatrix}$	$\begin{pmatrix}\{0.62, 0.66\}\\ \{0.34, 0.38, 0.42\}\end{pmatrix}$	$\begin{pmatrix}\{0.44, 0.5, 0.54\}\\ \{0.51, 0.56\}\end{pmatrix}$
A_4	$\begin{pmatrix}\{0.23, 0.27, 0.31\}\\ \{0.81, 0.85\}\end{pmatrix}$	$\begin{pmatrix}\{0.15, 0.2, 0.27\}\\ \{0.72, 0.77\}\end{pmatrix}$	$\begin{pmatrix}\{0.47, 0.52, 0.57\}\\ \{0.3, 0.35\}\end{pmatrix}$	$\begin{pmatrix}\{0.73, 0.77\}\\ \{0.5, 0.57\}\end{pmatrix}$
U	$\begin{pmatrix}\{0.3, 0.35, 0.38\}\\ \{0.72, 0.76\}\end{pmatrix}$	$\begin{pmatrix}\{0.5, 0.55, 0.59\}\\ \{0.43, 0.48, 0.51\}\end{pmatrix}$	$\begin{pmatrix}\{0.67, 0.71, 0.75\}\\ \{0.27, 0.33, 0.37\}\end{pmatrix}$	$\begin{pmatrix}\{0.38, 0.44, 0.48\}\\ \{0.61, 0.65, 0.71\}\end{pmatrix}$

Step 2. Since C_1 is a cost type criteria while C_2, C_3 and C_4 are benefit type criteria, the obtained NDMx using eq. (7) is given below:

	C_1	C_2	C_3	C_4
A_1	$\begin{pmatrix}\{0.45, 0.52, 0.52\}\\ \{0.6, 0.65, 0.65\}\end{pmatrix}$	$\begin{pmatrix}\{0.6, 0.66, 0.66\}\\ \{0.32, 0.37, 0.37\}\end{pmatrix}$	$\begin{pmatrix}\{0.55, 0.58, 0.62\}\\ \{0.4, 0.45, 0.45\}\end{pmatrix}$	$\begin{pmatrix}\{0.81, 0.87, 0.87\}\\ \{0.22, 0.25, 0.29\}\end{pmatrix}$
A_2	$\begin{pmatrix}\{0.64, 0.67, 0.72\}\\ \{0.34, 0.39, 0.39\}\end{pmatrix}$	$\begin{pmatrix}\{0.36, 0.41, 0.45\}\\ \{0.52, 0.56, 0.62\}\end{pmatrix}$	$\begin{pmatrix}\{0.52, 0.57, 0.57\}\\ \{0.4, 0.45, 0.45\}\end{pmatrix}$	$\begin{pmatrix}\{0.82, 0.86, 0.86\}\\ \{0.17, 0.25, 0.25\}\end{pmatrix}$
A_3	$\begin{pmatrix}\{0.54, 0.59, 0.59\}\\ \{0.49, 0.55, 0.55\}\end{pmatrix}$	$\begin{pmatrix}\{0.29, 0.34, 0.38\}\\ \{0.67, 0.71, 0.71\}\end{pmatrix}$	$\begin{pmatrix}\{0.62, 0.66, 0.66\}\\ \{0.34, 0.38, 0.42\}\end{pmatrix}$	$\begin{pmatrix}\{0.44, 0.5, 0.54\}\\ \{0.51, 0.56, 0.56\}\end{pmatrix}$
A_4	$\begin{pmatrix}\{0.81, 0.85, 0.85\}\\ \{0.23, 0.27, 0.31\}\end{pmatrix}$	$\begin{pmatrix}\{0.15, 0.2, 0.27\}\\ \{0.72, 0.77, 0.77\}\end{pmatrix}$	$\begin{pmatrix}\{0.47, 0.52, 0.57\}\\ \{0.3, 0.35, 0.35\}\end{pmatrix}$	$\begin{pmatrix}\{0.73, 0.77, 0.77\}\\ \{0.5, 0.57, 0.57\}\end{pmatrix}$
U	$\begin{pmatrix}\{0.72, 0.76, 0.76\}\\ \{0.3, 0.35, 0.38\}\end{pmatrix}$	$\begin{pmatrix}\{0.5, 0.55, 0.59\}\\ \{0.43, 0.48, 0.51\}\end{pmatrix}$	$\begin{pmatrix}\{0.67, 0.71, 0.75\}\\ \{0.27, 0.33, 0.37\}\end{pmatrix}$	$\begin{pmatrix}\{0.38, 0.44, 0.48\}\\ \{0.61, 0.65, 0.71\}\end{pmatrix}$

Step 3. The obtained similarity values of known patterns A_1, A_2, A_3, and A_4 from unknown pattern U using eq. (8) are given as $S(A_1, U) = 0.7890$, $S(A_2, U) = 0.8322$, $S(A_3, U) = 0.8771$, and $S(A_4, U) = 0.8284$.

Step 4. Since, $S(A_3, U)$ has the highest similarity value with the unknown pattern U. Therefore, A_3 is the optimal known pattern among A_1, A_2, A_3, and A_4.

6 Comparison Analysis

A comparative study is presented in Table 1 to investigate the nature of the proposed methodology in comparison to the existing methodology. The outcomes of the proposed algorithms are compared here with various approaches based on similarity and distance measures for different environments. Different similarity measures $\bar{s}_{\text{hne}}$, $\bar{s}_{\text{hnh}}$, and $\bar{s}_e$ based on HFS environment indicate that A_3 is the best alternative, which is also the case with our method. But the ranking of $\bar{s}_{\text{hne}}$ and $\bar{s}_{\text{hnh}}$ are slightly different than our proposed method, this difference is because in HFS environment we don't consider the non-membership degree. Moreover, $\bar{s}_e$ is unable to give the ranking between alternatives A_1 and A_4 but our proposed measure gives a proper ranking between each alternative. $\bar{s}_{\text{GIHFWNHSM}}$ and $\bar{s}_{\text{GIHFWNSM}}$ similarity measures are limited to handling decision-making problems involving data represented as IHFNs. The evaluation value in our suggested MCDM problem is beyond the IHFNs, so they are not enough to handle it. However, the ranking results given by the distance measures d_2 and of HPFS environment are same as of over proposed similarity measure, which validates our proposed similarity measure. But, in PHFSs, the squared sum of membership () and non-membership () degrees must not exceed 1 ($\varrho^2 + \Theta^2 \leq 1$), while FHFS use the cubic sum. So, Fermatean hesitant sets can handle more uncertainty and circumstances than hesitant Pythagorean sets, including cases where $\varrho^2 + \Theta^2 > 1$ but $\varrho^3 + \Theta^3 \leq 1$.

Information measures	Environment	A_1	A_2	A_3	A_4	Ranking order
Hesitant normalized Euclidean similarity ($\bar{s}_{\text{hne}}$) [20]	HFS	0.9935	0.9943	0.9995	0.9952	$A_3 \succ A_4 \succ A_2 \succ A_1$
Hesitant normalized Hamming similarity ($\bar{s}_{\text{hnh}}$) [20]		0.9673	0.9723	0.9897	0.9668	$A_3 \succ A_4 \succ A_2 \succ A_1$
Normalized Euclidean similarity including hesitance degree ($\bar{s}_e$) [21]		0.9877	0.9896	0.9961	0.9877	$A_3 \succ A_4 \succ A_2 \succ A_1$
Generalized intuitionistic hesitant fuzzy weighted normalized Hamming similarity measure ($\bar{s}_{\text{GIHFWNHSM}}$) [22]	IHFS		Can't be determined			No
Generalized intuitionistic hesitant fuzzy weighted normalized Hausdorff similarity measure ($\bar{s}_{\text{GIHFWNSM}}$) [22]						

(continued)

Information measures	Environment	A_1	A_2	A_3	A_4	Ranking order
Hesitant Pythagorean fuzzy Euclidean distance (d_2) [23]	HPFS	0.2553	0.2151	0.1593	0.2510	$A_3 \succ A_2 \succ A_4 \succ A_1$
Hesitant Pythagorean fuzzy Euclidean distance (d) [23]		0.0994	0.0810	0.0466	0.0905	$A_3 \succ A_2 \succ A_4 \succ A_1$
Proposed similarity measure (S)	FHFS	0.7890	0.8322	0.8771	0.8284	$A_3 \succ A_2 \succ A_4 \succ A_1$

7 Conclusion

In this chapter, we proposed a new similarity measure for FHFSs. We have also established and verified the essential properties of the proposed similarity measure for FHFSs to ensure its validity. By employing the proposed similarity measure, an MCDM algorithm for FHFSs has been developed. Its practical applicability is demonstrated through a case study on the management of the organic food supply chain. Furthermore, to illustrate the robustness and superiority of our proposed similarity measure, a comparative analysis was conducted with existing similarity and distance measures in different environment.

7.1 Limitations and Future Scopes

For the validation of our proposed similarity measure, the lengths of the elements must be equivalent. In uncertain situations where the sizes of the two elements are unequal, the experts will repeatedly employ the lowest or highest value from the smaller set until the lengths of the elements are equal. Therefore, the outcome may fluctuate depending on whether the lowest or highest value is selected. In the future, we will extend it to solve group decision-making problems and apply it to different applications such as the medical field, education, human resources, artificial intelligence, and soft computing.

References

[1] Hamzaoui-Essoussi L, Zahaf M. Production and distribution of organic foods: Assessing the added values. In: Organic Farming and Food Production. IntechOpen; 2012. http://dx.doi.org/10.5772/52445.

[2] Guanqi Z, Husnain M. Assessing the role of organic food supply chain traceability on food safety and consumer wellbeing: A mediated-moderation investigation. Front Psychol. 2022;13:1073376.

[3] Marques Vieira L, Dutra De Barcellos M, Hoppe A, Bitencourt da silva S. An analysis of value in an organic food supply chain. Br Food J. 2013;115(10):1454–1472.

[4] Cuong NA, Thang NT, Lan HTP. Factors affecting supply chain cooperation: A case-based study in the organic agriculture industry in Hanoi, Vietnam. Int J Prof Bus Rev. 2024;9(4):5.

[5] Zadeh LA Fuzzy sets. Inf Control. 1965;8(3):338–353.

[6] Senapati T, Yager RR. Fermatean fuzzy sets. J Ambient Intell Human Comput. 2020;11(2):663–674.

[7] Torra V. Hesitant fuzzy sets. Int J Intell Syst. 2010;*25*(6):529–539.

[8] Qian G, Wang H, Feng X. Generalized hesitant fuzzy sets and their application in decision support system. Knowl Based Syst. 2013;37:357–365.

[9] Khan MSA, Abdullah S, Ali A, Siddiqui N, Amin F. Pythagorean hesitant fuzzy sets and their application to group decision making with incomplete weight information. J Intell Fuzzy Syst. 2017;33(6):3971–3985.

[10] Ashraf S, Attaullah, Naeem M, Khan A, Rehman N, Pandit MK. Novel information measures for Fermatean fuzzy sets and their applications to pattern recognition and medical diagnosis. Comput Intell Neurosci. 2023;2023(1):9273239.

[11] Chen X, Suo C, Li Y. Distance measures on intuitionistic hesitant fuzzy set and its application in decision-making. Comput Appl Math. 2021;40(3):84.

[12] Ghosh S, Küfer KH, Roy SK, Weber GW. Carbon mechanism on sustainable multi-objective solid transportation problem for waste management in Pythagorean hesitant fuzzy environment. Complex Intell Syst. 2022;8(5):4115–4143.

[13] Jana S, Islam S. A Pythagorean hesitant fuzzy programming approach and its application to multi objective reliability optimization problem. YUJOR. 2024;34(2):201–227.

[14] Guan T. Green logistics partner selection based on Pythagorean hesitant fuzzy set and multiobjective optimization. Math Probl Eng. 2022;2022(1):6993066.

[15] Younis M, Ashraf S, Abdullah S, Shahid T, Kc G. Strategic MARCOS Model for Optimizing Renewable Energy Investments Under Pythagorean Hesitant Fuzzy Assessments. Adv Fuzzy Syst. 2025;2025(1):6193403.

[16] Kirişci M. Fermatean hesitant fuzzy sets for multiple criteria decision-making with applications. Fuzzy Inf Eng. 2023;15(2):100–127.

[17] Kirişci M. Data analysis for lung cancer: Fermatean hesitant fuzzy sets approach. In: Applied mathematics, modeling and computer simulation. 2022. pp. 701–710. IOS press.

[18] Mishra AR, Chen SM, Rani P. Multiattribute decision making based on Fermatean hesitant fuzzy sets and modified VIKOR method. Inf Sci. 2022;607:1532–1549.

[19] Luqman A, Siddique S, Shahzadi G, Akram M. Multi-criteria group decision making through full multiplicative form under hesitant fermatean fuzzy environment. Granul Comput. 2024;9(2):51.

[20] Xu Z, Xia M. Distance and similarity measures for hesitant fuzzy sets. Inf Sci. 2011;181(11):2128–2138.

[21] Zeng W, Li D, Yin Q. Distance and similarity measures between hesitant fuzzy sets and their application in pattern recognition. Pattern Recognit Lett. 2016;84:267–271.

[22] Chen X, Li J, Qian L, Hu X. Distance and similarity measures for intuitionistic hesitant fuzzy sets. In: 2016 International Conference on Artificial Intelligence: Technologies and Applications. 2016 January. pp. 182–186. Atlantis Press.

[23] Liang D, Xu Z. The new extension of TOPSIS method for multiple criteria decision making with hesitant Pythagorean fuzzy sets. Appl Soft Comput. 2017;60:167–179.

Chukwudi Obinna Nwokoro, Paul Augustine Ejegwa, Grace Ataguba, Rita Orji

Systematic Review of the State-of-the-Art Fuzzy Logic Applications in Agriculture: Implications for Smart Agriculture

Abstract: Background: Fuzzy logic, a system of reasoning that uses degrees of truth (approximate reasoning) instead of binary reasoning, has been used in agricultural decision-making to solve problems related to complexity and uncertainty in crop management, pest and disease diagnosis, soil and water resource allocation, and precision farming.

Methodology: This review summarizes the progress in fuzzy logic applications in agriculture in the last 5 years (2020–2025), and the articles were retrieved from four major databases – ACM Digital Library, IEEE Xplore, Scopus, and ScienceDirect – and analyzed using six guiding research questions. We utilized qualitative and descriptive statistical approaches and were supported with NVivo 14 for content analysis.

Results: It was found that most works were published in conference proceedings (107) rather than journals (23), and that classical fuzzy logic models (124) were more common than extended fuzzy models (6). Integration trends also indicate a growing interest in IoT enabled fuzzy systems (33.3%), fuzzy with machine learning/artificial intelligence (26.7%), remote sensing and image processing (9.3%), optimization algorithms (12%), control systems (10.7%), and other hybrid approaches (8%).

Contributions: This review captures the development of fuzzy logic in precision farming, points out areas of research that are not balanced, and shows the integration paths that improve real-time adaptability and operational efficiency. It discusses, among others, the limited use of extended fuzzy models, the limited database coverage, and the absence of large-scale, multisite validations as challenges. Future directions are to include q-rung orthopair fuzzy sets in order to boost decision-making precision, to broaden the database inclusion for thorough coverage, and to start large-scale, multisite validations.

Keywords: fuzzy logic, smart agriculture, uncertainty, IoT, decision support

Chukwudi Obinna Nwokoro, Department of Computer Science, Nigerian British University, Asia, Abia State, Nigeria, e-mail: nwokoro.chukwudi@nbu.edu.ng
Paul Augustine Ejegwa, Department of Mathematics, Joseph Sarwuan Tarka University, Makurdi, Nigeria, e-mail: ejegwa.augustine@uam.edu.ng
Grace Ataguba, Department of Computer Science, Faculty of Computer Science, Dalhousie University, Halifax, Canada, e-mail: grace.ataguba@dal.ca
Rita Orji, Department of Computer Science, Faculty of Computer Science, Dalhousie University, Halifax, Canada, e-mail: Rita.orji@dal.ca

 | https://doi.org/10.1515/9783112222249-011

1 Introduction

In recent years, the agricultural sector has witnessed more interest in fuzzy logic due to the growing complexity and uncertainty overtaking modern farming systems [1–3]. This is one of the fundamental driving forces that have sped up the evolution of precision agriculture. In our study, we use smart farming and agriculture synonymously to indicate the radical technological, decision-making, data-driven, and sustainable practices that combine to bring forth the modern agricultural systems for efficiency, productivity, and excellent resources management. Now, agricultural operations face multifaceted challenges, including uncertainties in weather variations, soil variability, pest pressures, and changing market prices; the classical decision-making methods cannot incorporate such unpredictability of agricultural variables. Then, FL came into existence, decisively addressing imprecision and uncertainty, to become the paradigms for modeling these complexities and thus enable the practice of agricultural activities in a more adaptive and data-driven manner [4], working with approximate values and degree of truth instead of binary options, and including various techniques ranging from classic fuzzy sets, intuitionistic fuzzy sets (IFS), fuzzy MCDM, and others. These models aggregate multiple criteria (e.g., soil quality, climate conditions, and expected market prices), where input values may be uncertain or based on expert judgments as in crop selection and yield estimation [5]. Also, in the diagnosis of pests and diseases, hybrid approximate decision-analytic frameworks enable experts to assess risk levels using qualitative levels such as "low," "moderate," and "high," which are mathematically processed for generating recommendations [6].

Making uncertainty in agricultural systems enter into the approximate reasoning phase has led to more accurate choice modeling for smart farming [7]. For example, fuzzy inference serves the purpose with nonlinear dynamics in animal monitoring and greenhouse control [8], whereas irrigation systems based on fuzzy logic use sensor data to optimize water use [9]. Moreover, fuzzy models integrated with IoT and remote sensing can greatly improve and complement sustainability and adaptability of digitalized agriculture.

1.1 Gaps in Knowledge

Despite these encouraging developments providing insight into the overall success of soft-based intelligent methods and highlighting potential future avenues for digital eco-farming, it is essential to review the progress made with artificial intelligence (AI) in transforming agriculture. This comprehensive review can help steer future applications and innovations in the field. Persistent contests, such as the need for standardized structures [5–8], harmonious amalgamation with big data analytics [12, 13], and the enlargement of user-friendly boundaries for farmers [14, 15], should be addressed meritoriously. Additionally, the potential of hybrid fuzzy systems, where fuzzy logic is combined with other computational intelligence techniques, warrants further exploration to understand their potential to drive the next generation of smart agriculture. Though existing reviews have explored this problem space, little or no work has uncovered these challenges and potential. For example, Widayat et al. [16] directly address fuzzy methods in smart farming, Spanaki et al. [17] cover various advanced technologies including fuzzy-based decision-making tools, Abbasi et al. [18] explore digital transformation in agriculture, touching on expert systems and soft computing, and also Elbasi et al. [19] considered soft computing and logic-based models under the AI umbrella. To understand the state-of-the-art trends, we reviewed relevant articles published between 2020 and 2025 in leading academic databases, including the Association for Computing Machinery Digital Library, IEEE Xplore, ScienceDirect, and Google Scholar. These databanks were selected because of their widespread coverage of high-impact research in agricultural engineering, computer science, and smart farming technologies, also have been widely used in related systematic reviews [11–15].

1.2 Objectives of the Chapter

This chapter aims to critically review the state-of-the-art fuzzy logic applications in agriculture, in specific terms, the objectives are to:

1. examine recent trends in approximate reasoning with eco-framing;
2. identify methodological strengths and limitations in recent investigations;
3. explore integration pathways with complementary technologies such as IoT, AI, and machine learning; and
4. highlight persistent gaps and propose future research directions for sustainable smart agro-industry.

1.3 Research Questions

The following research questions guide the review:

a) RQ1: Where and how are fuzzy logic methods being used in agriculture today, how has their use evolved over the last 5 years (2020–2025), and what evidence exists for impact under increasing complexity and uncertainty?
b) RQ2: What are the fundamental principles and classifications of fuzzy logic models used in agricultural decision-making, including extensions such as intuitionistic, Pythagorean, Fermatean, *q*-rung orthopair, type-2, hesitant, and picture fuzzy sets (PFSs)?
c) RQ3: How effective are fuzzy logic models in supporting decision-making in the context of uncertainty, vagueness, and inaccuracies within different sectors of agriculture?

d) RQ4: What ways are fuzzy logic methods combined with new technologies like IoT, remote sensing, AI, and machine learning to enhance smart farming practices?
e) RQ5: What are the predominant patterns in methodological strategies, implementation difficulties, and practical barriers encountered when applying fuzzy logic systems in real-world agricultural settings?
f) RQ6: What fundamental factors contribute to the existence of research gaps in the existing literature, and what future research avenues could be pursued to enhance the use of fuzzy logic in sustainable, data-driven smart farming?

1.4 Contribution of the Review

Findings from our review show that fuzzy logic has evolved from basic rule-based systems to hybrid, cross-domain adaptive frameworks that integrate sensing technologies, IoT, and fact-based decision-making across the agricultural price chain. By synthesizing the state-of-the-art, our review contributes to a growing interest in smart and sustainable agriculture in four ways:
1) It maps the evolution of fuzzy systems from static rule-based models to dynamic, multisource architectures.
2) It highlights the integration of fuzzy approaches with sensing, interconnected devices paradigm, and machine learning for crop, water, and energy management.
3) It identifies persistent gaps such as limited multisite validation, postharvest underrepresentation, and insufficient explainability.
4) It outlines future directions for interoperability, low-power rural deployment, and robust hybrid frameworks. Findings from this review will inform and support researchers, practitioners, and policymakers in leveraging fuzzy logic to build more resilient, efficient, and agroecosystem.

2 Related Works

It is well recognized that many academic reviews have dealt with different themes of innovative and precision agriculture such as sensor-based applications, irrigation management, greenhouse information for monitoring, and geotechnology. Most of these works provide salient perspectives about incorporation of technology under farm practices as enlisted in Table 1.

The adoption of systematic reviews in agriculture has contributed significantly to knowledge development. There remains a critical need to explore the role of fuzzy logic. In particular, the classification and advanced vague reasoning extensions application, such as intuitionistic, Pythagorean, type-2, and *q*-rung fuzzy sets are either ab-

Table 1: Historical view of systematic reviews carried out by scholars.

S. no.	Authors	Year	Database	Domain	Focus and strength	Review methodology/ approach	Limitation
1	Abou-Mehdi -Hassani et al. [21]	2025	Web of Science	Smart greenhouse monitoring	Comprehensive sensor and system monitoring review	Systematic review; greenhouse monitoring systems	Overlooks logic for decision-making under ambiguous sensor outputs; no fuzzy inference layer
2	Bhat and Wang [22]	2025	IEEE Xplore	Control systems in autonomous farm vehicles	Strong in vehicular automation	Systematic review; autonomous vehicle control	No focus on fuzzy controllers for adaptive terrain reasoning
3	Botero-Valencia et al. [23]	2025	Scopus and Web of Science	Machine learning (ML) in crops	Solid overview of ML applications across crops and sustainability metrics	Systematic review on ML methods and sustainability	Neglects the integration of fuzzy-based interpretability and uncertainty modeling with ML
4	Hossain et al. [24]	2025	IEEE Xplore and ScienceDirect	Agricultural machinery	Detailed analysis of automation (GPS, robotics, and navigation systems)	Systematic review; survey of autonomous agricultural machinery	No mention of fuzzy control; overlooks robust decision-making under uncertain field dynamics
5	Yilmaz et al. [25]	2025	Web of Science and IEEE Xplore	Plant disease detection and remote imaging	High-quality survey of vision techniques for disease detection	Systematic review and technology evaluation	Excludes fuzzy approaches for handling imprecise visual patterns or expert feedback loops
6	Ali et al. [26]	2024	Web of Science	Smart agriculture	Covers smart agri-cyber threats and AI defenses	Systematic review of AI in cybersecurity for agriculture	No treatment of fuzzy methods for handling uncertain attack patterns

(continued)

Table 1 (continued)

S. no.	Authors	Year	Database	Domain	Focus and strength	Review methodology/ approach	Limitation
7	Chicaiza et al. [27]	2024	IEEE Xplore and Scopus	Broader smart farming technologies overview	Broad survey of digital solutions in agriculture (GPS, drones, and analytic pipelines)	Methodological overview; comparative analysis of smart farming tools	Lacks specific taxonomy of fuzzy logic; overlooks experimental implementation
8	Hammouch et al. [28]	2024	IEEE Xplore	Irrigation systems	Covers irrigation methods, efficiency gains, and automation techniques	Systematic review and meta-analysis; intelligent irrigation	Missing analysis of fuzzy logic as a core decision methodology; lacks comparative performance metrics
9	Mohyuddin et al. [29]	2024	IEEE Xplore	Machine learning in precision farming	Detailed discussion on ML techniques	Comprehensive review of ML approaches in precision farming	They did not adopt interpretability or fuzzy-ML hybrids
10	Muhammed et al. [30]	2024	Web of Science, Scopus, and IEEE Xplore	Integrated smart farming platform	Covers architecture and IoT-AI integration in Agri-Tech	Structured review of AIoT architectures	Mentions fuzzy logic only in passing; lacks deployment case studies
11	Naseer et al. [31]	2024	IEEE Xplore, ScienceDirect, SpringerLink, and Scopus	Adopting cybersecurity and Internet of Things (IoT) in agriculture	Details on global connected devices network adoption and cybersecurity concerns	Systematic review; thematic and trend analysis	Fails to explore fuzzy logic integration; lacks discussion on uncertainty modeling
12	Scussi et al. [32]	2024	Web of Science and Scopus	Sensor technologies and data analytics	Focused review on sensor-driven data acquisition and preprocessing	Systematic review of sensor and data approaches	Does not integrate fuzzy logic into sensor fusion or interpretation stages

14	Widayat et al. [16]	2024	Web of Science	Smart farming	Comprehensive decade-long focus on rule-based expert systems in oil palm cultivation	Systematic review and thematic synthesis	Limited generalizability beyond oil palm; excludes fuzzy logic and advanced soft computing methods
15	Garg and Alam [33]	2023	Web of Science	Overview of smart agriculture technologies	General coverage of technologies and trends	Literature review; digital agriculture overview	Lacks analysis of fuzzy decision support frameworks
16	Raouhi et al. [34]	2023	Scopus and Web of Science	Drone and UAV applications in agriculture	Thorough review of aerial robotics-driven use in agriculture	Systematic review; UAV application domains	Does not explore fuzzy interpretation of sensor data or contextual rule extraction from UAV imagery
17	Radočaj et al. [35]	2023	Web of Science	Geospatial navigation in precision farming	Detailed analysis of GNSS-based navigation and tractor automation	Systematic review; GNSS applications	No coverage of decision-making models under uncertainty; lacks rule-based or intelligent logic layers
18	Abbasi et al. [18]	2022	Web of Science	Digitalized transformation in agriculture (Industry 4.0)	Broad overview on digitization	Systematic literature review and thematic mapping	Superficial treatment of fuzzy logic; treats it as peripheral without in-depth classification or implementation insights
19	Bwambale et al. [36]	2022	Web of Science	Smart irrigation	Detailed survey on irrigation control approaches (sensors, ML, and optimization	Systematic review; irrigation control strategies	Limited mention of fuzzy approaches to irrigation decision-making
20	Elbasi et al. [19]	2022	IEEE Xplore	Broad tech integration in agriculture	Covers diverse technological innovations	Systematic literature review of AI in agriculture	Does not detail fuzzy logic models or decision-making implications
21	Maraveas [37]	2022	Web of Science	Smart greenhouses	Extensive survey of AI tech: sensors, environmental control, and predictive models	Systematic review; AI in greenhouse systems	Fuzzy logic seldom discussed; lacks uncertainty modeling or explainable decision layers

(continued)

Table 1 (continued)

S. no.	Authors	Year	Database	Domain	Focus and strength	Review methodology/ approach	Limitation
22	Spanaki et al. [17]	2022	Scopus and Web of Science	AgriTech innovation and operational disruption	Broad AI-driven AgriTech perspective, including analytics and predictive control	Systematic review; disruptive AI technologies	Treats fuzzy logic as a subset; lacks depth in application or validation
23	Tan et al. [38]	2022	Web of Science	Oil palm agriculture	Longitudinal scope over a decade	Systematic review for crop, water, and soil management categories	Domain-restricted and uses only classical fuzzy sets
24	Omar [39]	2021	IEEE Xplore	Smart farming	Broad overview of IoT platforms, connectivity, and device architecture	Systematic review of IoT technologies in farming	Ignores decision intelligence layers such as fuzzy logic; lacks uncertainty handling discussion
25	Song and Dong [40]	2021	IEEE Xplore	Agricultural apparatuses	Evaluates AI-based recommendation tools for farmers	Systematic review; classification of recommendation systems	Omits fuzzy logic; focuses on crisp rule and ML systems
26	Farooq et al . [41]	2020	Scopus and ScienceDirect	IoT system adoption and deployment insights	Comprehensive survey of IoT deployment (sensors, networks, and platforms) in farming	Systematic review and thematic synthesis	Omits fuzzy logic integration; lacks discussion on decision-making under uncertainty
27	Navarro et al. [42]	2020	Web of Science	Sensor networks and precision data collection	Reviews sensors, monitoring, and connectivity solutions	Systematic review of IoT-enabled applications	Ignores fuzzy-based decision layers atop IoT; lacks evaluation of uncertainty modeling

sent or insufficiently addressed in managing uncertainty, imprecision, and vagueness in agricultural decision-making.

3 Methodology

We employed the Preferred Reporting Items for Systematic Reviews and Meta-Analyses (PRISMA) guidelines [43], which have been widely adopted by researchers because these guidelines provide evidence-based recommendations designed to improve the reporting of logical reviews and meta-analyses [44]. They ensure transparency, clarity, and rigor in presenting review findings. In addition to PRISMA, we also employed the Grant and Booth's [45] SALSA (Search, Appraisal, Synthesis, and Analysis) methodology for reporting our analysis.

3.1 Search Strategy

We explored related articles published within the last 5 years (January 2020 to July 2025) from four electronic databases, which include IEEE (Institute of Electrical and Electronics Engineers), ScienceDirect, ACM (Association for Computing Machinery) Digital Library, and Scopus. These four databases were chosen considering their related impact in the field. We considered search strings such as:

(Fuzzy logic AND Fuzzy control)
OR
(Fuzzy inference system AND Smart agriculture)
OR
(Precision farming OR intelligent agriculture OR agriculture)

These strings were used to sort for full text, titles, and abstracts related to the search keywords. A total of 11,823 papers were found, and distributed as follows: Scopus (920), ACM digital library (1,755), IEEE (6,166), and ScienceDirect (2,982). The search was completed on July 31, 2025. Other databases, such as PubMed and Web of Science, were not explored because they have limited contribution to the field. However, in subsequent reviews, other databases and additional search strings will be covered extensively.

3.2 Appraisal

Each database from the search results was converted into BibTeX format, which was uploaded to Rayyan. Rayyan is a free web and mobile application that supports the systematic review process [41, 42], which helps to detect multiple reviews or remove duplicate papers. This was achieved with the help of inclusion and exclusion criteria. Through a careful review of documents, we excluded 10 duplicates and removed 2,199 irrelevant papers based on the keywords used in this study. Therefore, 9,624 papers were selected after screening. More so, we applied the following inclusion and exclusion criteria to select relevant papers.

3.2.1 Inclusion Criteria

a) Full papers on fuzzy logic, inference system, precision, smart, and intelligent agriculture
b) Studies involved in implementing farming

It was closely selected to address the core aspects of this review, focusing on the adoption of fuzzy logic for smart agriculture, for proper scrutiny of research that reported and experimented and implemented their findings with an empirical presentation.

3.2.2 Exclusion Criteria

a) Systematic review, books, and survey (Criterion 1)
b) Studies not related to agriculture (Criterion 2)
c) Machine learning (Criterion 3)
d) No fuzzy (Criterion 4)

To follow a methodological rigor, we selected relevant papers that sufficiently answer our research questions using the different inclusion/exclusion criteria. Hence, papers that failed to meet the inclusion criteria were excluded. Overall, we excluded:
a) 4,878 publications comprising systematic reviews, books, and surveys based on Criterion 1;
b) 2,652 studies due to their lack of relevance to agriculture, as required by Criterion 2;
c) 1,749 studies that focused specifically on machine learning, meeting Criterion 3; and
d) 161 studies that failed to incorporate fuzzy logic, as required by Criterion 4.

In total, 9,440 papers were excluded and 130 eligible papers were included in the final analysis. In order to maintain the validity and internal consistency of our article selection, we used four researchers who screened articles in collaboration. In general, the four researchers agreed in 92% of the cases and settled differences in 8%. Figure 1 illustrates the flow of different procedures in the PRISMA diagram.

3.3 Synthesis

We developed a coding sheet to extract the necessary data from the articles after filtering them. Our coding sheet contains 13 themes, including metadata, fuzzy logic models and extensions, methodological approaches and model design, and future research directions and gaps. For example, metadata covers bibliometric information about the studies we included, such as year and type of publication, venue, DOI, system architecture, and the country where the publication was carried out. Table 2 covers the information about the themes, data extracted, and the associated research questions.

3.4 Analysis

We followed a series of steps in cleaning our coding sheet by eliminating unnecessary information and standardizing the data in preparation for data analysis. First, we carefully excluded data that did not support our research topics and summarized the items where additional language created ambiguity. Next, we standardized our coding sheet to address missing data points by double-checking entries from the relevant journals. These steps enhanced the quality of our dataset and the validity of our synthesis. For our analysis, we employed both content and descriptive analyses. Our descriptive analysis covered research questions RQ1, RQ2, RQ3, RQ4, RQ5, and RQ6 using Microsoft Excel. Furthermore, we employed content analysis in answering research questions RQ3 and RQ4 using NVivo 14. In addition, we generated graphs and charts in association with our descriptive analysis using Microsoft Excel.

We adopted NVivo 14 for the content analysis [47] since it is a qualitative data analysis tool. Basically, it was used to code and analyze the textual data for answering various research questions (RQI, RQ2, RQ3, RQ4, and RQ5), which helps to reveal patterns and identification of themes and relationships with the exploratory data.

4 Results and Discussion

This section presents the findings of our exploration, which were obtained by methodically examining the information gathered. The results are arranged according to the order of the research questions. First, we report the results of our bibliometric analysis, showing the publications we gathered for this scoping study, which will help to illustrate the distribution of papers by year, venue of the manuscript, and methodology employed. Second, we present results of RQ1, which elaborates the extent to which fuzzy methods were adopted in the agricultural domain. Third, we examine the application areas of fuzzy approaches in the agricultural domain (RQ2). Fourth, we assess whether

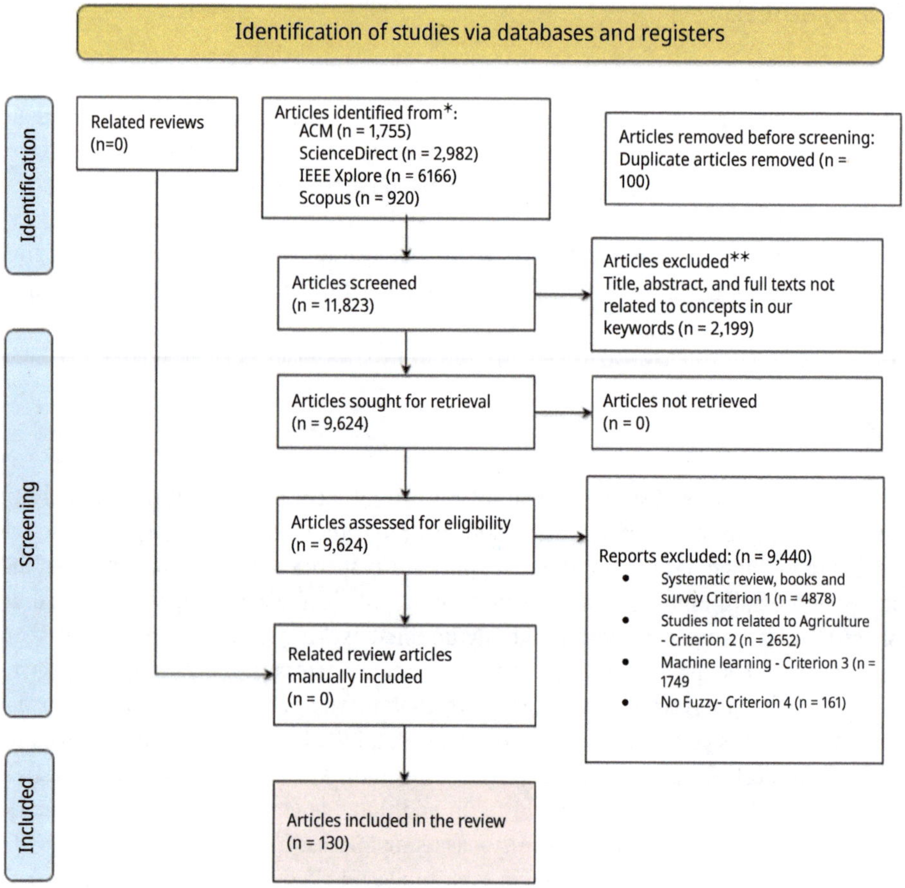

Figure 1: Adapted PRISMA diagram (Pageetal, 2021) showing our scoping review search and filtering processes. The asterisk symbol is typically used to depict noted data points.

the developed models support decision-making under uncertainty (RQ3). Fifth, we analyze the extent of fuzzy logic integration with complementary technologies for smart agriculture (RQ4). Sixth, we identify common procedural patterns and useful associations within fuzzy logic systems (RQ5). Finally, we synthesize existing gaps and propose future directions to advance smart agriculture (RQ6). Representative examples from the included studies are shown in Figures 2-12

4.1 Bibliometric Analysis

This section presents results, showing different scholarly contributions to the field from 2015 to 2025, which focuses on research trends and patterns aimed at improving

Table 2: Summary of themes, explanations, and relevant research questions from the literature review.

S. no.	Coding theme	Description	Related research questions
1	**Metadata**	This reveals information about selected article's in term of the following fields: year and type of publication, venue, DOI, system architecture, and country were the publication was carried out	Bibliometric analysis
2	**Fuzzy logic models and extensions**	Focuses on types and theoretical underpinnings of fuzzy logic models used in agriculture, including extensions such as intuitionistic, type-2, hesitant, *q*-rung, etc.	RQ1
3	**Application domains in agriculture**	Describes specific agricultural areas where fuzzy logic is applied: crop production, animal farming, irrigation, pest management, yield prediction, etc.	RQ2 and RQ3
4	**Decision-making under uncertainty**	Evaluates how fuzzy models handle imprecision, vagueness, and lack of crisp data in diverse agricultural tasks, enhancing decision quality.	RQ3
5	**Integration with smart technologies**	Explores how fuzzy models are integrated with IoT, AI, ML, remote sensing, edge computing, etc., to enable precision and real-time decisions in smart farming.	RQ4
6	**Methodological approaches and model design**	Examines fuzzy model architectures, algorithms, data processing techniques, and design choices adopted in the reviewed studies.	RQ1 and RQ5
7	**Model evaluation and accuracy**	Describes how the proficiency of fuzzy replicas is assessed, including metrics used and comparison with other techniques (ML, statistical models).	RQ3 and RQ5
8	**Implementation challenges and limitations**	Identifies key technical, practical, and contextual limitations in real-world deployment of fuzzy systems in agriculture.	RQ5
9	**Recommendations for practice and policy**	Highlights recommendations provided by authors for future applications or improvements in fuzzy-based agricultural systems.	RQ5 and RQ6
10	**Future research directions and gaps**	Summarizes identified research gaps and areas that require further exploration to advance the field of fuzzy decision support in smart agriculture.	RQ6

Table 2 (continued)

S. no.	Coding theme	Description	Related research questions
11	**Data characteristics and collection**	Discusses types of data (sensor, satellite, and survey), sources, and dataset characteristics used in fuzzy modeling for agriculture.	RQ2 and RQ3
12	**System architecture and smart agriculture design**	Reviews how fuzzy logic is incorporated into broader system architectures (hardware/software stack, mobile/web apps, edge/cloud platforms) for smart farming.	RQ4, RQ5, and RQ6
13	**Implications for sustainable agriculture**	Connects fuzzy logic applications to goals of sustainable, resource-efficient, and climate-resilient agriculture.	RQ2 and RQ6

smart agriculture. Based on our analysis of the publication years, we found that more papers were published in the year 2022 to 2024 (Figure 2). Although, for 2025, we completed data collection on August 6, 2025, while more research was still gaining ground, a total of 160 papers were collected from the IEEE database. This is followed by ScienceDirect with eight papers, while Scopus had four, and then ACM is the least with one paper (Figure 3). Furthermore, we found that within 2020 and 2025, most works were published as conference papers (107) compared to journals (22) (Figure 4).

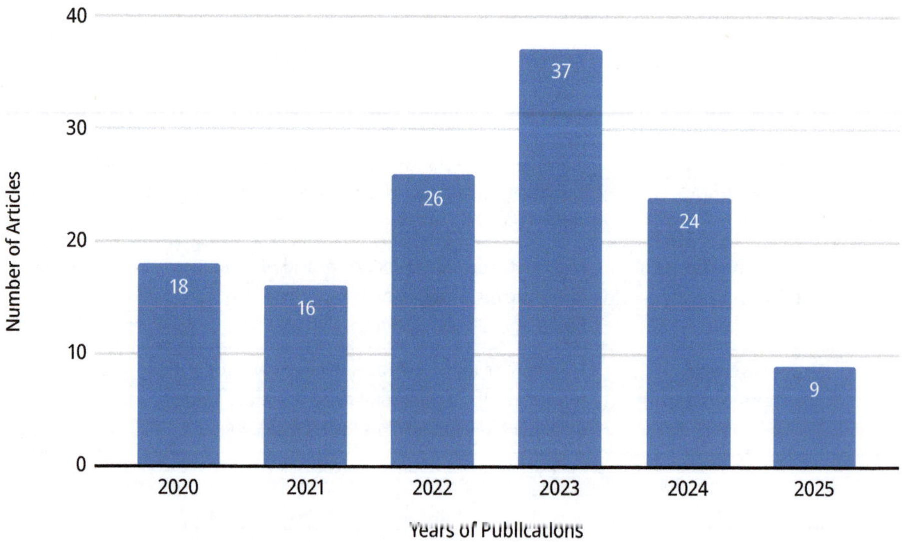

Figure 2: Number of publications.

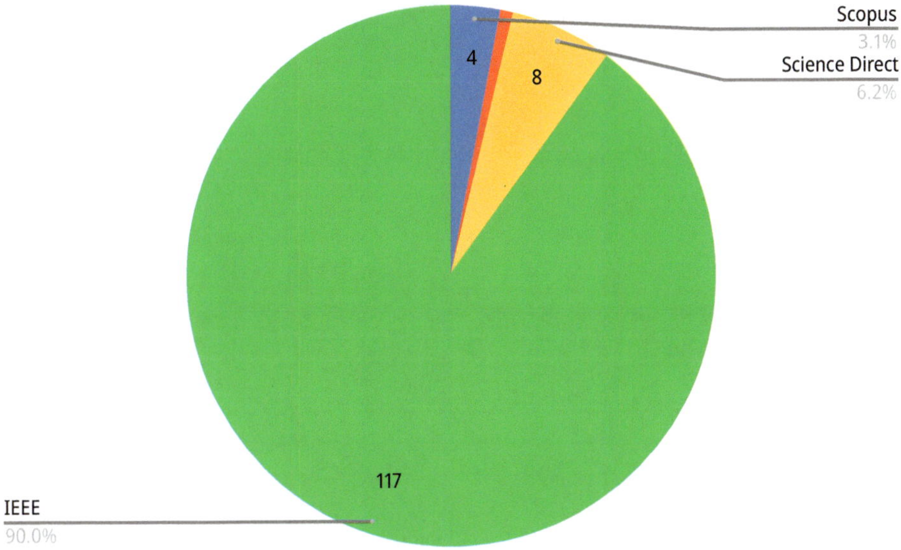

Figure 3: Number database used in this study.

In addition, we analyzed the distribution of smart agriculture research across different geographical locations and we found that this is predominantly concentrated in a few countries (Figure 5), notably India (38), Indonesia (16), Bangladesh (9), China (8), and both Nigeria and the Philippines (7 each), with other countries contributing less significantly.

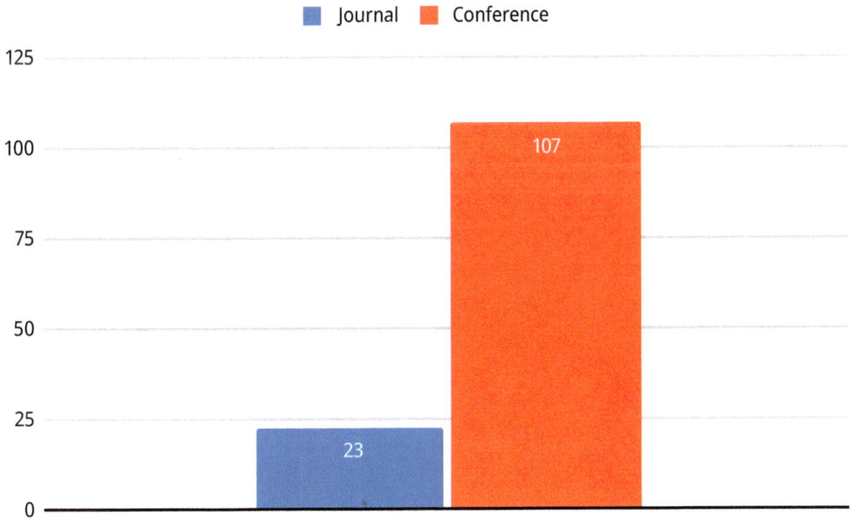

Figure 4: Distribution of smart agriculture research publications in journals and conference papers.

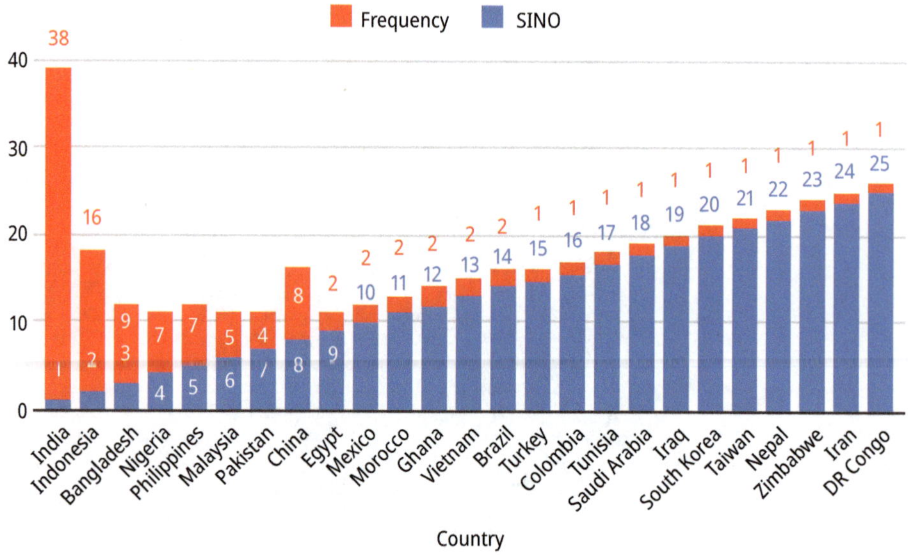

Figure 5: Global geographical concentration of research in smart agriculture.

The bibliometric analysis reveals a consistent increase in the number of publications between 2022 and 2025, indicating that the research community has made progressive advances in the field of smart agriculture during this period. Interestingly, the overwhelming value of conference proceedings as compared to the corrected journal publications highlights the incipient and dynamic character of fuzzy logic applications in this field. This pattern mirrors the predisposition of academics to share early results and novel methods in fast-track outlets. Furthermore, most of the cited papers are included in the IEEE database, proving that technology-based outlets are key drivers in this intellectual discourse. Geographically, the pattern of papers presents evidence that research effort is focused in regions where agriculture plays a central role in driving economic and societal development, i.e., Bangladesh, China, Nigeria, India, Indonesia, and the Philippines. These examples demonstrate the increasing pace of smart farming research and, at the same time, the imbalances in the global dispersion of it, which stresses the importance of more inclusive participation and the greater involvement of underrepresented regions besides the ones with already notable academic activities.

4.2 RQ1: Where and How Are Fuzzy Logic Methods Being Used in Agriculture Today, How Has Their Use Evolved Over the Last 5 Years (2020–2025), and What Evidence Exists for Impact Under Increasing Complexity and Uncertainty?

To answer this research question, we analyzed data collected under the "Application Domains in Agriculture" coding theme. Hence, we report our results in two Sections 4.2.1 and 4.2.2: In Section 4.2, we present the different application domains also found from our review, Section 4.2.2 observed some major milestones in the field. The evolution began with early rule-based systems and later expanded through the integration of sensing technologies and IoT. This was followed by the development of hybrid intelligent systems that combine AI with optimization, and more recently, cross-domain adaptive systems that utilize multisource data across agricultural subfields.

4.2.1 Applications of Fuzzy Logic in Smart Agriculture

Based on our findings from this review, we found the different areas of application of fuzzy logic. As shown in Table 3, crop production, irrigation and water resource management, livestock, aquaculture, insect farming, controlled environment agriculture, intelligent systems, IoT integration, and sustainability initiatives are all included in these fields.

Table 3: Thematic classification of smart agriculture domains with references.

S. no.	Theme	Subdomains/ focus areas	Representative fuzzy methods	Representative authors	Application areas/ examples
1	Crop production and management	Precision farming; crop monitoring; yield optimization; disease detection	Type-1 (Mamdani/ Sugeno); ANFIS; neuro-fuzzy; fuzzy clustering (FCM); fuzzy MCDM (TOPSIS); fuzzy + ML (RF, CNN/RNN); fuzzy + IoT; fuzzy inference/expert/ DSS; Fuzzy + PCA/ RGB/ILC; fuzzy deep forest; fuzzy + satellite imagery	Souza et al. [48]; Ravindranath et al. [49]; Clotz et al. [50]; Gupta [12]; Singhal and Haripriya [51]; Mendonca et al. [52]; Mathew et al. [53]; Rani et al. [54]; Suwastika et al. [55]; Ahmad et al. [56]; Hadi et al. [57]; Aditya Pai H et al. [58];	Maize growth monitoring; sweet pepper production; crop classification; disease detection; plant protection; postharvest processing; smart gardening; yield prediction; crop advisory; leaf image classification; viticulture

Table 3 (continued)

S. no.	Theme	Subdomains/ focus areas	Representative fuzzy methods	Representative authors	Application areas/ examples
				Ciptaningtyas et al. [9]; Aghaseyedabdollah et al. [59] Hosseinzadeh et al. [60]; Ming et al. [15]; Koshariya et al. [61]; Gopi and Karthikeyan [62]; Kamble and Jadhav [63]; Rajeswari et al. [3], Vats et al. [64]; Macatangay et al. [65]; Aziz et al. [66]; Chen et al. [67]; Shankar and Jayaparvathy [68]; Zheng et al. [69]; Savakar, Talawar and Gowtham [70]; Bilokonska et al. [71]	processing; soil quality and crop health monitoring; plant pathology; smart cultivation
2	Irrigation and water resource management	Smart irrigation; soil moisture sensing; water optimization; resource conservation	Type-1 (basic fuzzy); fuzzy + IoT; FIS; fuzzy PID/ control; fuzzy clustering; fuzzy + remote sensing; fuzzy + trend analysis; fuzzy PID + swarm intelligence	Nandi and Mahmood [72]; Farooq et al. [73]; Fadhilah Rama et al. [74]; Pareja and Bandala [75]; Adnan [76]; Hartono et al. [77]; Suguna [78]; Awalludin et al. [79]; Bukhari et al. [80]; Agarwal et al. [81]; Yu and Yanxia [82]; Hua et al. [83]; Liu [84]; Mendes et al. [85]; Falahi et al. [86]	Automated irrigation control; precision irrigation; water resource management; soil moisture monitoring; irrigation automation

Table 3 (continued)

S. no.	Theme	Subdomains/ focus areas	Representative fuzzy methods	Representative authors	Application areas/ examples
3	Controlled and indoor environment agriculture	Greenhouse automation; hydroponics; vertical farming; aeroponics; climate control	Type-1 (basic fuzzy); fuzzy + IoT; FIS; fuzzy control; fuzzy PID + genetic algorithm; CNN + fuzzy control	Silvestre et al. [87]; Puno et al. [88]; Jayalakshmi et al. [89]; Tolis et al. [90]; Bolivar et al. [91]; Alaviyan et al. [92]; Yao et al. [93]; Hua et al. [83]	Greenhouse control and management; hydroponics (NFT); indoor farming; aeroponics
4	Livestock, aquaculture, and insect Farming	Livestock monitoring; aquaculture automation; diagnostics; environmental control	Type-1 (basic fuzzy); Mamdani and Sugeno; Fuzzy + IoT/related tech; fuzzy + solar energy + IoT; *C*-means fuzzy + deep NN	Zou et al. [94]; Mendoza et al. [95]; Arshad et al. [96]; Chua et al. [97]; Nibir et al. [98]; Shitsukane [99]; Shahana et al. [100]; Saravanan et al. [101]	Smart fishponds; livestock tracking; insect farming; animal health diagnostics; water quality and habitat monitoring
5	Smart systems, robotics, and IoT integration	Edge AI; autonomous control; real-time decision-making	Type-1 controllers; ANFIS; fuzzy + artificial bee colony; fuzzy + ESP-32; fuzzy + decoupling control; super-resolution + fuzzy deep learning; Tsukamoto fuzzy; self-tuning fuzzy PID; fuzzy logic controller; broad learning + fuzzy fusion; fuzzy clustering + PSO; fuzzy gain scheduling + APF; fuzzy *K*-means + IoT ; fuzzy + autonomous robotics; fuzzy + IoT; granular computing + fuzzy sets; fuzzy NN + spectrophotometry	Sharma et al. [7]; Salsabila et al. [102]; Liu [84]; Suwirmayanti et al. [103]; Fadlo et al. [104]; Magsumbol et al. [105]; Biswas et al. [106]; Revathi and Santhi [107]; Salsabila et al. [102]; Khan et al. [108]; Pradana et al. [109]; Seo et al. [110]; Yu et al. [82].; Yuebo et al. [111]; Vahedi et al. [112]; Kashyap et al. [113]; Hayat et al. [114]; Rasyid et al. [115]; Waluyo et al. [116]; Puri et al. [117]; Silva et al. [118]; Ogurtsov et al. [119]; Chen et al. [67]	Agricultural services; farm robotics; sensor network; remote sensing; UAV operations; soil diagnostics; environmental monitoring; WSN for smart farms; energy-aware smart farming; communication infrastructure

Table 3 (continued)

S. no.	Theme	Subdomains/ focus areas	Representative fuzzy methods	Representative authors	Application areas/ examples
6	Sustainability, energy, and decision support	Renewable energy; energy-efficient systems; forecasting; socioeconomics; decision support	Fuzzy AHP; fuzzy Bayesian imputation; cross-former (big data); fuzzy *C*-means; MapReduce; fuzzy clustering; fuzzy inference/dynamic analysis; fuzzy logic control; fuzzy optimization; fuzzy PI/STATCOM; fuzzy subset/data fusion; deep RL + fuzzy; neuro-fuzzy; fuzzy + power systems; fuzzy + IoT	Lv and Liu [120]; Kong et al. [121]; Sathyamoorthy et al. [122]; Shrivastava et al. [123]; Falana et al. [124]; Sharma et al. [125]; Rezaei and Uddin [126]; Chalamalasetty et al. [127]; Herabad and Afshar [128]; Sumalatha and Muneender [129]; Rajesh and Devi [130]; Pan et al. [131]; Rahman et al. [132]; Xu and Li [133]; Ding et al. [134]; Achroufene and Moula [135]; Balakrishnan et al. [136]; Li et al. [137]; Waluyo et al. [116]	Irrigation using solar energy, forecasting agricultural risks, farming with energy-conscious methods, environmental surveillance, wireless sensor networks for smart farms, communication systems infrastructure

The crop production and management category (Figure 6), which comprises 30 papers, denotes the most significant area of research activity in smart agriculture. This emphasis reflects its central focus on yield optimization, food security, and sustainable farming. The reviewed studies clearly indicate a wide range of methodologies, from traditional fuzzy approaches to advanced hybrid techniques that combine AI with fuzzy methods. For example, the research of Souza et al. applied Mamdani fuzzy logic to soil and air sensors that were then combined with a weather forecast for sweet pepper production support. Similarly, Hosseinzadeh et al. [60] combined a fuzzy RNN and nonconvex activation functions to improve the accuracy of crop disease detection. Multitechniques are particularly remarkable: Koshariya et al. [61] integrated deep learning, CNN, and neuro-fuzzy models for plant pathology identification, while Gopi and Karthikeyan [62] contributed by proposing an ANFIS framework reinforced by moth flame optimization for crop classification. The category still boasts some very powerful decision-support applications. Rajeswari et al. [138] devised a fuzzy decision support system for crop recommendation; on the other hand, Vedanth et al. [70] applied Fuzzy TOPSIS to aid in production-related decision-making.

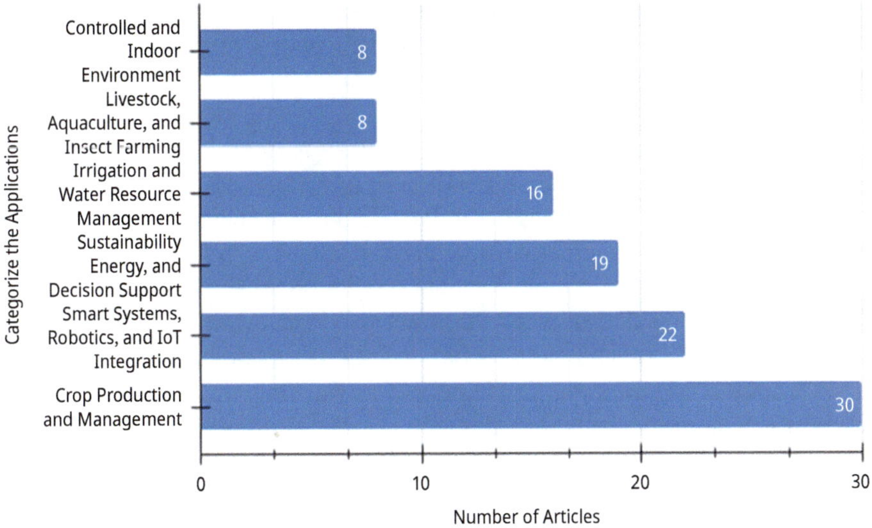

Figure 6: Categories where fuzzy logic has been applied.

Fuzzy systems are increasingly integrated with IoT and sensor technologies, resulting in a shift from static, rule-based models to real-time, data-driven decision-making throughout the agricultural supply chain. For example, fuzzy inference has been combined with in-field sensors for smart gardening, cultivation, and crop-health monitoring [9, 49, 55, 57, 58, 139], and scaled to regional assessments using satellite imagery [71]. Along the agricultural value chain, fuzzy logic has been fused with RGB indices and direct inference for disease detection [12, 66], with ILC for plant protection [67], with RF hybrids for yield prediction [51, 68], and with FCM/expert systems for postharvest classification and viticulture processing [50, 63].

Despite these advances, persistent gaps limit the generalization and large-scale adoption of fuzzy systems. Many solutions remain crop-specific [63], [83], [140]. IoT-based monitoring is common but rarely extended to closed-loop control [57, 58, 139]. Advanced hybrid systems also remain underexplored in terms of explainability [61, 69]. Furthermore, postharvest applications are underrepresented, and much of the evidence relies on pilot or laboratory datasets with limited multisite validation [64–65].

Beyond crops, fuzzy and hybrid intelligence extend to robotics, WSN/IoT, UAVs, and edge computing for automation, navigation, and environmental monitoring. These applications employ a wide range of approaches, including ANFIS, self-tuning fuzzy PID, fuzzy controllers, fuzzy *K*-means with IoT, clustering with PSO, fuzzy deep learning granular computing, and decoupling control [81, 103–108, 110–116, 118, 124, 140–143]. However, these archetypes often remain simulation-heavy or pilot-scale [102, 125], tussle with incomplete topography or environmental data [106, 145], lack interoperability, and continue to omit explainable AI [82, 108].

In the areas of sustainability, energy, and decision support, studies have applied fuzzy AHP/MCDM, Bayesian imputation, optimization, neuro-fuzzy systems, and deep RL-fuzzy methodologies to support energonomics [3, 115, 119, 120–123, 125–128, 130, 131, 134, 135, 145]. Irrigation-related work has employed fuzzy PID, ANFIS, IoT integration, and remote sensing for smart water management [19, 30, 50, 51, 63, 122, 132, 147, 148]. Controlled-environment agriculture has leveraged fuzzy PID with genetic algorithms, FIS with IoT, and CNN-fuzzy hybrids [86–92, 148]. In the case of livestock, aquaculture, and insect farming, fuzzy approaches have been combined with IoT, solar energy, as well as deep learning techniques such as C-means with DNNs [94–100, 149].

Across all domains, several cross-cutting needs remain, which include the implementation of multiseason and multilocation trials, integration of agrohydrothermal data, development of low-power or offline IoT systems for rural contexts, coupling with renewable energy, adaptability to multiple crops and species, interoperability standards (e.g., FIWARE and ROS-I), and the embedding of explainable AI for trustworthy, scalable deployment.

4.2.2 Evolutionary Phases

4.2.2.1 Phase 1: Rule-Based Fuzzy Logic

As shown in Figure 7, the earliest adoption of fuzzy logic in agriculture can be traced to Phase 1 (2020–2021), which centered on rule-based systems and IoT-enabled watering schedules. These approaches relied on expert-defined fuzzy rules to manage specific tasks such as greenhouse climate control and irrigation timing, as demonstrated in greenhouse automation studies by Sharma et al. [125], Arshad et al. [96], and Alaviyan et al. [92]. Although the systems effectively handled the unpredictability associated with biological and environmental factors, they remained limited due to their dependence on a large set of manually created rules and their inability to adapt to changing field conditions.

4.2.2.2 Phase 2: Integration with Sensing Technologies

In 2022, the second phase saw the emergence of intelligent farming, as fuzzy logic was integrated into embedded devices, sensor networks, and environmental monitoring systems, enabling real-time, data-driven decision processes. Many researches, such as Suwirmayanti et al. [103] in smart farming, Shitsukane [99] in aquaculture, and Tolis et al. [90] in greenhouse automation, showed that the application of sensors led to quicker responses from the systems, less manual work, and better control of the environment. Notwithstanding, some implementations like those by Rasyid et al. [115] and Salsabila et al. [102], pointed out the still-existing issues, especially concerning long-term deployment, scalability, energy efficiency, and field-level validation.

4.2.2.3 Phase 3: Hybrid Intelligent Systems

The third phase (2023) has been characterized by a move toward hybrid systems that utilize fuzzy logic in conjunction with advanced AI and optimization techniques, thereby allowing more flexible and sophisticated decisions to be made. For instance, Balakrishnan et al. [136] fused fuzzy logic with deep reinforcement learning to construct a decision-support system, whereas Hosseinzadeh et al. [60] made use of fuzzy logic in conjunction with CNNs for cultivating crops. The use of hybrid methods helped to overcome some of the drawbacks associated with the use of a single technique by providing improved predictability, as well as flexibility. However, their combination also brought about new problems, such as the increased demand for computation and the requirement for meticulous model tuning. This has been highlighted in the studies of Koshariya et al. [61] (neuro-fuzzy) and Vahedi et al. (fuzzy gain scheduling with artificial potential fields).

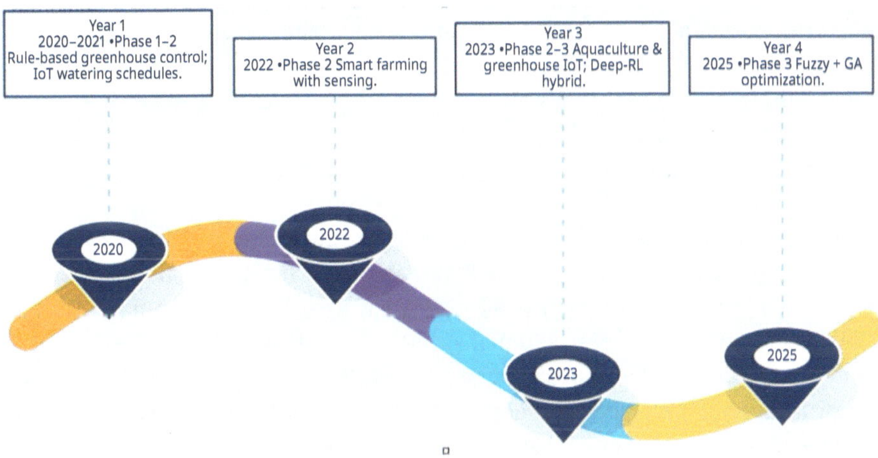

Figure 7: Evolution trend toward agricultural innovation.

4.2.2.4 Phase 4: Cross-Domain Adaptive Systems

The fourth phase (2024) focused on cross-domain adaptability, where systems based on fuzzy logic will be able to integrate data from different sources, will be working in various agricultural subfields, and will change their way of operation according to the conditions (see Figure 7). This transformation can be seen in the work of Kong et al. [121], who used fuzzy Bayesian imputation in big data settings; Herabad and Afshar [128], who combined fuzzy logic with deep reinforcement learning and IoT; and Ogurtsov et al. [119], who made use of granular computing for UAV-based precision agriculture. All these methods are leading us to a point where there is interoperability among crop management, livestock monitoring, water management, and energy optimization. Nonetheless, the complete cross-domain adaptability is still limited by the absence of interoperability standards, restricted dataset diversity, and the requirement of real-time, large-scale validation. These results reveal that fuzzy logic has not only influenced but also covered multiple agricultural domains, thereby providing powerful instruments to cope with complexity and uncertainty, which are the characteristics of modern farming systems.

4.3 RQ2: What Are the Fundamental Principles and Classifications of Fuzzy Logic Models Used in Agricultural Decision-Making, Including Extensions Such as Intuitionistic, Pythagorean, Fermatean, *q*-Rung Orthopair, Type-2, Hesitant, and Picture Fuzzy Sets?

To answer RQ2, we analyzed the data we collected under the coding theme "Fuzzy Logic Models and Extensions." Our analysis covered classical fuzzy models, extended fuzzy models, and hybrid/empirical validation approaches. In Sections 4.3.1 and 4.3.2, we report the results of our analysis.

4.3.1 Overview of Classical Fuzzy Logic Models in Agriculture

Our analysis identified several domains within agricultural environments where uncertainty and variability are prevalent, and where classical fuzzy logic models have been adopted to address these challenges through smart agriculture applications. These domains include crop production and management, irrigation and water resource management, controlled and indoor environment agriculture, livestock, aquaculture and insect farming, smart systems and robotics, and IoT integration, as well as sustainability, energy, and decision support. Table 4 summarizes the extent of research efforts across these areas, highlighting how conventional fuzzy reasoning prototypes have been implemented to improve decision-making and system performance in farming contexts.

The classical fuzzy logic systems adaptations in Table 4 have led to considerable agricultural management improvements in several areas, and the application of such systems to smart farming is a prospect still to come. Fuzzy logic methods in crop production systems have provided benefits that can be measured. For example, Aditya et al. [58] mentioned that the irrigation water use was reduced by 18% while the plants were kept in the optimal condition. Reliable disease detection of more than 90% accuracy was reported by Ciptaningtyas et al. [9] for the major crop pathogens. The effectiveness of fuzzy models was also demonstrated in yield prediction with margins of error being less than 8.5% [125], and in resource optimization showing 22–25% improvements in nutrient uptake efficiency [84]. The Mamdani-type inference systems are especially applicable to noisy agricultural environments. The resilience of the systems was shown to be $p > 0.8$ even at 25% noise levels by Dinçer et al. [151].

Table 4: Thematic comparative summary of fuzzy models in agriculture.

Model family	Uncertainty representation	Typical inference/ defuzzification	Learning/ hybridization	Fits when . . .	Notes/examples
Type-1 (Mamdani)	Single-grade membership; linguistic rules	Mamdani inference; centroid/bisector/MOM	Expert tuning; offline/ online rule/membership adjustment	Interpretability and edge-feasibility are priorities; actuator control; modest data	Robust to sensor noise; widely used in irrigation, greenhouse, robotics control
Type-1 (Sugeno/ Takagi–Sugeno)	Single-grade membership; functional consequents	Weighted average of rule outputs (no defuzz needed)	Parameter regression; adaptive tuning	Faster inference and easier optimization needed; controller synthesis	Common in control loops with continuous outputs and compact rule bases
Type-2 (interval/ general)	Footprint of uncertainty around membership grades	Type-reduction + defuzzification	Computational optimization; sometimes hybrid with ML	High epistemic uncertainty (multisensor fusion; conflicting sources)	Greater robustness to uncertainty; higher compute cost
Type-3/Higher-order	Higher-order uncertainty layers	Extended type-reduction; advanced schemes	Adaptive learning required	Complex, volatile subsystems (e.g., agri-energy faults)	Early evidence in adjacent domains; promising for agriculture with validation
Intuitionistic fuzzy sets (IFS)	Membership + nonmembership (with hesitation)	Aggregation operators; score/accuracy functions	Weight learning; expert calibration	Group decisions with conflicting appraisals; MCDM	Useful in sustainability assessment; needs data-driven calibration
Pythagorean/ Fermatean	Stronger constraints on membership pairs	Aggregation/score functions	Weight tuning; sensitivity analysis	Linguistic uncertainty with stricter bounds	Expressive for stakeholder aggregation under conflict
***q*-Rung orthopair (*q*-ROFS)**	Flexible membership-nonmembership powered by *q*	Aggregation + score/ accuracy	*q* selection; weight/ threshold learning	Need to widen acceptance range for evidence fusion	Used with AHP/TOPSIS in sustainability decisions

(continued)

Table 4 (continued)

Model family	Uncertainty representation	Typical inference/ defuzzification	Learning/ hybridization	Fits when . . .	Notes/examples
Hesitant/ picture fuzzy	Multiple candidate memberships/ additional neutral class	Aggregation of hesitancy; 4-valued logic for picture sets	Rule construction with hesitancy	Ambiguous labels; crowd/expert disagreement	Relevant for noisy annotations and weak supervision
ANFIS/neuro-fuzzy	Type-1 rules with trainable parameters	Sugeno-style inference	Gradient-based learning (supervised); hybrid training	Nonlinear mappings with enough labeled data	Bridges interpretability and accuracy; common in prediction tasks
Fuzzy PID/ control	Rule-based tuning of gains; linguistic error terms	Rule-based gain scheduling	Metaheuristics (GA and PSO) or online adaptation	Nonlinear control under disturbances and delays	Greenhouse climate, pump control, actuator compensation
Fuzzy MCDM (AHP/TOPSIS)	Linguistic pairwise comparisons/fuzzy ratings	Fuzzy aggregation; ranking	Weight elicitation; scenario analysis	Multicriteria planning under uncertainty	Cultivar selection, input allocation, and sustainability ranking
Fuzzy clustering (FCM)	Soft memberships to clusters	Objective minimization with fuzzifier m	m tuning; feature engineering	Noisy multimodal sensor data; preprocessing for downstream models	Water quality, disease grouping, soil-type segmentation

This points to the robustness of the systems in situations where precise measurements are hard to get.

While these achievements are noteworthy, there are still considerable constraints that have to be faced. The reported computational complexity ($O(n^2)$), accompanied by time limitations for processing in edge deployments, has been a major factor that restricts the extent to which these methods can be applied [13]. Also, the performance differences between indoor and outdoor usage have been noted to be very significant ($\Delta > 30\%$) [63], and the methods have also been considered unstable when the field conditions are changing rapidly. In addition, most of the research works are quite similar in that they have only a very limited validation area. That is, 68% of them are confined to one single region (e.g., soil moisture in India, greenhouse climate in China using Type-1 Mamdani/Sugeno models). Moreover, the quality of the dataset has been found to be another issue, as 37% of the studies that were reviewed did not mention the size of the dataset, thus, creating uncertainty regarding generalizability. Most importantly, 82% of the cases were not evaluated with respect to economic viability, thereby raising concerns over practical adoption.

The review revealed three main research avenues:

- Algorithmic development – hybrid neuro-fuzzy architectures and integration of advanced methods such as deep reinforcement learning and quantum-inspired fuzzy controllers for yield prediction are the focus here.
- Implementation research – optimization for edge computing and standardized benchmarking, including deployments of fuzzy irrigation controllers on Raspberry Pi/Arduino with LoRa-IoT integration are the focus of this area.
- Policy integration – application of fuzzy multicriteria decision models in national smart agriculture frameworks and sustainability assessments.

These scientific works taken together depict the potential of fuzzy systems in precision agriculture. However, taking the technology to its full benefits will still be a long journey that requires overcoming computational, methodological, and implementation challenges bridging the gap between theoretical advances and large-scale practical deployment working together.

4.3.2 Empirical Outcomes of Extended Fuzzy Models

Through the examination of the studies, we managed to spot just three publications where the major area of application of extended fuzzy models in agricultural science, although such models were supposed to allow for another opportunity for fuzziness, multicriteria trade-offs, and dynamic decision-making in complex environments. He and Jiang [152] were the first to propose the use of IFS in a multiple-attribute group decision-making framework for smart manufacturing quality control. Their IFS-based model is likely to be adapted to agricultural quality control processes, but empirical research is required to ascertain its overall applicability to farming systems.

Zhou et al. [153] explored Type-3 fuzzy logic together with adaptive learning mechanisms and succeeded in obtaining 83.3% accurate results in detecting and isolating faults in wind turbines. Their work, although technically promising for the introduction of renewable energy in agricultural grids, did not address the robustness requirements that are very important in agri-energy areas, thus limiting its agricultural applicability. Dinçer et al. [151] implemented q-rung orthopair fuzzy sets along with AHP-TOPSIS to prioritize environmental investments in agricultural projects. Though this method opens the door to using a practical linguistic framework for sustainability assessments, the heavy dependence on expert opinion seems to weaken it and also brings the issue of subjectivity and the need for more methodological refinement into the picture.

In particular, there is limited research on extended fuzzy models, which may be of theoretical and practical value for smart agriculture, as current industrial fuzzy models are not sufficiently adapted to agricultural contexts, technical applications do not always account for the specific complexities of agricultural systems, and expert judgment often dominates without sufficient data-driven validation. We therefore propose the use of hybrid validation frameworks to fill the gap between theoretical sophistication and real-world applicability in precision agriculture; future research can focus on developing agriculturally adapted fuzzy models that stay mathematically rigorous yet face the dynamic challenges of farming systems.

4.4 RQ3: How Effective Are Fuzzy Logic Models in Supporting Decision-Making in the Context of Uncertainty, Vagueness, and Inaccuracies Within Different Sectors of Agriculture?

The effectiveness of fuzzy logic models in agricultural decision-making may be evaluated in a multidimensional framework measuring its predictive simplicity, robustness, interpretability, and impact on actual decision processes to say if fuzzy systems in practice know how to deal with uncertainty, vagueness, and imprecision. Table 5 outlines the different branches of evaluation methods, together with their common measures, appropriate baselines for comparison, and the specific insights each approach provides in assessing the effectiveness of fuzzy logic models in agricultural applications.

As shown in Table 5 and Figure 8, classification-based assessments are more prevalent (69 out of 130, 53.1%), where accuracy, precision, recall, F_1-score, and AUC are used to evaluate discrete judgments (e.g., crop disease detection, pest identification, and soil condition classification), while regression-focused evaluations are less common (25 studies, 19.2%), using RMSE, MAE, MAPE, and nRMSE to measure predictive accuracy of continuous outputs (e.g., yield estimation, soil moisture forecasting, and irrigation scheduling). This distribution reflects the more extensive application of fuzzy logic to categorical decision-making as opposed to continuous prediction problems.

Table 5: Evaluation measures adapted by researchers.

S. no.	Evaluation family	Typical metrics	Number of studies	Percentage of studies (%)
1	Classification	Accuracy, precision, recall, F_1, and AUC	69	53.08
2	Regression	RMSE, MAE, MAPE, and nRMSE	25	19.23
3	Probabilistic	Brier, CRPS, and calibration	8	6.15
4	Decision impact	Economic gain, cost saved, and net benefit	5	3.85
5	Robustness	Performance versus noise/missingness	4	3.08
6	Interpretability	Rule count and expert Likert score	16	12.31
7	Not specifed	Not stated	3	2.31

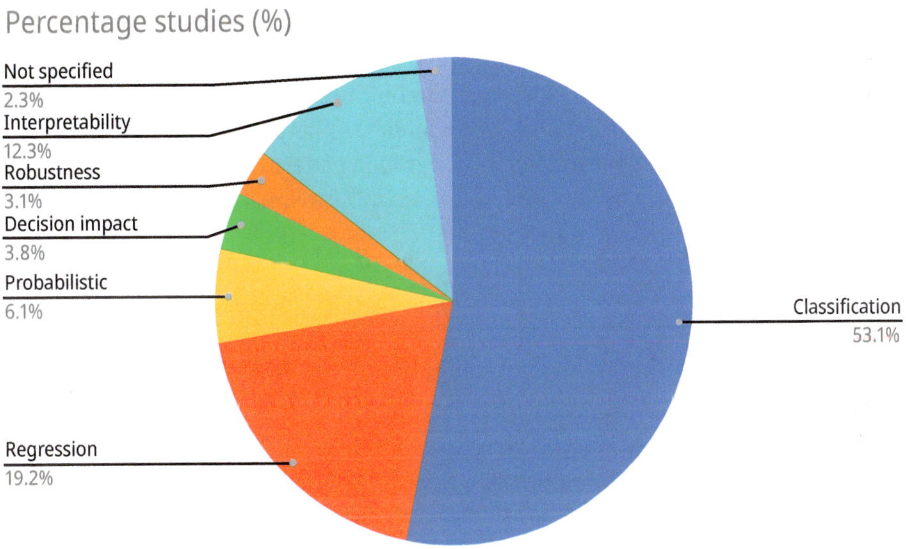

Figure 8: Evaluation distribution.

A total of eight studies (6.2%) used probabilistic metrics (e.g., Brier score, CRPS, and calibration measures) to explicitly validate uncertainty, a relatively weak form of validation given that uncertainty management is a core strength of fuzzy logic. Even less common were decision-impact evaluations, with only five studies (3.8%) showing that fuzzy logic outputs could be mapped to tangible outcomes in agriculture, such as economic gain, cost savings, or net benefit. Only four studies (3.1%) reported robustness testing, where performance is assessed under noise, missing data, or environmental variability, which is low given the variability inherent to agricultural environments.

Interpretability metrics were discussed in 16 studies, accounting for 12.3%, covering aspects such as rule number, linguistic clarity, and usability for experts. Although by this split fuzzy logic's best advantage – the generation of interpretable rules – looking into such interpretability factors, the low percentage still shows that formal interpretability assessment is far from becoming commonplace. In conclusion, three studies (2.3%) did not include any information about evaluation, which reduces transparency and complicates comparisons with other fields.

4.5 RQ4: In What Ways Are Fuzzy Logic Methods Combined with New Technologies, Such as IoT, Remote Sensing, AI, and Machine Learning, to Enhance Smart Farming Practices?

In RQ4, we examined the studies to understand how fuzzy logic is incorporated into emerging smart agricultural technologies, such as computational intelligence and cloud or edge computing, and to identify the scope and characteristics of this integration. The examined aspects included the different integration types associated with each technology and their frequency, the relationship between integration types and agricultural sectors, and the evolution of integration patterns over time. The synthesis of these integration patterns and their characteristics is summarized in Table 6.

Table 6: Integration approaches of fuzzy logic in smart agriculture literature examples and study counts.

Integration category	Examples from literature	Description of integration	Number of studies
IoT-enabled fuzzy systems	Tian et al. [154], Ravindranath et al. [49], Rasyid et al. [115], Hartono et al. [77], Suwastika et al. [55], Mendes et al. [85], Suguna [78], Hayat et al. [114], Rasyid et al. [115], Waluyo et al. [116], Tolis et al. [90], Awalludin et al. [79], Ahmad et al. [139], Puri et al. [117], Bolivar et al. [91], Hadi et al. [57], Aditya et al. [58], Ciptaningtyas et al. [9], Alaviyan et al. [92], Silva et al. [118], Shitsukane [99]	Uses IoT devices, wireless sensor networks, and real-time monitoring integrated with fuzzy inference for precision agriculture decisions.	**25**
Fuzzy + machine learning/AI	Prabakaran et al. [155] (SVM), Remya and Sasikala [156] (ANFIS + Adam), Shah et al. [157] (deep shallow NN), Liang et al. [158] (fuzzy TOPSIS + LSSVM + FWA), Hosseinzadeh et al. [60] (CNN), Hua et al. [149] (fuzzy RNN),	Combines fuzzy systems with AI/ML algorithms for prediction, classification, and decision optimization.	**20**

Table 6 (continued)

Integration category	Examples from literature	Description of integration	Number of studies
	Ming et al. [15] (FNN), Hang et al. [159] (Choquet fuzzy ensemble), Kumar and Kukreja [160] (PSPNET), Miao and Li [161] (type-2 fuzzy + whale optimization), Boonyopakorn and Bualeard [162] (neuro-fuzzy), Gopi and Karthikeyan [62] (ANFIS + MFO), Li et al. [163] (neuro-fuzzy), Balakrishnan et al. [136] (deep RL), Koshariya et al. [61] (CNN neuro-fuzzy), Rani et al. [54] (NN + fuzzy control), Herabad and Afshar [128] (deep RL + IoT)		
Fuzzy + remote sensing/image processing	Mendes et al. [85] (remote sensing), Bilokonska et al. [71] (satellite imagery), Aziz et al. [66] (RGB indices), Yao et al. [93] (CNN + fuzzy control), Peng et al. [164] (3D-CNN + fuzzy C-means), Kong et al. [121] (big data imputation + fuzzy), Villaseñor-Aguilar et al. [165] (fuzzy classification + Raspberry Pi)	Uses satellite/drone imagery or spectral data with fuzzy logic for crop health, land mapping, and yield estimation.	**7**
Fuzzy + optimization algorithms	Erdoğdu et al. [166] (NSGA-II), Liang et al. [158] (FWA), Hua et al. [149] (GA), Revathi and Santhi [107] (ABC), Miao and Li [161] (whale optimization), Gopi and Karthikeyan [62] (MFO), Pan et al. [131] (GA), Vahedi et al. [112] (artificial potential fields), Alaviyan et al. [92] (Optimization + IoT)	Uses fuzzy systems alongside optimization for control tuning, multiobjective decision-making, and scheduling.	**9**
Fuzzy + control systems	Bukhari et al. [80] (FIS + MPC + PI), Zhou et al. [153] (predictive control), Souza et al. [140] (Mamdani with Sensors), Khan et al. [108] (fuzzy DL), Ding et al. [134] (fuzzy PI), Rahman et al. [132] (fuzzy logic control), Vats et al. [64] (fuzzy inference model), Hua et al. [149] (fuzzy PID + swarm intelligence)	Applies fuzzy logic to control irrigation, machinery, energy, or greenhouse environments.	**8**

Table 6 (continued)

Integration category	Examples from literature	Description of integration	Number of studies
Other hybrid approaches	Brainya et al. [167] (BWM, MULTIMOORA), Alamoodi et al. [168] (MCDM), He and Jiang [152] (MAGDM), Ogurtsov et al. [119] (granular computing), Yu et al. [82] (spectrophotometry), and Shemim and Witkowski [142] (network coding)	Diverse integrations that do not fall neatly into the above categories.	**6**

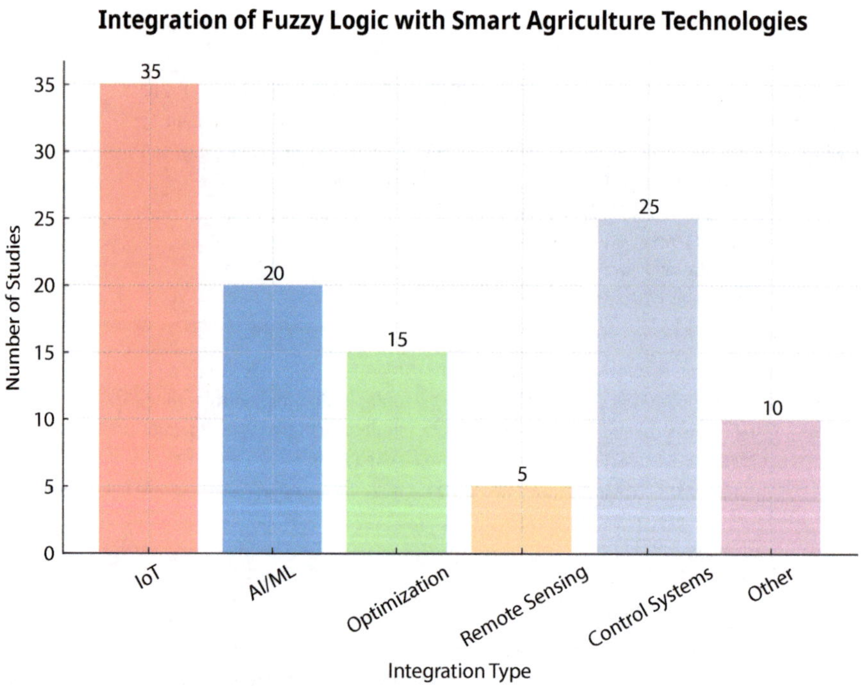

Figure 9: Fuzzy logic integration with smart agriculture technologies.

The stacked bar chart (Figure 9) demonstrates that IoT is the most prevalent technology integrated with fuzzy logic in precision agriculture, accounting for nearly half of the described cases. The second largest group involves machine learning and AI methods (e.g., neural networks, CNNs, and clustering), followed by optimization algorithms such as genetic algorithms and swarm intelligence. Control systems, including fuzzy PID and MPC, are also common. On the other hand, the application of remote sensing and more niche-type integrations (such as hardware implementations and decision-support schemes) are in fewer numbers. These distributions somewhat accentuate the IoT-enabled fuzzy system for data collection and real-time decision-making, letting AI/ML methods provide prediction and adaptation.

4.6 RQ5: What Are the Predominant Patterns in Methodological Strategies, Implementation Difficulties, and Practical Barriers Encountered When Applying Fuzzy Logic Systems in Real-World Agricultural Settings?

This study compares PFSs with other nonfuzzy methods to address RQ5. More specifically, this question seeks to ascertain the extent to which PFSs have been used, modified, or extended into real-world problem domains. The study thus attempts to close the gap between theoretical development and practical implementation by analyzing evidence from the field, distinguishing domain-specific changes, and investigating factors affecting the use and scalability of PFS-based models in various industries.

Table 7: Distribution of fuzzy logic methodologies across agricultural application areas.

Methodology /application area →	Crop production and management	Irrigation and water management	Livestock and aquaculture	Disease detection and monitoring	Cross-domain and hybrid systems
Rule-based fuzzy logic	☑ Sharma et al. [125] – *Challenge:* Defining precise rules; *Limitation:* Poor adaptability	☑ Falahi et al. [86] – *Challenge:* Scalability; *Limitation:* Sensitive to sensor faults	☑ Agarwal et al. [81] – *Challenge:* Membership function tuning; *Limitation:* Limited automation scope	☑ Adnan [169]. – *Challenge:* Expert bias; *Limitation:* Hard to generalize	✗
Fuzzy logic + sensing tech	☑ Bukhari et al. [80] – *Challenge:* Sensor calibration; *Limitation:* High maintenance costs	☑ Farooq et al. [73] – *Challenge:* Real-time data processing; *Limitation:* Network dependency	☑ Liu [84] – *Challenge:* Data noise; *Limitation:* Reduced accuracy in extreme weather	☑ Nandi and Mahmood [72] – *Challenge:* Connectivity delays; *Limitation:* Vulnerable to sensor failure	☑ Rama et al. [74] – *Challenge:* Data integration; *Limitation:* High implementation cost

Table 7 (continued)

Methodology ↓/application area →	**Crop production and management**	**Irrigation and water management**	**Livestock and aquaculture**	**Disease detection and monitoring**	**Cross-domain and hybrid systems**
Hybrid intelligent systems	☑ Revathi and Santhi [107] – *Challenge:* Model complexity; *Limitation:* Requires skilled staff	✕	☑ Haider et al. [170] – *Challenge:* Data heterogeneity; *Limitation:* Long training time	✕	☑ Salsabila et al. [102] – *Challenge:* Computational load; *Limitation:* Energy consumption
Cross-domain adaptive systems	✕	☑ Xu and Li [146] – *Challenge:* Domain interoperability; *Limitation:* Low transferability	✕	☑ Liu et al. [84] – *Challenge:* Multidata fusion; *Limitation:* Difficult replication	☑ Sathyamoorthy et al. [122] – *Challenge:* Standardization gaps; *Limitation:* High data dependency

The methodological trajectory evolves from the simple rule-based systems to hybrid intelligent systems integrating sensing technologies, AI, and IoT (Table 7). Representative studies have spanned applications in various domains, such as crop protection, irrigation management, and postharvest processing (e.g., Agarwal et al. [81], Bukhari et al. [80], Nandi and Mahmood [72], Lv and Liu [120], and Farooq et al. [73]). Yet, despite these innovations, several implementation challenges exist, such as the complexity in integrating heterogeneous datasets, limited scalability, and the need for domain experts in formulating rules. Other issues are budget constraints, maintenance requirements, and the contrast between lab-tested accuracy and that actually obtained in the field.

The heatmap in Figure 10 indicates irrigation and water management as the most intensively focused application areas to date, showing maturity in both methodology and application. Crop protection and classification appear as moderately warm zones, driven by the increasing integration of sensing and imaging technologies. By contrast, postharvest processing remains a cooler zone, with relatively fewer contributions but a growing niche of specialized interest. This warm-to-cool gradient illustrates the balance between established applications and underexplored areas, pointing to clear opportunities for innovation in emerging agricultural domains.

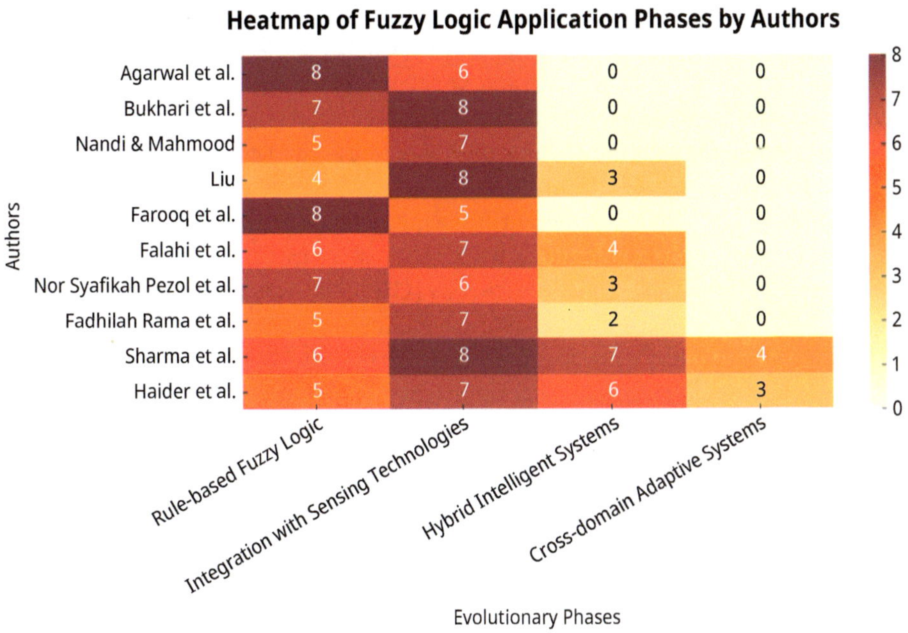

Figure 10: Heatmap of fuzzy logic application phases by authors.

4.6 RQ6: What Fundamental Factors Contribute to the Existence of Research Gaps in the Existing Literature, and What Future Research Avenues Could Be Pursued to Enhance the Use of Fuzzy Logic in Sustainable, Data-Driven Smart Farming?

Based on the studies we were able to access, fuzzy logic is widely proposed and combined with many other techniques (ANFIS, type-2/3 fuzzy, CNNs, GATs, optimization algorithms, IoT/FPGA deployments, MCDM, etc.), showing clear promise for uncertainty-tolerant decision support, resource-efficient control, and explainable rules for farmers. However, most work stops at lab or simulation scale, uses limited/single-crop datasets, omits standard performance metrics or economic evaluation, depends on static/expert-crafted rule sets, and rarely demonstrates long-term, multisite, real-world deployments, which together block reliable adoption and broad scalability of fuzzy solutions in sustainable, data-driven smart agriculture.

One major drawback is the absence of multisite, real-world field validation; as demonstrated by Bin et al. [2], Souza et al. [48], and Rasyid et al. [115], many studies are limited to a particular crop and region or to laboratory or simulation settings. Evaluation practices are also inconsistent, with several studies failing to report essential performance metrics such as accuracy, IoU, percentage improvement, latency, energy consumption, or cost efficiency, exemplified by Prabakaran et al. [155], Peng et al. [164], Agarwal [81], and Mendes [85]. Dataset limitations further restrict generalizability, as many works rely on small, proprietary, or single-crop datasets [82, 159, 164], without public benchmarks or cross-regional validation. Methodologically, many fuzzy systems depend on fixed, physically crafted membership functions and rules that cannot adapt to new conditions [78, 51, 109].

Computational and deployment constraints remain underexplored, with hybrid deep-fuzzy models often computationally heavy and few optimizations for edge or low-power environments. Prabakaran et al.'s [155] FPGA implementations are rare exceptions. Although advanced uncertainty-handling approaches such as type-2/type-3 fuzzy systems and *q*-ROFN are proposed, their complexity [171–173], tuning challenges, and real-time feasibility remain underinvestigated [153, 161] . Apart from that, the primary challenge lies in integrating different system layers, and there are very few practical examples that combine IoT, GIS, cloud-based energy management, and economic analysis within a single cohesive framework. Many of these are without any sustainability factors, down to including solar power or CO_2 reduction measures, whereas the others do not have economic analyses-based thinking for the opposite side of the arguments. In contrast, the decision-making problems analyzed under the different fuzzy-MCDM approaches mostly use subjective weights [70, 168].

Human factors are another gap, with minimal evaluation of explainability, farmer-in-the-loop processes, user interface design, or sociotechnical adoption, despite work such as Shahana et al. [100] noting differences in interpretability between Mamdani and Sugeno systems. Security, resilience, and standardization are rarely addressed, leaving agri-IoT systems vulnerable to network or cyber disruptions [24, 86]. Finally, benchmarking and reproducibility remain weak, as few studies provide public datasets, open-source implementations, or standardized evaluation protocols [159, 164], limiting comparability and slowing practical uptake. However, Figure 11 shows cluster distribution of various domains considered in this study.

In general, Figure 11 illustrates a network visualization showing four main clusters of research areas in sustainable, data-driven smart agriculture, each represented by distinct colors and labeled as crop production (purple), soil health (yellow), water management (yellow), pest and disease control (red), and market forecasting (red, adjacent to pest control). Larger nodes suggest more central or influential themes, whereas smaller nodes imply less linked ones. Each node inside these areas indicates a different study topic. Modern research is characterized by its compartmentalized structure, with grey connecting lines illustrating the relationships or overlaps between themes; these connections are stronger within clusters and weaker between different clusters. The layout hence points to an urgent need for interdisciplinary permeation to enable fuzzy logic and its extensions to achieve greater utility in agriculture. The intersection in the center of the layout indicates the shared dependencies among these clusters, which collectively imply how crop production, water management, pest and malady control, and market prediction are interconnected components that strengthen the overall system of smart agriculture.

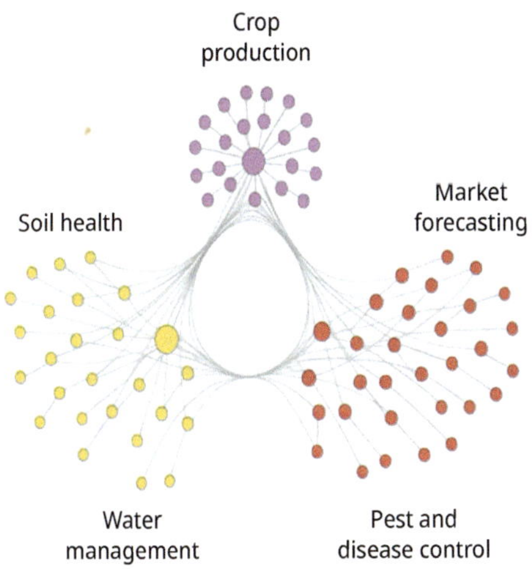

Figure 11: Research clusters and fragmentation in fuzzy logic applications for smart agriculture.

4.7 Multicriteria Sentence Ranking Framework for Smart Agriculture

Our suggested system will foster a smart agricultural environment by applying MCDM methods to prioritize sentences or statements according to their significance, relevance, and dependability within the context. There are things to consider-misfits of information inadequacy to compromise particular local agro-ecological conditions with emphasis upon the sustainability of impact as wherein the framework would have worked toward the rising demand for an evidence-based, context-driven support system for decision making in the application of agricultural science and research. These limitations highlight the mismatch between available information and the specific agro-ecological conditions required for sustainable agricultural impact. The proposed framework therefore aims to support evidence-based, context-driven decision-making in agricultural research and practice, while improving the interpretation of complex agricultural data and enhancing transparency for researchers and practitioners.

To integrate the major objectives of SLR, an approach has been put forth for the automatic organization and synthesis of key findings for automatic extractionas shown in Figure 12:

1. Topic rank extractor: Find sentences dealing with crop management, irrigation, pest control, and climate adaptation, whereas the perspective is that the topic model may be domain-specific (e.g., a special LDA designed for agricultural datasets).
2. In a semantic rank extractor approach, one may employ agricultural transformer-based models (SciBERT/AgriBERT) to compute semantic similarities of study objectives, methods, and results.
3. In the keyword rank extractor of a graph-based approach, one uses keyword co-occurrence networks (e.g., based on abstracts and conclusions) to rank concepts

according to their influence and are considered to cover concepts such as precision irrigation, fuzzy logic, or IoT sensor networks.

4. At position rank extractor, sentences in Conclusion, Discussions, and Results will be given more weightage.
5. Rank fusion: Combine the rankings with adjustable weights to create structured and concise summaries for each paper.

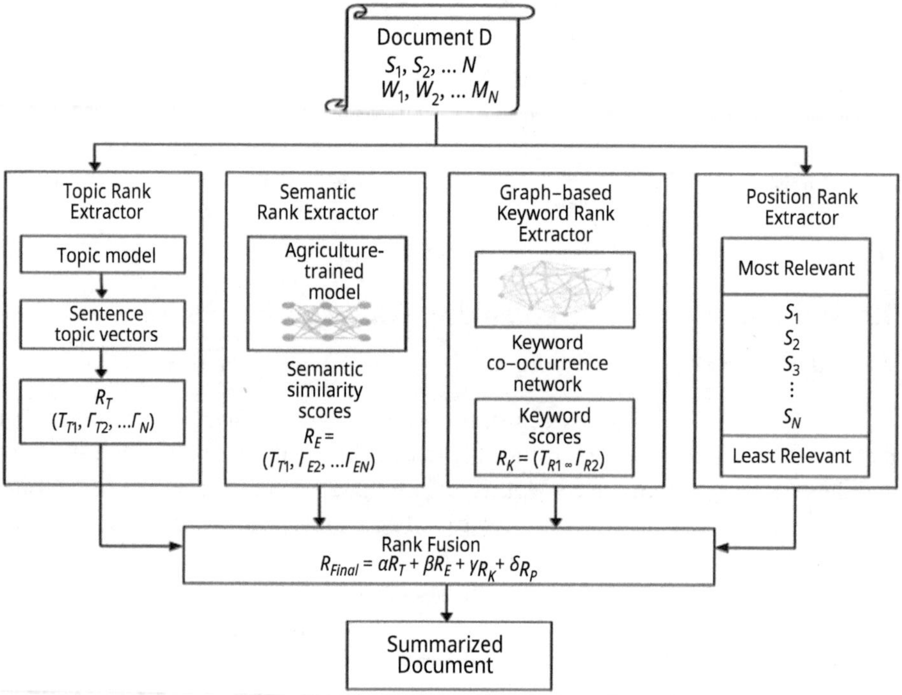

Figure 12: Proposed framework for smart agriculture.

5 Limitations

During this review, all proposed research questions were answered, and the current state-of-the-art and best practices, gaps, and opportunities in smart agriculture were identified. Some minor constraints still become evident. First, our search was limited to four major research databases: IEEE Xplore, Scopus, ScienceDirect, and ACM Digital Library. Although we believe that the resulting dataset should provide a robust and representative view of the research in this domain, we propose that future work should look at extending the list. Second, from preliminary searches with several other databases, we had a high rate of irrelevant results. That is a clear indication of one limitation of the search terms: other databases may use different terms for what are some key terms in smart agriculture, for instance, "precision cultivation," "digital farming," "crop analytics," or "agro-technological innovation." Future study should, therefore, consider broadening keyword searches so as to capture these variations and yield more insights. Even beyond, the present review, which is concerned with computational intelligence, IoT, and decision-support systems, could be extended to focus on other contemporary farming technologies.

6 Implications for Future Work

This review underscores the importance of advancing fuzzy logic applications in agriculture through the adoption of more expressive and adaptable fuzzy set frameworks. Although IFS and standard fuzzy logic have shown their usefulness, they frequently struggle to manage the high levels of ambiguity, reluctance, and multidimensional preference information present in intricate agricultural situations. Future research should concentrate on incorporating q-rung orthopair fuzzy sets, which offer greater flexibility in representing uncertainty and reluctance by accommodating bigger degrees of membership and nonmembership, to increase decision accuracy in multicriteria agricultural scenarios. The enhanced capacity of PFS to represent positive, neutral, and negative preferences makes them especially appropriate for complex agricultural decision-making situations such as crop selection in the face of climate variability, irrigation scheduling, and pest management tactics. Future agricultural decision-support frameworks can attain greater robustness, adaptability, and interpretability by combining these sophisticated fuzzy models with contemporary sensing technologies, hybrid intelligent systems, and domain-specific data analytics; this may ultimately result in more accurate and sustainable farming practices.

7 Conclusion

In this chapter, we presented an exploratory review that examined different studies published within the last 5 years on the role of fuzzy logic systems in smart agriculture. Based on our analysis, we found that, besides fuzzy logic, there were other variations and advanced extensions such as q-rung orthopair and picture fuzzy, that were limitedly covered in our review. These restricted fuzzy logic systems' capacity to demonstrate complex uncertainties. In addition, there were quite a few number of evaluations establishing the accuracy of fuzzy logic systems, and most of these studies overlooked their robustness, probabilistic reliability, and practical applicability. Also, there is a need to address how many fuzzy systems are tailored to specific tasks such as crop yield prediction, irrigation scheduling, or livestock monitoring, and their integration with smart technologies. Overall, we recommend that future work should explore methodological innovations, leveraging fuzzy logic's theoretical underpinnings to make them scalable and sustainable in the real world.

References

[1] Zihad SMRA, Bin L, Shahzad M, Khan H, Bashir MM, Ullah A et al. Fuzzy logic, geostatistics, and multiple linear models to evaluate irrigation metrics and their influencing factors in a drought-prone agricultural region. Environ Res. 2023;234:116509. doi: 10.1016/j.envres.2023.116509.

[2] Bin L, Shahzad M, Khan H, Bashir MM, Ullah A, Siddique M. Sustainable smart agriculture farming for cotton crop: A fuzzy logic rule-based methodology. Sustainability. 2023;15(18):13874. doi: 10.3390/su151813874.

[3] Rajesh J, Dommeti D, Nallapati SR, Lokesh C, Bhuvanesh SP, Padyala V, Srinivas PVVS. Application of fuzzy logic in smart agriculture to recognise tomato fruit ripeness.

[4] Hanilçi F, Çakır E, Kuyu YÇ. Navigating the technological frontier: Machine learning infused with fuzzy logic for control system advancements. In: Kahraman C, Cevik Onar, S, Cebi, S, Oztaysi, B, Tolga, AC, Ucal Sari, I(eds). Intelligent and fuzzy systems. Cham: Springer Nature Switzerland. 2024. pp. 37–42. doi: 10.1007/978-3-031-67192-0_5.

[5] Kumar A, Bewerwal A, Vikas, Srivastava D, Jain S. An intelligent yield prediction of crops in smart agriculture management using enhanced fuzzy-based AI framework. In: 2024 2nd International Conference on Disruptive Technologies (ICDT). 2024 Mar. pp. 1068–1073. doi: 10.1109/ICDT61202.2024.10489101.

[6] Mokarram M, Pourghasemi HR, Pham TM. Identification of suitable location to cultivate grape based on disease infestation using multi-criteria decision-making (MCDM) and remote sensing. Ecol Inform. 2023;76:102142. doi: 10.1016/j.ecoinf.2023.102142.

[7] Sharma MK, Dhiman N, Mishra LN, Mishra VN, Sahani SK. Mediative fuzzy extension technique and its consistent measurement in the decision making of medical application. Math Probl Eng. 2021;2021(1):5530681. doi: 10.1155/2021/5530681.

[8] Pai VR, Pai SG, Suhasi PM, Rekha PM Identification and classification of skin diseases using deep learning techniques [Internet]. 2023 Dec 12 [cited 2023 Dec 12]. Available from: https://www.researchsquare.com

[9] Ciptaningtyas HT, Sabilla IA, Haqqi MI. IoT-based smart irrigation for Allium ascalonicum using fuzzy logic. In: 2023 Eighth International Conference on Informatics and Computing (ICIC). 2023 Dec. Manado, Indonesia: IEEE. 2023, pp. 1–6. doi: 10.1109/ICIC60109.2023.10382022.

[10] Abdullah N, Durani NAB, Shari MFB, Siong KS, Hau VKW, Siong WN, Ahmad IKA. Towards smart agriculture monitoring using fuzzy systems. IEEE Access. 2021;9:4097–4111. doi: 10.1109/ACCESS.2020.3041597.

[11] Ferraro DO, De paula R. A fuzzy knowledge-based model for assessing risk of pesticides into the air in cropping systems. Sci Total Environ. 2022;820:153158. doi: 10.1016/j.scitotenv.2022.153158.

[12] Gupta T. Detection of rice leaf diseases by using fuzzy logic application. In: 2024 IEEE International Conference on Computing, Power and Communication Technologies (IC2PCT); 2024 Feb. Greater Noida, India: IEEE. 2024. pp. 1056–1059. doi: 10.1109/IC2PCT60090.2024.10486263.

[13] Patil BD, Gupta S, Sheikh AI, Kumar SS, Pilli LK, Bhupal Raj G. IoT and Big Data Integration for Real-Time Agricultural Monitoring. J Adv Zool. 2023;44(S-2):3079–3089. doi: 10.53555/jaz.v44iS2.1542.

[14] Bissadu K, Hossain G, Velagala LP. Enhancing cybersecurity resilience for low-income farmers in developing nations: A fuzzy cognitive mapping approach. In: 2024 IEEE International Conference on Consumer Electronics (ICCE). 2024Jan. Las Vegas, NV, USA. IEEE. 2024, pp. 1–6. doi: 10.1109/ICCE59016.2024.10444324.

[15] Ming F, Mou J, Cui Y. Application of fuzzy neural networks for intelligent irrigation of wheat. In: 2023 IEEE 6th Information Technology, Networking, Electronic and Automation Control Conference (ITNEC). 2023 Feb. Chongqing, China: IEEE. 2023, pp. 148–152. doi: 10.1109/ITNEC56291.2023.10082703.

[16] Widayat IW, Arsyad AA, Mantau AJ, Adhitya Y, Köppen M. Fuzzy methods in smart farming: A systematic review. Informatica. 2024;36(2):453–489. doi: 10.15388/24-INFOR579.

[17] Spanaki K, Sivarajah U, Fakhimi M, Despoudi S, Irani Z. Disruptive technologies in agricultural operations: A systematic review of AI-driven AgriTech research. Ann Oper Res. 2022;308(1):491–524. doi: 10.1007/s10479-020-03922-z.

[18] Abbasi R, Martinez P, Ahmad R. The digitization of agricultural industry: A systematic literature review on agriculture 4.0. Smart Agric Technol. 2022;2:100042. doi: 10.1016/j.atech.2022.100042.

[19] Elbasi E, Mostafa N, AlArnaout Z, Zreikat AI, Cina E, Varghese G, Shdefat A, Topcu AE, Abdelbaki W, Mathew S, Zaki C. Artificial intelligence technology in the agricultural sector: A systematic literature review. IEEE Access. 2023;11:171–202. doi: 10.1109/ACCESS.2022.3232485.

[20] Akkem Y, Biswas SK, Varanasi A. Smart farming using artificial intelligence: A review. Eng Appl Artif Intell. 2023;120:105899. doi: 10.1016/j.engappai.2023.105899.

[21] Abou-Mehdi-Hassani F, Zaguia A, Ait Bouh H, Mkhida A. Systematic literature review of smart greenhouse monitoring. SN Comput Sci. 2025;6(2):95. doi: 10.1007/s42979-024-03640-4.

[22] Bhat VS, Wang Y. Revisiting the control systems of autonomous vehicles in the agricultural sector: A systematic literature review. IEEE Access. 2025;13:54686–54721. doi: 10.1109/ACCESS.2025.3555142.

[23] Botero-Valencia J et al. Machine learning in sustainable agriculture: Systematic review and research perspectives. Agriculture. 2025;15(4):377. doi: 10.3390/agriculture15040377.

[24] Hossain MS, Rahman M, Rahman A, Kabir MM, Mridha MF, Huang J, Shin J. Automatic navigation and self-driving technology in agricultural machinery: A state-of-the-art systematic review [Internet]. IEEE Access. 2025 Jul 7 [cited 2025 Jul 7]. Available from. https://ieeexplore.ieee.org/abstract/document/11015488.

[25] Yilmaz E, Bocekci SC, Safak C, Yildiz K. Advancements in smart agriculture: A systematic literature review on state-of-the-art plant disease detection with computer vision. doi: 10.1049/cvi2.70004.

[26] Ali G, Mijwil MM, Buruga BA, Abotaleb M, Adamopoulos I. A survey on artificial intelligence in cybersecurity for smart agriculture: State-of-the-art, cyber threats, artificial intelligence applications, and ethical concerns. Mesopotamian J Comput Sci. 2024;2024:53–103. doi: 10.58496/MJCSC/2024/007.

[27] Chicaiza K, Paredes RX, Sarzosa IM, Yoo SG, Zang N. Smart farming technologies: A methodological overview and analysis. IEEE Access. 2024;12:164922–164941. doi: 10.1109/ACCESS.2024.3487497.

[28] Hammouch H, El-Yacoubi MA, Qin H, Berbia H. A systematic review and meta-analysis of intelligent irrigation systems. IEEE Access. 2024;12:128285–128304. doi: 10.1109/ACCESS.2024.3421322.

[29] Mohyuddin G, Khan MA, Haseeb A, Mahpara S, Waseem M, Saleh AM. Evaluation of machine learning approaches for precision farming in smart agriculture system: A comprehensive review. IEEE Access. 2024;12:60155–60184. doi: 10.1109/ACCESS.2024.3390581.

[30] Muhammed D, Ahvar E, Ahvar S, Trocan M, Montpetit MJ, Ehsani R. Artificial intelligence of things (AIoT) for smart agriculture: A review of architectures, technologies and solutions. J Netw Comput Appl. 2024;228:103905. doi: 10.1016/j.jnca.2024.103905.

[31] Naseer A, Shmoon M, Shakeel T, Ur Rehman S, Ahmad A, Gruhn V. A systematic literature review of the IoT in agriculture – global adoption, innovations, security, and privacy challenges. IEEE Access. 2024;12:60986–61021. doi: 10.1109/ACCESS.2024.3394617.

[32] Soussi A, Zero E, Sacile R, Trinchero D, Fossa M. Smart sensors and smart data for precision agriculture: A review. Sensors. 2024;24(8):2647. doi: 10.3390/s24082647.

[33] Garg D, Alam M Smart agriculture: a literature review [Internet]. 2023 Apr [cited 2025 Jul 07]. Available from: https://www.tandfonline.com/doi/abs/10.1080/23270012.2023.2207184

[34] Raouhi EM, Lachgar M, Hrimech H, Kartit A. Unmanned aerial vehicle-based applications in smart farming: A systematic review. Int J Adv Comput Sci Appl. 2023;14(6). doi: 10.14569/IJACSA.2023.01406123.

[35] Radočaj D, Plaščak I, Jurišić M. Global navigation satellite systems as state-of-the-art solutions in precision agriculture: A review of studies indexed in the Web of Science. Agriculture. 2023;13(7):1417. doi: 10.3390/agriculture13071417.

[36] Bwambale E, Abagale FK, Anornu GK. Smart irrigation monitoring and control strategies for improving water use efficiency in precision agriculture: A review. Agric Water Manag. 2022;260:107324. doi: 10.1016/j.agwat.2021.107324.

[37] Maraveas C. Incorporating artificial intelligence technology in smart greenhouses: Current state of the art. Appl Sci. 2023;13(1):14. doi: 10.3390/app13010014.

[38] Tan XJ, Cheor WL, Yeo KS, Leow WZ. Expert systems in oil palm precision agriculture: A decade systematic review. J King Saud Univ Comput Inf Sci. 2022;34(4):1569–1594. doi: 10.1016/j.jksuci.2022.02.006.

[39] Omar S. Internet of things (IoT) for smart farming: A systematic review. Int J Comput Appl. 2021;174(27):47–54. doi: 10.5120/ijca2021921182.

[40] Song C, Dong H. Application of intelligent recommendation for agricultural information: A systematic literature review. IEEE Access. 2021;9:153616–153632. doi: 10.1109/ACCESS.2021.3127201.

[41] Farooq MS, Riaz S, Abid A, Umer T, Zikria YB. Role of IoT technology in agriculture: A systematic literature review. Electronics. 2020;9(2):319. doi: 10.3390/electronics9020319.

[42] Navarro E, Costa N, Pereira A. A systematic review of IoT solutions for smart farming. Sensors. 2020;20(15):4231. doi: 10.3390/s20154231.

[43] Page MJ, McKenzie JE, Bossuyt PM, Boutron I, Hoffmann TC, Mulrow CD, Shamseer L, Tetzlaff JM, Akl EA, Brennan SE, Chou R, Glanville J, Grimshaw JM, Hróbjartsson A, Lalu MM, Li T, Loder EW, Mayo-Wilson E, McDonald S, McGuinness LA, Stewart LA, Thomas J, Tricco AC, Welch VA, Whiting P, Moher D. The PRISMA 2020 statement: An updated guideline for reporting systematic reviews. BMJ. 2021;372:n71. doi: 10.1136/bmj.n71.

[44] Gough D, Thomas J, Oliver S. An introduction to systematic reviews. 2017; 1–352.

[45] Grant MJ, Booth A. A typology of reviews: An analysis of 14 review types and associated methodologies. Health Inf Libr J. 2009;26(2):91–108. doi: 10.1111/j.1471-1842.2009.00848.x.

[46] Ouzzani M, Hammady H, Fedorowicz Z, Elmagarmid A. Rayyan – a web and mobile app for systematic reviews. Syst Rev. 2016;5(1):210. doi: 10.1186/s13643-016-0384-4.

[47] Ataguba G, Chan G, Orji R. Exploring persuasive games for emotion regulation: A state-of-the-art scoping review. Int J Hum Comput Interact. 2025;0(0):1–41. doi: 10.1080/10447318.2025.2483863.

[48] Souza RT, Costa ML, Mayrink J, Feitosa FE, Rocha Filho EA, Leite DF, Vettorazzi J, Calderon IM, Sousa MH, Passini R Jr, Baker PN, Kenny L, Cecatti JG. Perinatal outcomes from preterm and early term births in a multicenter cohort of low risk nulliparous women. Sci Rep. 2020;10(1):8508. doi: 10.1038/s41598-020-65022-z.

[49] RC PG, Salimath K, SK KV, M SS, Manjunath TC. Design and implementation of a type-1 fuzzy controller driven IoT-integrated automated plant watering system for gardening. In: 2024 4th International Conference on Pervasive Computing and Social Networking (ICPCSN). 2024 May. Salem, India: IEEE. 2024, pp. 790–796. doi: 10.1109/ICPCSN62568.2024.00133.

[50] Clotz C, Adlam F, Reddy K. Design and implementation of a fuzzy logic-based automated food dryer. In: 2025 33rd Southern African Universities Power Engineering Conference (SAUPEC). 2025 Jan. Pretoria, South Africa: IEEE. 2025, pp. 1–6. doi: 10.1109/SAUPEC65723.2025.10944389.

[51] Singhal P, V H. Detection of rice leaf diseases to increase its yield by using fuzzy logic approach. In: 2023 3rd International Conference on Advancement in Electronics & Communication Engineering (AECE). 2023 Nov. Ghaziabad, India: IEEE. 2023, pp. 703–706. doi: 10.1109/AECE59614.2023.10428281.
[52] Mendonça V, Santos M, Muñoz DM, Llanos CH. Fuzzy logic for assessing the health of a corn crop within family farming. In: 2024 IEEE Latin American Conference on Computational Intelligence (LA-CCI). 2024 Nov. Bogota D.C., Colombia: IEEE. 2024, pp. 1–6. doi: 10.1109/LA-CCI62337.2024.10814764.
[53] Mathew GA, Jotwani V, Singh AK. Fuzzy logic based soil nutrient categorization system for decision making in soil nutrient mitigation. In: 2025 IEEE 14th International Conference on Communication Systems and Network Technologies (CSNT). 2025 Mar. Bhopal, India: IEEE. 2025, pp. 765–769. doi: 10.1109/CSNT64827.2025.10968282.
[54] Rani VA, Guna G, Manonmani A, Yuvaraj P, Priya JS. Intelligent decision making in molasses extraction through neural network modeling and fuzzy logic control. In: 2024 International Conference on Power, Energy, Control and Transmission Systems (ICPECTS). 2024 Oct. Chennai, India: IEEE. 2024, pp. 1–8. doi: 10.1109/ICPECTS62210.2024.10780239.
[55] Suwastika NA, Helmi M, Aulia MMS, Wardana AA. Design of a fogponics farming system based on the Internet of Things and fuzzy logic. In: 2022 2nd International Conference on Intelligent Cybernetics Technology & Applications (ICICyTA). 2022 Dec. Bandung, Indonesia: IEEE. 2022, pp. 99–104. doi: 10.1109/ICICyTA57421.2022.10038174.
[56] Ahmad I, Shariffudin SE, Ramli AF, Maharum SMM, Mansor Z, Kadir KA. Intelligent plant monitoring system via IoT and fuzzy system. In: 2021 IEEE 7th International Conference on Smart Instrumentation, Measurement and Applications (ICSIMA). 2021 Aug. Bandung, Indonesia: IEEE. 2021, pp. 123–127. doi: 10.1109/ICSIMA50015.2021.9526312.
[57] Hadi MS, Kusuma AUPY, Mizar MA, Lestari D, Witjoro A, Irvan M. IoT and fuzzy logic based smart mushroom cultivation technology. In: 2021 7th International Conference on Electrical, Electronics and Information Engineering (ICEEIE). 2021 Oct. Malang, Indonesia: IEEE. 2021, pp. 346–350. doi: 10.1109/ICEEIE52663.2021.9616842.
[58] AP H, TR M, Kumar AS, Agarwal J, Abhinav K, Ms G. IoT based smart hydroculture control system using fuzzy computing. In: 2023 International Conference on Computer Science and Emerging Technologies (CSET). 2023 Oct. Bangalore, India: IEEE. 2023, pp. 1–8. doi: 10.1109/CSET58993.2023.10346948.
[59] Aghaseyedabdollah M, Alaviyan Y, Yazdizadeh A. IoT based smart greenhouse design with an intelligent supervisory fuzzy optimized controller. In: 2021 7th International Conference on Web Research (ICWR). 2021 May. Tehran, Iran: IEEE. 2021, pp. 311–317. doi: 10.1109/ICWR51868.2021.9443022.
[60] Hosseinzadeh M, Khoramdel J, Borhani Y, Najafi E. A new fuzzy logic based learning rate scheduling method for crop classification with convolutional neural network. In: 2022 8th International Conference on Control, Instrumentation and Automation (ICCIA). 2022 Mar. Tehran, Iran: IEEE. 2022, pp. 1–5. doi: 10.1109/ICCIA54998.2022.9737183.
[61] Koshariya AK, Ashraf MS, Nigam R, Kumar TS, Kumar D, Awasthi G, Elavarasan G. Implementation of deep convolution neuro-fuzzy network to plant disease detection, risk assessment, and classification. In: 2023 International Conference on Sustainable Computing and Data Communication Systems (ICSCDS). 2023 Mar. Erode, India: IEEE. 2023, pp. 575–581. doi: 10.1109/ICSCDS56580.2023.10104792.
[62] Gopi PSS, Karthikeyan M. Automated crop type classification using moth flame optimization with adaptive neuro fuzzy inference systems. In: 2022 International Conference on Applied Artificial Intelligence and Computing (ICAAIC). 2022 May. Salem, India: IEEE. 2022, pp. 867–872. doi: 10.1109/ICAAIC53929.2022.9792985.
[63] Kamble P, Jadhav BT. Fuzzy expert system for acidification and deacidification process in red wine grape juice. In: 2022 IEEE Bombay Section Signature Conference (IBSSC). 2022 Dec. Mumbai, India: IEEE. 2022, pp. 1–6. doi: 10.1109/IBSSC56953.2022.10037276.

[64] Vats S, Shankar VG, Mohapatra S, Anshika, Singh H. Enhancing crop yield through recommendation system using fuzzy inference model. In: 2023 4th IEEE Global Conference for Advancement in Technology (GCAT). 2023 Oct. Bangalore, India: IEEE. 2023, pp. 1–5. doi: 10.1109/GCAT59970.2023.10353412.

[65] Macatangay EC, Dela Cruz CAC, Villaverde JF. Determination of tomato fruit stages using principal component analysis and fuzzy logic algorithm. In: 2022 IEEE International Conference on Artificial Intelligence in Engineering and Technology (IICAIET); 2022 Sept. Kota Kinabalu, Malaysia: IEEE. 2022. pp. 1–6. doi: 10.1109/IICAIET55139.2022.9936870.

[66] Aziz NH, Narashid RH, Razak TR, Anshah SA, Talib N, Abd Latif Z, Hashim N, Zainuddin K. Detection of bacterial leaf blight disease using RGB-based vegetation indices and fuzzy logic. In: 2023 19th IEEE International Colloquium on Signal Processing & Its Applications (CSPA); 2023 Mar. Kedah, Malaysia: IEEE. 2023. pp. 134–139. doi: 10.1109/CSPA57446.2023.10087429.

[67] Chen H, Xie J, Xu L, Feng Q, Lin Q, Cai K. Feature selection for portable spectral sensing data of soil using broad learning network in fusion with fuzzy technique. IEEE Sens J. 2024;24(5):5644–5653. doi: 10.1109/JSEN.2023.3331026.

[68] Gowri Shankar D, Jayaparvathy R. Fuzzy–random forest predictive models for improvement of sugarcane yield and quality. In: 2025 Fifth International Conference on Advances in Electrical, Computing, Communication and Sustainable Technologies (ICAECT). 2025 Jan. Bhilai, India: IEEE. 2025, pp. 1–7. doi: 10.1109/ICAECT63952.2025.10958839.

[69] Zheng W, Yan L, Gou C, Wang FY. Fuzzy deep forest with deep contours feature for leaf cultivar classification. IEEE Trans Fuzzy Syst. 2022;30(12):5431–5444. doi: 10.1109/TFUZZ.2022.3177764.

[70] Vedanth GS, Kodipalli S, BR A, GN, Soma B. Intelligent crop selection for maximizing agricultural productivity: A fuzzy TOPSIS perspective. In: 2023 International Conference on Recent Advances in Science and Engineering Technology (ICRASET). 2023 Nov. B G Nagara, India: IEEE. 2023, pp. 1–5. doi: 10.1109/ICRASET59632.2023.10419956.

[71] Bilokonska Y, Yailymova H, Yailymov B, Shelestov A, Shumilo L, Lavreniuk M. Losses assessment for winter crops based on satellite data and fuzzy logic. In: 2020 IEEE 5th International Symposium on Smart and Wireless Systems within the Conferences on Intelligent Data Acquisition and Advanced Computing Systems (IDAACS-SWS). 2020 Sept. IEEE. 2020. pp. 1–5. doi: 10.1109/IDAACS-SWS50031.2020.9297059.

[72] Nandi PK, Mahmood MA. An automated irrigation and fertilization management system using fuzzy logic. In: 2021 5th International Conference on Electrical Information and Communication Technology (EICT). 2021 Dec. Khulna, Bangladesh: IEEE. 2021, pp. 1–5. doi: 10.1109/EICT54103.2021.9733472.

[73] Farooq M, Hussain A, Hashim S, Yang L, Ali M. Automated irrigation system based on irrigation gates using fuzzy logic. In: 2020 International Conference on Internet of Things and Intelligent Applications (ITIA). 2020 Nov. Zhenjiang, China: IEEE. 2020, pp. 1–5. doi: 10.1109/ITIA50152.2020.9312344.

[74] Rama F, Mulyana E, Mardiati R, Fathonih A. Design of automatic watergate system using ESP32 microcontroller based on fuzzy logic method. In: 2021 7th International Conference on Wireless and Telematics (ICWT). 2021 Aug. Bandung, Indonesia: IEEE. 2021, pp. 1–6. doi: 10.1109/ICWT52862.2021.9678410.

[75] Pareja M, Bandala A. Fuzzy irrigation system with rain detection and fertilizer control. In: 2020 IEEE Region 10 Conference (TENCON). 2020 Nov. Osaka, Japan: IEEE. 2020, pp. 1357–1362. doi: 10.1109/TENCON50793.2020.9293841.

[76] Adnan R. Institute of Electrical and Electronics Engineers, IEEE Malaysia Control System Society. In: (ed). 2020 IEEE International Conference on Automatic Control and Intelligent Systems (I2CACIS 2020): proceedings: virtual conference. 2020 June 20Shah Alam. Malaysia. Piscataway, NJ: IEEE. 2020. doi: 10.1109/I2CACIS49202.2020.

[77] Hartono A, Hirani Putri ST, Dzaki N, Kumila BN, Sanjaya E, Hakiem N. Developing sensor of a water level control prototype using fuzzy logic based on soil moisture levels with Internet of Things integration. In: 2024 12th International Conference on Cyber and IT Service Management (CITSM). 2024 Oct. Batam, Indonesia: IEEE. 2024, pp. 1–4. doi: 10.1109/CITSM64103.2024.10775887.
[78] Suguna M. Fuzzy logic controller based IoT for smart irrigation system. In: 2021 International Conference on Advancements in Electrical, Electronics, Communication, Computing and Automation (ICAECA). 2021 Oct. Coimbatore, India: IEEE. 2021, pp. 1–5. doi: 10.1109/ICAECA52838.2021.9675725.
[79] Awalludin N, Gerhana YA, Mulyana E, Lukman N, Dauni P, Dzikrayah F. Implementation of fuzzy logic method to measure soil moisture and environmental temperature in automatic watering system based on Internet of Things. In: 2022 8th International Conference on Wireless and Telematics (ICWT). 2022 July. Yogyakarta, Indonesia: IEEE. 2022, pp. 1–5. doi: 10.1109/ICWT55831.2022.9935141.
[80] Bukhari M, Athar SO, Ullah M, Aman MN. A fuzzy-logic-based smart irrigation controller for precision agriculture. IEEE Internet Things J. 2024;11(22):37257–37268. doi: 10.1109/JIOT.2024.3440200.
[81] Agarwal A, Arya YDS, Agarwal G, Agarwal S, Gola KK. A fuzzy based decision support system for irrigation process in precision agriculture. In: 2020 9th International Conference System Modeling and Advancement in Research Trends (SMART). 2020 Dec. Moradabad, India: IEEE. 2020, pp. 448–452. doi: 10.1109/SMART50582.2020.9337080.
[82] Yu Y, Qu Y, Zhang H, Zhang D. Indicator analysis of lake water quality based on fuzzy neural network and spectrophotometry. In: 2022 IEEE 10th Joint International Information Technology and Artificial Intelligence Conference (ITAIC). 2022 June. Chongqing, China: IEEE. 2022, pp. 1543–1546. doi: 10.1109/ITAIC54216.2022.9836751.
[83] Hua S, Xu M, Li S, Han K, Xu Z, Zhao Z. Intelligent water and fertilizer control strategy of fuzzy PID optimized by swarm intelligence algorithm. In: 2023 2nd International Conference on Artificial Intelligence and Intelligent Information Processing (AIIIP). 2023 Oct. Hangzhou, China: IEEE. 2023, pp. 262–267. doi: 10.1109/AIIIP61647.2023.00056.
[84] Liu X. Analysis of integrated water resources scheduling system in the Yellow River Basin based on fuzzy cluster cyclic iteration algorithm. In: 2022 8th International Conference on Hydraulic and Civil Engineering: Deep Space Intelligent Development and Utilization Forum (ICHCE). 2022 Nov. Xi'an, China: IEEE. 2022, pp. 1202–1205. doi: 10.1109/ICHCE57331.2022.10042610.
[85] Mendes WR, Videira AME, Er-Raki S, Heeren DM, Dutta R, Araújo FMU. Development of a fuzzy variable rate irrigation control system based on remote sensing data to fully automate center pivots. IEEE Trans Autom Sci Eng. 2024;21(4):6109–6125. doi: 10.1109/TASE.2023.3322120.
[86] Al Falahi MS, Canlas FQ, Nair S. Water control system for Al Nakheel tree using fuzzy logic with trend analysis. In: 2020 International Conference on Computing. Electronics & Communications Engineering (iCCECE)2020 AugSouthend. United Kingdom: IEEE. 2020. pp. 143–149. doi: 10.1109/iCCECE49321.2020.9231256.
[87] Silvestre EOT, Bados JSF, Villanueva JMA, Rojo CAC, Aburto NJM. Design of a fuzzy controller for the calculation of atmospheric comfort in a greenhouse. In: 2022 3rd International Conference on Artificial Intelligence, Robotics and Control (AIRC). 2022 May. Cairo, Egypt: IEEE. 2022, pp. 46–51. doi: 10.1109/AIRC56195.2022.9836977.
[88] Puno JCV, Haban JJI, Alejandrino JD, Bandala AA, Dadios EP. Design of a nutrient film technique hydroponics system with fuzzy logic control. In: 2020 IEEE Region 10 Conference (TENCON). 2020 Nov. IEEE. 2020. pp. 403–408. doi: 10.1109/TENCON50793.2020.9293749.
[89] Jayalakshmi B, Haritha H, Maniyeri A, Arjun N. Fuzzy based irrigation and lighting systems for indoor farming. In: 2020 s International Conference on Inventive Research in Computing Applications (ICIRCA). 2020 July. Coimbatore, India: IEEE. 2020, pp. 1183–1186. doi: 10.1109/ICIRCA48905.2020.9183154.

[90] Tolis F, Blounas TF, Tsipianitis D. Implementation of a small scale smart greenhouse structure using fuzzy logic and IoT. In: 2023 14th International Conference on Information, Intelligence, Systems & Applications (IISA). 2023 July. Volos, Greece: IEEE. 2023, pp. 1–8. doi: 10.1109/IISA59645.2023.10345847.

[91] Bolivar PBN et al. IoT-based aeroponic system for seasonal plants using fuzzy logic. In: TENCON 2022 – 2022 IEEE Region 10 Conference (TENCON); 2022 Nov; Hong Kong, Hong Kong. IEEE. 2022. pp. 1–6. doi: 10.1109/TENCON55691.2022.9977457.

[92] Alaviyan Y, Aghaseyedabdollah M, Sadafi M, Yazdizade A. Design and manufacture of a smart greenhouse with supervisory control of environmental parameters using fuzzy inference controller. In: 2020 6th Iranian Conference on Signal Processing and Intelligent Systems (ICSPIS). 2020 Dec. Mashhad, Iran: IEEE. 2020, pp. 1–6. doi: 10.1109/ICSPIS51611.2020.9349619.

[93] Yao L, Du C, Dong F. Design of plant growth lamp control system based on convolutional neural network and fuzzy control. In: 2024 5th International Conference on Computer Engineering and Application (ICCEA). 2024 Apr. Hangzhou, China: IEEE. 2024, pp. 1795–1798. doi: 10.1109/ICCEA62105.2024.10604274.

[94] Zou C, Zhang Y, Yuan Z. An intelligent adverse delivery outcomes prediction model based on the fusion of multiple obstetric clinical data. Comput Methods Biomech Biomed Engin. 2023;0(0):1–15. doi: 10.1080/10255842.2023.2262663.

[95] Mendoza CDP, Caguimbal RGR, Mandocdoc JB, Sarmiento JS. Developing fuzzy rules for small scale rearing of black soldier fly eggs. In: 2023 International Conference on Disruptive Technologies (ICDT); 2023 May. Greater Noida, India: IEEE. 2023. pp. 708–713. doi: 10.1109/ICDT57929.2023.10151393.

[96] Arshad NID, Kemalasari, Puspita E. Early detection of cow pregnancy based on volume control of cow urine using fuzzy logic method. In: 2021 International Electronics Symposium (IES); 2021 Sept. Surabaya, Indonesia: IEEE. 2021. pp. 126–130. doi: 10.1109/IES53407.2021.9594012.

[97] Chua A, Imperial ED, Mababangloob GR, Tinhay JB, Concepcion R. Intelligent water quality management in maturation tanks for Penaeus monodon using fuzzy logic. In: 2023 8th International Conference on Business and Industrial Research (ICBIR); 2023 May. Bangkok, Thailand: IEEE. 2023. pp. 675–681. doi: 10.1109/ICBIR57571.2023.10147655.

[98] Nibir R, Hassan J, Bokani A, Ray B. Fuzzy logic based trajectory optimization for livestock health monitoring using Internet of Drones (IoD). In: 2023 13th International Conference on Cloud Computing. Data Science & Engineering (Confluence)2023 Jan. Noida, India. IEEE. 2023, pp. 692–697. doi: 10.1109/Confluence56041.2023.10048844.

[99] Shitsukane A. Integrating fuzzy logic, solar power and IoT for real-time water quality monitoring in cage fish culture. SSRN Electron J. 2025. doi: 10.2139/ssrn.5236846.

[100] Shahana G, Ezhilarasi P, RajeshKannan S. Comparison of Mamdani and Sugeno fuzzy based data aggregation models for developing smart aquaculture system. In: 2022 International Conference on Innovative Computing, Intelligent Communication and Smart Electrical Systems (ICSES). 2022 July. Chennai, India: IEEE. 2022, pp. 1–8. doi: 10.1109/ICSES55317.2022.9914378.

[101] Saravanan H, Rajesh M, Vincent R. Fusion of C-means fuzzy logic and deep neural networks for enhanced fish disease detection. In: 2024 15th International Conference on Computing Communication and Networking Technologies (ICCCNT). 2024 June. Kamand, India: IEEE. 2024, pp. 1–5. doi: 10.1109/ICCCNT61001.2024.10724765.

[102] Salsabila AL, Astawa IGP, Budikarso A, Aswoyo B. Design and implementation of intelligent systems using ESP-32 based fuzzy logic on smart agriculture system for mustard plants. In: 2023 International Electronics Symposium (IES). 2023 Aug. Denpasar, Indonesia: IEEE. 2023, pp. 224–229. doi: 10.1109/IES59143.2023.10242499.

[103] Suwirmayanti NLG, Diaz RAN, Aryanto IKAA, Pradipta GA, Prasada IBMI. Development of papaya plant automation systems with the Internet of Things concept using fuzzy logic. In: 2022 4th

International Conference on Cybernetics and Intelligent System (ICORIS). 2022 Oct. Prapat: Indonesia. IEEE. 2022, pp. 1–6. doi: 10.1109/ICORIS56080.2022.10031515.

[104] Fadlo S, Elmahjoub AA, Rabbah N. Energy performance analysis of a differential wheeled mobile robot with fuzzy logic controller. In: 2021 IEEE International IoT, Electronics and Mechatronics Conference (IEMTRONICS). 2021 Apr. Toronto, ON, Canada: IEEE. 2021, pp. 1–5. doi: 10.1109/IEMTRONICS52119.2021.9422631.

[105] Magsumbol J-AV et al. FLi-BMS: A fuzzy logic-based intelligent battery management system for smart farm. In: 2022 IEEE 14th International Conference on Humanoid, Nanotechnology, Information Technology, Communication and Control, Environment, and Management (HNICEM) 2022 DecBoracay Island. Philippines: IEEE. 2022. pp. 1–5. doi: 10.1109/HNICEM57413.2022.10109388.

[106] Biswas B, Haider MR, Islam MR. Fuzzy logic-based precision foliar application amount determination using digital image processing of maize canopy. Int J Sci Res Multidiscip Stud.

[107] Revathi A, Santhi SG. Coverage optimization using fuzzy and artificial bee colony algorithm in wireless sensor networks. In: 2023 First International Conference on Advances in Electrical, Electronics and Computational Intelligence (ICAEECI). 2023 Oct. Tiruchengode, India: IEEE. 2023, pp. 1–7. doi: 10.1109/ICAEECI58247.2023.10370941.

[108] Khan JA, Khan MA, Al-Khalidi M, AlHammadi DA, Alasiry A, Marzougui M, Zhang Y, Khan F. Design of super resolution and fuzzy deep learning architecture for the classification of land cover and landsliding using aerial remote sensing data. IEEE J Sel Top Appl Earth Obs Remote Sens. 2025;18:337–351. doi: 10.1109/JSTARS.2024.3490775.

[109] Putra Pradana FG, Sarno R, Triarjo S. Development of smart farming control system based on Tsukamoto fuzzy algorithm. In: 2023 International Conference on Computer Science. Information Technology and Engineering (ICCoSITE)2023 Feb. Jakarta, Indonesia. IEEE. 2023, pp. 348–353. doi: 10.1109/ICCoSITE57641.2023.10127754.

[110] Seo H-S, Chung IS, Kim BM, Kwou YR. The design and implementation of automata-based testing environment for Java multi-thread programs. In: Proceedings Eighth Asia-Pacific Software Engineering Conference. 2001 Dec. pp. 221–228. doi: 10.1109/APSEC.2001.991480.

[111] Yuebo L, Haitao Y, Hongyan L, Qingxue L. Fuzzy clustering and routing protocol with rules tuned by improved particle swarm optimization for wireless sensor networks. IEEE Access. 2023;11:128784–128800. doi: 10.1109/ACCESS.2023.3332914.

[112] Vahedi AM, Nobahari H, Alizad M. Fuzzy gain scheduling of artificial potential fields for online path planning and obstacle avoidance of an aerial robot. In: 2022 10th RSI International Conference on Robotics and Mechatronics (ICRoM). 2022 Nov. Tehran, Iran: IEEE. 2022, pp. 309–316. doi: 10.1109/ICRoM57054.2022.10025308.

[113] Kashyap S, Spielman AF, Ramnarayan N, Sahana SD, Pant R, Kaur B, Rajkumar N, Premkumar R, Alam S, Murthy S. Family-centered postnatal training and adherence to newborn care practices in district hospitals in two Indian states: a quasi-experimental intervention study. medRxiv [Preprint. 2021 Mar 24 [cited 2025 Aug 30. doi: 10.1101/2021.03.15.21252605.

[114] Hayat AD, Hassan A, Hajar M, Ali EA. Fuzzy logic controller for 4WD/4WS autonomous agricultural robotic. In: 2022 IEEE 3rd International Conference on Electronics, Control, Optimization and Computer Science (ICECOCS). 2022 Dec. Fez, Morocco: IEEE. 2022, pp. 1–6. doi: 10.1109/ICECOCS55148.2022.9983035.

[115] Harun Al Rasyid MU, Syauqi Ahsan A, Najaa MF. Fuzzy logic for automatic watering system of smart agriculture with IoT. In: 2020 International Conference on Applied Science and Technology (iCAST). 2020 Oct. Padang, Indonesia: IEEE. 2020, pp. 1–7. doi: 10.1109/iCAST51016.2020.9557707.

[116] Waluyo W, Widura A, Hadiatna F, Anugerah D. Comparative power and energy consumptions between scheduled and fuzzy controlling on an IoT-based vertical farming. In: 2021 3rd International Conference on High Voltage Engineering and Power Systems (ICHVEPS). 2021 Oct. Bandung, Indonesia: IEEE. 2021, pp. 278–282. doi: 10.1109/ICHVEPS53178.2021.9601038.

[117] Puri V, Chandramouli M, Le CV, Hoa TH. Internet of Things and fuzzy logic based hybrid approach for the prediction of smart farming system. In: 2020 International Conference on Computer Science, Engineering and Applications (ICCSEA). 2020 Mar. Gunupur, India: IEEE. 2020, pp. 1–5. doi: 10.1109/ICCSEA49143.2020.9132933.
[118] Da silva sg, da Silveira MF, Bertoldi AD, Domingues MR, dos Santos IS. Maternal and child-health outcomes in pregnancies following assisted reproductive technology (ART): a prospective cohort study. BMC Pregnancy Childbirth. 2020;20(1):106. doi: 10.1186/s12884-020-2755-z.
[119] Ogurtsov M, Korolyov V, Khodzinskyi O. Improving the productivity of UAV operations based on granular computing and fuzzy sets. In: 2021 IEEE 6th International Conference on Actual Problems of Unmanned Aerial Vehicles Development (APUAVD). 2021 Oct. Kyiv, Ukraine: IEEE. 2021, pp. 33–36. doi: 10.1109/APUAVD53804.2021.9615419.
[120] Lv L, Liu RZ. Credit rating modeling of small and micro enterprises based on fuzzy analytic hierarchy process. In: 2022 14th International Conference on Measuring Technology and Mechatronics Automation (ICMTMA). 2022 Jan. Changsha, China: IEEE. 2022, pp. 732–737. doi: 10.1109/ICMTMA54903.2022.00151.
[121] Kong J, Fan X, Zuo M, Yan W, Jin X. FICformer: A multi-factor fuzzy Bayesian imputation cross-former for big data-driven agricultural decision support systems. IEEE Trans Fuzzy Syst. 2025;33(1):69–81. doi: 10.1109/TFUZZ.2024.3363213.
[122] Sathyamoorthy M, Kumar R, Vanitha CN, Sharma B, Syamraj V, Chowdhury S. An efficient integrated approach of fuzzy C-means map reduce for weather forecasting data collection. In: 2023 6th International Conference on Information Systems and Computer Networks (ISCON). 2023 Mar. pp. 1–7. doi: 10.1109/ISCON57294.2023.10112170.
[123] Shrivastava H, Gopalakrishnan S, Al-Fatlawy MH, Sivasubramanian S, Sheela MS, Zhu K. Construction of evaluation model of regional economic development level based on fuzzy clustering algorithm. In: 2023 s International Conference On Smart Technologies For Smart Nation (SmartTechCon). 2023 Aug. Singapore, Singapore: IEEE. 2023, pp. 678–682. doi: 10.1109/SmartTechCon57526.2023.10391095.
[124] Falana OJ, Ebo IO, Tinubu CO, Adejimi OA, Ntuk A. Detection of cross-site scripting attacks using dynamic analysis and fuzzy inference system. In: 2020 International Conference in Mathematics, Computer Engineering and Computer Science (ICMCECS). 2020 Mar. Ayobo, Nigeria: IEEE. 2020, pp. 1–6. doi: 10.1109/ICMCECS47690.2020.240871.
[125] Sharma S, Kumar R, Pathak BK. Analyzing the impact of uncertainties with fuzzy logic on service composition in smart agriculture. In: 2024 International Conference on Emerging Smart Computing and Informatics (ESCI). 2024 Mar. Pune, India: IEEE. 2024, pp. 1–5. doi: 10.1109/ESCI59607.2024.10497423.
[126] Rezaei N, Uddin MN. Fuzzy logic based adaptive overcurrent protection for wind farms. In: 2021 2nd International Conference on Robotics, Electrical and Signal Processing Techniques (ICREST). 2021 Jan. Dhaka, Bangladesh: IEEE. 2021, pp. 768–771. doi: 10.1109/ICREST51555.2021.9331046.
[127] Chalamalasetty S, Bikku T, Mukherjee P, Tyagi N, Kesarwani J, Kesharwani A. Envisioning revolutionary technological changes: The future terrain of fuzzy logic. In: 2024 2nd International Conference on Disruptive Technologies (ICDT). 2024 Mar. Greater Noida, India: IEEE. 2024, pp. 232–237. doi: 10.1109/ICDT61202.2024.10489074.
[128] Herabad MG, Afshar NP. Fuzzy-based deep reinforcement learning for frost forecasting in IoT edge-enabled agriculture. In: 2022 8th Iranian Conference on Signal Processing and Intelligent Systems (ICSPIS). 2022 Dec. Behshahr, Iran: IEEE. 2022, pp. 1–5. doi: 10.1109/ICSPIS56952.2022.10044063.
[129] Sumalatha K, Muneender E. Fuzzy logic controller strategy for ameliorating DC link voltage for grid connected hybrid PV-wind generation systems. In: 2021 s International Conference on Smart Technologies in Computing, Electrical and Electronics (ICSTCEE). 2021 Dec. Bengaluru, India: IEEE. 2021, pp. 1–7. doi: 10.1109/ICSTCEE54422.2021.9708549.

[130] Rajesh M, Devi AL. Design and development of a fuzzy-based single-phase STATCOM operator for non-linear loads in standalone PV systems to improve power quality. In: 2022 OPJU International Technology Conference on Emerging Technologies for Sustainable Development (OTCON). 2023 Feb. Raigarh, Chhattisgarh, India: IEEE. 2023, pp. 1–6. doi: 10.1109/OTCON56053.2023.10114000.
[131] Pan Z, Zhang S, Huang L, Qin H, He C, Wu Z. Wind power smoothing with genetic algorithm based fuzzy control of energy storage in wind farm. In: 2025 2nd International Conference on Smart Grid and Artificial Intelligence (SGAI). 2025 Mar. Changsha, China: I EE. 2025, pp. 303–308. doi: 10.1109/SGAI64825.2025.11009312.
[132] Rahman MT, Shib SK, Dipto DR, Shufian A, Hossain MS. Enhanced dynamic voltage restorer (DVR) with fuzzy logic control for voltage regulation. In: 2025 Fourth International Symposium on Instrumentation, Control, Artificial Intelligence, and Robotics (ICA-SYMP). 2025 Jan. pp. 42–47. doi: 10.1109/ICA-SYMP63674.2025.10876470.
[133] Xu X, Li M. An intelligent fuzzy system based on the optimization of ecological agriculture areas. IEEE Trans Emerg Top Comput Intell. 2024;8(5):3251–3262. doi: 10.1109/TETCI.2023.3243909.
[134] Ding C, You H, Guo Y. Low voltage ride through of doubly-fed wind farm based on STATCOM with fuzzy PI control. In: 2023 6th International Conference on Energy, Electrical and Power Engineering (CEEPE). 2023 May. pp. 378–382. doi: 10.1109/CEEPE58418.2023.10166798.
[135] Achroufene A, Moula BA. Flood prediction based on the fuzzy subset data fusion approach. In: 2024 International Conference on Information and Communication Technologies for Disaster Management (ICT-DM). 2024 Nov. Setif, Algeria: IEEE. 2024, pp. 1–7. doi: 10.1109/ICT-DM62768.2024.10798950.
[136] Balakrishnan D, Mariappan U, Chowdary PM, Krishna MH, Kumar MVNJ, Abhishekh P. Deep reinforcement learning with fuzzy inference system for prediction of crop sowing windows. In: 2023 2nd International Conference on Futuristic Technologies (INCOFT). 2023 Nov. Belagavi, Karnataka, India: IEEE. 2023, pp. 1–6. doi: 10.1109/INCOFT60753.2023.10425119.
[137] Li J, Jia L, Peng D, Hou R. Auxiliary modelling error and probability density function based neuro-fuzzy short-term wind power prediction. In: 2023 8th International Conference on Power and Renewable Energy (ICPRE); 2023 Sept. Shanghai, China. IEEE. 2023. pp. 1264–1268. doi: 10.1109/ICPRE59655.2023.10353702.
[138] Rajeswari AM, Anushiya AS, Fathima KSA, Priya SS, Mathumithaa N Fuzzy decision support system for recommendation of crop cultivation based on soil type [Internet]. 2025 Aug 28 [cited 2025 Aug 28]. Available from: https://ieeexplore.ieee.org/abstract/document/9142899
[139] Ahmad I, Shariffudin SE, Ramli AF, Maharum SMM, Mansor Z, Kadir KA. Intelligent plant monitoring system via IoT and fuzzy system. In: 2021 IEEE 7th International Conference on Smart Instrumentation, Measurement and Applications (ICSIMA). 2021 Aug. Bandung, Indonesia: IEEE. 2021, pp. 123–127. doi: 10.1109/ICSIMA50015.2021.9526312.
[140] Souza G, Aquino PT, Filev Maia R, Kamienski C, Soininen J-P. A fuzzy irrigation control system. In: 2020 IEEE Global Humanitarian Technology Conference (GHTC). 2020 Oct. Seattle, WA, USA: IEEE. 2020, pp. 1–6. doi: 10.1109/GHTC46280.2020.9342960.
[141] Kashyap P, Jaiswal A, Kumar M. Fuzzy K-means clustering (FKmC) to maximize the Energy Efficiency in Sensor-Enabled Internet of Things. In: 2021 IEEE 4th International Conference on Computing, Power and Communication Technologies (GUCON). Kuala Lumpur, Malaysia: IEEE. Sept. 2021. pp. 1–6. doi: 10.1109/GUCON50781.2021.9573937.
[142] S FSK, Witkowski U. Integrative Approach to Enhance Energy-Efficiency of Wireless Sensor Networks: Evaluating the Impact of Fuzzy-Logic Clustering and Network Coding in Routing Protocol. In: 2024 Advances in Science and Engineering Technology International Conferences (ASET). Abu Dhabi, United Arab Emirates: IEEE. June. 2024. pp. 1–6. doi: 10.1109/ASET60340.2024.10708647.

[143] Dela Flora Da Silveira J, Martins TR, Da Silva CN, Dos Reis JV. New Solution based on Fuzzy System for Planning IoT Communication Infrastructure for Rural Areas. In: 2021 IEEE International Conference on Fuzzy Systems (FUZZ-IEEE). Luxembourg, Luxembourg: IEEE. July. 2021. pp. 1–6. doi: 10.1109/FUZZ45933.2021.9494578.

[144] Chen T, Zhao R, Chen J, Zhang Z. Data-Driven Active Disturbance Rejection Control of Plant-Protection Unmanned Ground Vehicle Prototype: A Fuzzy Indirect Iterative Learning Approach. IEEECAA J Autom Sin. Aug. 2024;11(8):1892–1894. doi: 10.1109/JAS.2023.124158.

[145] Seo J, Cha M, Oh K, Park Y-J, Kwon TJ. Development of Steering Control Algorithms with Self-tuning Fuzzy PID for All-terrain Cranes. In: 2020 20th International Conference on Control, Automation and Systems (ICCAS). Oct. 2020. pp. 727–730. doi: 10.23919/ICCAS50221.2020.9268346.

[146] Xu X, Li M. An Intelligent Fuzzy System Based on the Optimization of Ecological Agriculture Areas. IEEE Trans Emerg Top Comput Intell. Oct. 2024;8(5):3251–3262. doi: 10.1109/TETCI.2023.3243909.

[147] Izahar NAZ, Derahman MN, Mohamed MA, Imas SS. Smart fertigation system with mobile application and fuzzy logic optimization. Int J Adv Technol Eng Explor. Dec. 2023;10(109). doi: 10.19101/IJATEE.2023.10102045.

[148] Bressane A, da AJ, Garcia S, De castro MV, Xerfan SD, Ruas G, Negri RG. Fuzzy Machine Learning Applications in Environmental Engineering: Does the Ability to Deal with Uncertainty Really Matter? Sustainability. Jan. 2024;16(11, Art. no. 11). doi: 10.3390/su16114525.

[149] Hua X, Yunde S, Yidong W, Haofeng Q, Mingge W. Application of an improved genetic algorithm fuzzy PID control algorithm for greenhouse. In: 2024 4th Asia-Pacific Conference on Communications Technology and Computer Science (ACCTCS). Shenyang, China: IEEE. Feb. 2024. pp. 36–41. doi: 10.1109/ACCTCS61748.2024.00013.

[150] Zou Z, Liu R, Wang Y, Huang P, Xu L. Design of aquaculture intelligent system based on fuzzy theory. In: 2023 IEEE 7th Information Technology and Mechatronics Engineering Conference (ITOEC). Chongqing, China: IEEE. Sept. 2023. pp. 464–468. doi: 10.1109/ITOEC57671.2023.10291527.

[151] Dincer H, Yuksel S, Mikhaylov A, Barykin SE, Aksoy T, Hacioglu U. Analysis of Environmental Priorities for Green Project Investments Using an Integrated q-Rung Orthopair Fuzzy Modeling. IEEE Access. 2022;10:50996–51007. doi: 10.1109/ACCESS.2022.3174058.

[152] He C, Jiang J. Improving Quality Evaluation for Control Systems of Machining Line Product Processing With Smart Algorithmic Techniques in Intuitionistic Fuzzy Multiple-Attribute Group Decision-Making. IEEE Access. 2025;13:18024–18039. doi: 10.1109/ACCESS.2025.3532961.

[153] Zhou A et al. Fault Detection and Isolation in Wind Turbines: Type 3 Fuzzy Logic Systems and Adaptive Random Search Learning. IEEE Access. 2024;12:129347–129361. doi: 10.1109/ACCESS.2024.3456818.

[154] Tian J, Li D, Jia X. IoT Smart Agriculture and Agricultural Product Income Insurance Participant Behavior Based on Fuzzy Neural Network. Comput Intell Neurosci. May 2022;2022:1–12. doi: 10.1155/2022/4778975.

[155] Prabakaran G, Vaithiyanathan D, Ganesan M. FPGA based effective agriculture productivity prediction system using fuzzy support vector machine. Math Comput Simul. July 2021;185:1–16. doi: 10.1016/j.matcom.2020.12.011.

[156] Remya S, Sasikala R. Performance evaluation of optimized and adaptive neuro fuzzy inference system for predictive modeling in agriculture. Comput Electr Eng. Sept 2020;86:106718. doi: 10.1016/j.compeleceng.2020.106718.

[157] Shah A et al. FuzzyShallow: A framework of deep shallow neural networks and modified tree growth optimization for agriculture land cover and fruit disease recognition from remote sensing and digital imaging. Measurement. Sept. 2024;237:115224. doi: 10.1016/j.measurement.2024.115224.

[158] Liang D, Fu Y, Xu Z. Time-Varying Intuitionistic Fuzzy Integral for Emergency Materials Demand Prediction With Case-Based Reasoning. IEEE Trans Fuzzy Syst. Sept. 2022;30(9):3617–3632. doi: 10.1109/TFUZZ.2021.3119427.

[159] Hang Y, Meng X, Wu Q. Application of Improved Lightweight Network and Choquet Fuzzy Ensemble Technology for Soybean Disease Identification. IEEE Access. 2024;12:25146–25163. doi: 10.1109/ACCESS.2024.3365829.

[160] Kumar D, Kukreja V. Application of PSPNET and Fuzzy Logic for Wheat Leaf Rust Disease and its Severity. In: 2022 International Conference on Data Analytics for Business and Industry (ICDABI), Sakhir. Bahrain: IEEE. Oct. 2022. pp. 547–551. doi: 10.1109/ICDABI56818.2022.10041575.

[161] Miao M, Li J. Application of WOA-based interval type-2 fuzzy logic system in wind power prediction. In: 2023 3rd International Conference on Frontiers of Electronics, Information and Computation Technologies (ICFEICT). 2023 May. Yangzhou, China: IEEE. 2023, pp. 318–323. doi: 10.1109/ICFEICT59519.2023.00060.

[162] Boonyopakorn P, Bualeard P. Applying neuro fuzzy system to analyze durian minerals within soil for precision agriculture. In: 2020 17th International Conference on Electrical Engineering/Electronics, Computer, Telecommunications and Information Technology (ECTI-CON). 2020 June. IEEE. 2020. pp. 135–138. doi: 10.1109/ECTI-CON49241.2020.9158093.

[163] Li J, Jia L, Peng D, Hou R. Auxiliary modelling error and probability density function based neuro-fuzzy short-term wind power prediction. In: 2023 8th International Conference on Power and Renewable Energy (ICPRE); 2023 Sept. Shanghai, China: IEEE. 2023. pp. 1264–1268. doi: 10.1109/ICPRE59655.2023.10353702.

[164] Peng M et al. Advanced image segmentation for precision agriculture using CNN-GAT fusion and fuzzy C-means clustering. Comput Electron Agric. 2024;226:109431. doi: 10.1016/j.compag.2024.109431.

[165] Villaseñor-Aguilar MJ, Padilla-Medina JA, Prado-Olivarez J, Martinez-Diaz S, Méndez-Gurrola -I-I, Barranco-Gutiérrez AI. Fuzzy classification of color carrots (Dacus Carota) using Raspberry Pi towards Farming 4.0. In: 2023 IEEE 13th Annual Computing and Communication Workshop and Conference (CCWC). 2023 Mar. Las Vegas, NV, USA: IEEE. 2023, pp. 1174–1178. doi: 10.1109/CCWC57344.2023.10099149.

[166] Erdoğdu A, Dayi F, Yildiz F, Yanik A, Ganji F. Combining fuzzy logic and genetic algorithms to optimize cost, time and quality in modern agriculture. Sustainability. 2025;17(7):2829. doi: 10.3390/su17072829.

[167] Brainy JRVJ, Suganthi KDN, Narayanamoorthy S, Ilakiya U, Innab N, Alshammari A, Ahmadian A, Jeon J. A perspective study for the assessment of field robots in agriculture: An enhanced fuzzy MADM approach. Comput Electron Agric. 2023;214:108296. doi: 10.1016/j.compag.2023.108296.

[168] Alamoodi AH, Garfan S, Deveci M, Albahri OS, Albahri AS, Yussof S, Homod RZ, Sharaf IM, Moslem S. Evaluating agriculture 4.0 decision support systems based on hyperbolic fuzzy-weighted zero-inconsistency combined with combinative distance-based assessment. Comput Electron Agric. 2024;227:109618. doi: 10.1016/j.compag.2024.109618.

[169] Adnan R. Institute of Electrical and Electronics Engineers, IEEE Malaysia Control System Society. In: (ed). 2020 IEEE International Conference on Automatic Control and Intelligent Systems (I2CACIS 2020): 20th June 2020, Shah Alam, Malaysia: proceedings: virtual conference. Piscataway, NJ: IEEE. 2020. doi: 10.1109/I2CACIS49202.2020.

[170] Haider MH, Wang Z, Khan AA, Ali H, Zheng H, Usman S, Kumar R, Bhutta MUM, Zhi P. Robust mobile robot navigation in cluttered environments based on hybrid adaptive neuro-fuzzy inference and sensor fusion. J King Saud Univ Comput Inf Sci. 2022;34(10):9060–9070. doi: 10.1016/j.jksuci.2022.08.031.

[171] Nwokoro CO, Ejegwa PA. A Decade of Picture Fuzzy Sets in Multi-Criteria Decision-Making: A Comprehensive Review of Trends, Gaps, and Future Directions. Knowl Decis Syst Appl. 2025 May 10 1:145164.

[172] Ejegwa PA, Anum MT, Kausar N, Nwokoro CO, Cagin TA. A Novel q-Rung Orthopair Fuzzy Distance Metric for Energy Source Selection. J Appl Res Ind Eng. 2025 Jul 29. e226106.

[173] Nwokoro CO, Inyang UG, Eyoh IJ, Uzoka FM, Akinnuwesi BA, Ejegwa PA. Meta-analysis of predictive modelling approaches and systematic reviews for maternal healthcare outcomes. Asian J Sci Appl Technol. 2024 Apr 15;13(1):19–35.

Parul Tomar*

Modified Arithmetic Operations on Trapezoidal Fuzzy Numbers and Its Applications in Fuzzy Transportation Problems

Abstract: Decisions made under uncertainty are often resolved with the use of fuzzy numbers to represent data that cannot be modeled exactly. This chapter offers a review of the arithmetic operations of trapezoidal fuzzy numbers (TrFNs) around fuzzy transportation problems. In the previous multiplication expression suggested, this resulted in an inconsistency by not allowing the multiplication of all TrFNs – only non-negative TrFNs. Upon exploring a situation with no restrictions over fuzzy numbers, the consequences of this approach generally resulted in mathematically invalid outputs causing inaccurate total transportation costs. Thus, this chapter offers a generalized expression of the additive and multiplicative operations on TrFNs to resolve this issue. The new method produces accurate results with non-negative TrFNs on unrestricted TrFNs as well – producing corrected total trapezoidal fuzzy transportation costs (TrFTCs). Overall, the new TrFTCs provided more valid and accurate outputs than previous solutions. Therefore, the proposed modification of the optimization process under uncertainty is more valid and reliable.

Keywords: TrFNs, transportation problem, TrFTCs

1 Introduction

The transportation problem serves as a model of general importance in operations research and decision-making. The parameters of the real world – such as supply, demand, and cost – are often not known exactly, and fuzzy modeling represents one way of handling uncertainty. Of several fuzzy representations, trapezoidal fuzzy numbers (TrFNs) have been extensively used because, in addition to being simple, they represent imprecision more flexibly.

In fuzzy optimization problems, the arithmetic operations with trapezoidal fuzzy quantities are necessary. Several scholars have studied operations with trapezoidal fuzzy quantities, with Raj et al. [1] recently developing a new formula for the multiplication of TrFNs and using this in an application involving trapezoidal fuzzy transportation

***Corresponding Author: Parul Tomar,** Department of Applied Sciences and Humanities, Dronacharya Group of Institutions, Greater Noida, Uttar Pradesh, e-mail: parul1003tomar@gmail.com

 | https://doi.org/10.1515/9783112222249-012

problems (TrFTPs). The multiplication produced smaller spreads and lower defuzzified values for total trapezoidal fuzzy transportation costs (TrFTCs) compared to other existing research in the literature [2–4].

However, a detailed analysis reveals that the expression proposed by Raj et al. holds valid only when both fuzzy numbers are non-negative. In scenarios where multiplication involves a non-negative fuzzy number and a fuzzy number with negative support, the expression fails to satisfy the necessary conditions of trapezoidal membership. Consequently, the resulting values and total TrFTCs reported in are mathematically incorrect.

The intent of this chapter is to overcome this limitation by developing a generalized multiplication operation for TrFNs, valid both in a non-negative fuzzy situation and in unlimited fuzzy conditions. The operation will maintain the trapezoidal structure and correctly compute fuzzy arithmetic operations. The new expression will also be used in the TrFTP, which will be presented by Raj et al. [1], and total TrFTCs will be calculated to provide validation for the new operation.

The remainder of this chapter has the following structure. Section 2 presents the preliminaries, definitions and concepts from which the four following Sections are derived from. Section 3 provides the multiplication approach strategy of Raj et al. [1]. Section 4 looks at the limitations of the existing multiplication expression. Section 5 demonstrates with analytical examples that the existing results were indeed incorrect. Section 6 discusses the modified multiplication method to address these limitations. Section 7 then utilizes the original methodology corrected after the feedback provided in Section 5 to provide computed results for the TrFTP. Finally, Section 8 brings us to a conclusion with a summary of findings and future directions for research.

2 Preliminaries

Definition 2.1 ([5]). A TrFN $\tilde{\wp}$ is defined by four real parameters $[\varsigma, \varrho, \tau, \kappa]$, where $\varsigma \leq \varrho \leq \tau \leq \kappa$. Its membership function is:

$$\mu_{\tilde{\wp}}(\varkappa) = \begin{cases} 0, & \varkappa < \varsigma, \\ \frac{\varkappa - \varsigma}{\varrho - \varsigma}, & \varsigma \leq \varkappa < \varrho, \\ 1, & \varrho \leq \varkappa \leq \tau, \\ \frac{\kappa - \varkappa}{\kappa - \tau}, & \tau < \varkappa \leq \kappa, \\ 0, & X > \kappa. \end{cases}$$

This representation is widely used in fuzzy transportation and optimization problems for modeling uncertain quantities.

Definition 2.2 ([5]). For TrFNs $\tilde{\wp} = [\varsigma_1, \varrho_1, \tau_1, \kappa_1]$ and $\tilde{\mathfrak{H}} = [\varsigma_2, \ell_2, \tau_2, k_2]$, the basic arithmetic operations are defined as:

(a) Addition: $\tilde{\wp} + \tilde{\mathfrak{H}} = [\varsigma_1 + \varsigma_2, \varrho_1 + \varrho_2, \tau_1 + \tau_2, \kappa_1 + \kappa_2]$
(b) Subtraction: $\tilde{\wp} - \tilde{\mathfrak{H}} = [\varsigma_1 - \kappa_2, \varrho_1 - \tau_2, \tau_1 - \varrho_2, \kappa_1 - \varsigma_2]$
(c) Multiplication: The multiplication operation is often defined only for non-negative fuzzy numbers as: $\tilde{\wp} \cdot \tilde{\mathfrak{H}} = [\varsigma_1\varsigma_2, \varrho_1\varrho_2, \tau_1\tau_2, \kappa_1\kappa_2]$

Definition 2.3 ([5]). The defuzzified value or crisp representative of a TrFN $\tilde{\wp} = [\varsigma, \varrho, \tau, \kappa]$ is obtained by the centroid method, defined as:

$$\text{Defuzz}(\tilde{\wp}) = \frac{\varsigma + 2\varrho + 2\tau + \kappa}{6}$$

3 Raj et al.'s Method

Raj et al. [1, Definition 3.1, Section 3, p. 1072] proposed eq. (1) to evaluate the multiplication of two TrFNs.

Let $\tilde{\wp} = [\varsigma_1, \varrho_1, \tau_1, \kappa_1]$ and $\tilde{\mathfrak{H}} = [\varsigma_2, \varrho_2, \tau_2, \kappa_2]$ be two TrFNs. Then:

$$\tilde{\wp} \cdot \tilde{\mathfrak{H}} = \left[\varsigma_1\varsigma_2, \varrho_1\varrho_2, \frac{\varrho_1\tau_2 + \varrho_2\tau_1 + \tau_1\tau_2 - \varrho_1\varrho_2}{2}, \frac{\varsigma_1\kappa_2 + \varsigma_2\kappa_1 + \kappa_1\kappa_2 - \varsigma_1\varsigma_2}{2}\right] \quad (1)$$

Raj et al. [1] considered the TrFTP, represented by Table 1, to show the advantage of their proposed expression (1) over the existing multiplication [2–4] of two TrFNs.

Table 1: TrFTP [1, Section 4, Example 4.2, p. 1074].

Destination		D_1	D_2	D_3	Supply
Origin	O_1	[1, 4, 9, 19]	[1, 2, 5, 9]	[2, 5, 8, 18]	[1, 5, 7, 9]
	O_2	[8, 9, 12, 26]	[3, 5, 8, 12]	[7, 9, 13, 28]	[4, 7, 8, 10]
	O_3	[11, 12, 20, 27]	[0, 5, 10, 15]	[4, 5, 8, 11]	[4, 5, 8, 11]
Demand		[3, 5, 8, 12]	[4, 8, 9, 10]	[2, 4, 6, 8]	

Raj et al. [1] pointed out that on solving the TrFTP, represented by Table 1, the results, shown in Table 2, are obtained.

In this chapter, it is pointed out that to evaluate the total TrFTC, there is a need to evaluate multiplication of a non-negative TrFN with an unrestricted TrFN. While, Raj et al.'s [1] expression (1) can be used only to evaluate multiplication of non-negative TrFNs.

Table 2: Existing results [1].

Method	TrFTP represented by Table 1 [1, Section 4, Example 4.2, p. 1074]	
	Fuzzy feasible solutions (FFSs)	**Total TrFTC**
Fuzzy north-west corner method	$\tilde{x}_{11} = [1, 5, 7, 9], \tilde{x}_{21} = [-6, -2, 3, 11], \tilde{x}_{22} = [-7, 4, 10, 16], \tilde{x}_{32} = [-12, -2, 5, 17], \tilde{x}_{33} = [2, 4, 6, 8]$	$[-60, 32, 235, 425]$
Fuzzy least cost method	$\tilde{x}_{12} = [1, 5, 7, 9], \tilde{x}_{21} = [-5, 3, 7, 15], \tilde{x}_{22} = [-5, 1, 4, 9], \tilde{x}_{31} = [-4, -1, 4, 9], \tilde{x}_{33} = [2, 4, 6, 8]$	$[-90, 50, 242.5, 510]$
Fuzzy Vogel's approximation method	$\tilde{x}_{11} = [1, 5, 7, 9], \tilde{x}_{22} = [4, 7, 8, 10], \tilde{x}_{31} = [-6, -2, 3, 11], \tilde{x}_{32} = [-6, 0, 2, 6], \tilde{x}_{33} = [2, 4, 6, 8]$	$[-45, 51, 220.5, 420]$

However, to evaluate the total TrFTC, presented in Table 2, Raj et al. [1] have used their proposed eq. (1) to evaluate multiplication of a non-negative TrFN with an unrestricted TrFN, which is mathematically incorrect. Due to the same reason, the total TrFTC, presented in Table 2, is not correct. Also, Raj et al.'s [1] eq. (1) is generalized to overcome its limitation. Finally, the correct total TrFTC of the TrFTP, represented by Table 1, is obtained by the generalized expression.

4 Limitation of Raj et al.'s Multiplication

It is pertinent to mention that Raj et al.'s [1] eq. (1) will be valid only if $\left[\varsigma_1\varsigma_2, \varrho_1\varrho_2, \frac{\varrho_1\tau_2+\varrho_2\tau_1+\tau_1\tau_2-\varrho_1\varrho_2}{2}, \frac{\varsigma_1\kappa_2+\varsigma_2\kappa_1+\kappa_1\kappa_2-\varsigma_1\varsigma_2}{2}\right]$ is a TrFN i.e., the condition (2) will be satisfied:

$$\varsigma_1\varsigma_2 \le \varrho_1\varrho_2 \le \frac{\varrho_1\tau_2 + \varrho_2\tau_1 + \tau_1\tau_2 - \varrho_1\varrho_2}{2} \le \frac{\varsigma_1\kappa_2 + \varsigma_2\kappa_1 + \kappa_1\kappa_2 - \varsigma_1\varsigma_2}{2} \tag{2}$$

It can be easily verified that if $\tilde{\wp} = [\varsigma_1, \varrho_1, \tau_1, \kappa_1]$ and $\tilde{\mathfrak{H}} = [\varsigma_2, \varrho_2, \tau_2, \kappa_2]$ are two non-negative TrFNs, i.e., $\varsigma_1 \ge 0$ and $\varsigma_2 \ge 0$ then, the condition (2) will always be satisfied. However, if either $\tilde{\wp} = [\varsigma_1, \varrho_1, \tau_1, \kappa_1]$ or $\tilde{\mathfrak{H}} = [\varsigma_2, \varrho_2, \tau_2, \kappa_2]$ is not a non-negative TrFN, i.e., either the condition $\varsigma_1 \ge 0$ or $\varsigma_2 \ge 0$ will not be satisfied. Then, the condition (2) will not necessarily be satisfied. Hence, Raj et al.'s [1] expression (1) cannot be used if

(i) $\varsigma_1 < 0$ and $\varsigma_2 < 0$
(ii) $\varsigma_1 \ge 0$ and $\varsigma_2 < 0$
(iii) $\varsigma_1 < 0$ and $\varsigma_2 \ge 0$

The following example clearly indicates that if for the TrFNs $\tilde{\wp} = [\varsigma_1, \varrho_1, \tau_1, \kappa_1]$ and $\tilde{\mathfrak{H}} = [\varsigma_2, \varrho_2, \tau_2, \kappa_2]$ either the condition $\varsigma_1 \ge 0$ or the condition $\varsigma_2 \ge 0$ will not be satisfied. Then, the TrFN, obtained by Raj et al.'s [1] expression (1), will not necessarily be TrFN.

Let $\tilde{\wp} = [0, 5, 10, 15]$ and $\tilde{\mathfrak{H}} = [-12, -2, 5, 17]$ be two TrFNs. Comparing $\tilde{\wp} = [0, 5, 10, 15]$ and $\tilde{\mathfrak{H}} = [-12, -2, 5, 17]$ with $\tilde{\wp} = [\varsigma_1, \varrho_1, \tau_1, \kappa_1]$ and $\tilde{\mathfrak{H}} = [\varsigma_2, \varrho_2, \tau_2, \kappa_2]$, $\varsigma_1 = 0$ and $\varsigma_2 = -12$.

Since, the condition $\varsigma_1 < 0$ and $\varsigma_2 < 0$ is satisfying. So, the expression (1) cannot be used to evaluate the multiplication of $\tilde{\wp}$ and $\tilde{\mathfrak{H}}$. However, if the expression (1) is used then $\tilde{\wp} \cdot \tilde{\mathfrak{H}} = [0, -10, 32.5, 37.5]$ which is not a TrFN as the condition $0 \leq -10 \leq 32.5 \leq 37.5$ is not satisfying.

5 Incorrectness of the Existing Total TrFTC

Raj et al. [1] have claimed that on considering their proposed expression (1), the total TrFTC of the TrFTP, represented by Table 1, corresponding to the FFS obtained by the following methods:

(i) Fuzzy north-west corner method is $[-60, 32, 235, 425]$.
(ii) Fuzzy least cost method is $[-90, 50, 242.5, 510]$.
(iii) Fuzzy Vogel's approximation method is $[-45, 51, 220.5, 420]$.

In actual case, these total TrFTCs are not correct as a mathematically incorrect result is considered to evaluate it. To validate this claim, it is shown that the first total TrFTC, i.e., $[-60, 32, 235, 425]$ is incorrect. On the same direction, one may easily verify that other total TrFTCs, presented in the third column of Table 2, are also not correct.

The first total TrFTC, i.e., $[-60, 32, 235, 425]$ is evaluated as follows:

It is obvious from Table 2 that the following FFS is obtained on solving the TrFTP, represented by Table 1, by the fuzzy north-west corner method:

$$\tilde{\varkappa}_{11} = [1, 5, 7, 9], \tilde{\varkappa}_{21} = [-6, -2, 3, 11], \tilde{\varkappa}_{22} = [-7, 4, 10, 16], \varkappa_{32} = [-12, -2, 5, 17], \tilde{\varkappa}_{33} = [2, 4, 6, 8]$$

The total TrFTC corresponding to this FFS is $[1, 4, 9, 19] \cdot [1, 5, 7, 9] + [8, 9, 12, 26] \cdot [-6, -2, 3, 11] + [3, 5, 8, 12] \cdot [-7, 4, 10, 16] + [0, 5, 10, 15] \cdot [-12, -2, 5, 17] + [4, 5, 8, 11] \cdot [2, 4, 6, 8]$.

Since, according to Raj et al.'s [1] expression (1):

$$[8, 9, 12, 26] \cdot [-6, -2, 3, 11] = [-48, -18, 28.5, 133], \quad [1, 4, 9, 19] \cdot [1, 5, 7, 9] = [1, 20, 58, 99],$$
$$[0, 5, 10, 15] \cdot [-12, -2, 5, 17] = [0, -10, 32.5, 37.5], \quad [4, 5, 8, 11] \cdot [2, 4, 6, 8] = [8, 20, 45, 67],$$
$$[3, 5, 8, 12] \cdot [-7, 4, 10, 16] = [-21, 20, 71, 88.5]$$

So, $[1, 4, 9, 19] \cdot [1, 5, 7, 9] + [8, 9, 12, 26] \cdot [-6, -2, 3, 11] + [3, 5, 8, 12] \cdot [-7, 4, 10, 16] + [0, 5, 10, 15] \cdot [-12, -2, 5, 17] + [4, 5, 8, 11] \cdot [2, 4, 6, 8] = [1, 20, 58, 99] + [-48, -18, 28.5, 133] + [-21, 20, 71, 88.5] + [0, -10, 32.5, 37.5] + [8, 20, 45, 67] = [-60, 32, 235, 425]$.

It is obvious that to find the total TrFTC, it is assumed that $[0, -10, 32.5, 37.5]$, obtained by multiplying the TrFNs $[0, 5, 10, 15]$ and $[-12, -2, 5, 17]$ using expression (1), is a TrFN. While, it is obvious that $[0, -10, 32.5, 37.5]$ is not a TrFN as the necessary condition (2) is not satisfying. Hence, the first total TrFTC i.e., $[-60, 32, 235, 425]$ is incorrect.

6 Modified Raj et al.'s Multiplication

It is obvious from Section 3 that to find the correct total TrFTC of the TrFTP, represented by Table 1, there is a need to use:

(i) An expression to evaluate the multiplication of two non-negative TrFNs
(ii) An expression to evaluate multiplication of two such TrFNs $\tilde{\wp} = [\varsigma_1, \varrho_1, \tau_1, \kappa_1]$ and $\tilde{\mathfrak{H}} = [\varsigma_2, \varrho_2, \tau_2, \kappa_2]$ for which the condition $\varsigma_1 \geq 0$ and $\varsigma_2 < 0$ will be satisfied

Keeping the same in mind, in this section, the following expressions are proposed to evaluate the multiplication of two TrFNs.

Let $\tilde{\wp} = [\varsigma_1, \varrho_1, \tau_1, \kappa_1]$ and $\tilde{\mathfrak{H}} = [\varsigma_2, \varrho_2, \tau_2, \kappa_2]$ be two TrFNs such that $\varsigma_1 \geq 0$ and ς_2 is any real number. Then on following the method, used by Raj et al. [1] to derive the expression (1), we have:

(i) $\tilde{\wp}.\tilde{\mathfrak{H}} = \left[\left(\frac{\varsigma_1+\kappa_1}{2}\right)\varsigma_2 - \left(\frac{\kappa_1-\varsigma_1}{2}\right)|\varsigma_2|, \left(\frac{\varrho_1+\tau_1}{2}\right)\varrho_2 - \left(\frac{\tau_1-\varrho_1}{2}\right)|\varrho_2|, \left(\frac{\varrho_1+\tau_1}{2}\right)\tau_2 + \left(\frac{\tau_1-\varrho_1}{2}\right)|\varrho_2|, \left(\frac{\varsigma_1+\kappa_1}{2}\right)\kappa_2 + \left(\frac{\kappa_1-\varsigma_1}{2}\right)|\varsigma_2|\right]$, if $|\varsigma_2| < |\kappa_2|$ and $|\varrho_2| < |\tau_2|$

(ii) $\tilde{\wp}.\tilde{\mathfrak{H}} = \left[\left(\frac{\varsigma_1+\kappa_1}{2}\right)\varsigma_2 - \left(\frac{\kappa_1-\varsigma_1}{2}\right)|\varsigma_2|, \left(\frac{\varrho_1+\tau_1}{2}\right)\varrho_2 - \left(\frac{\tau_1-\varrho_1}{2}\right)|\tau_2|, \left(\frac{\varrho_1+c_1}{2}\right)\tau_2 + \left(\frac{c_1-\varrho_1}{2}\right)|\tau_2|, \left(\frac{\varsigma_1+\kappa_1}{2}\right)\kappa_2 + \left(\frac{\kappa_1-\varsigma_1}{2}\right)|\varsigma_2|\right]$, if $|\varsigma_2| < |\kappa_2|$ and $|\varrho_2| > |\tau_2|$

(iii) $\tilde{\wp}.\tilde{\mathfrak{H}} = \left[\left(\frac{\varsigma_1+\kappa_1}{2}\right)\varsigma_2 - \left(\frac{\kappa_1-\varsigma_1}{2}\right)|\kappa_2|, \left(\frac{\varrho_1+\tau_1}{2}\right)\varrho_2 - \left(\frac{\tau_1-\varrho_1}{2}\right)|\varrho_2|, \left(\frac{\varrho_1+\tau_1}{2}\right)\tau_2 + \left(\frac{\tau_1-\varrho_1}{2}\right)|\varrho_2|, \left(\frac{\varsigma_1+\kappa_1}{2}\right)\kappa_2 + \left(\frac{\kappa_1-\varsigma_1}{2}\right)|\kappa_2|\right]$, if $|\varsigma_2| > |\kappa_2|$ and $|\varrho_2| < |\tau_2|$

(iv) $\tilde{\wp}.\tilde{\mathfrak{H}} = \left[\left(\frac{\varsigma_1+\kappa_1}{2}\right)\varsigma_2 - \left(\frac{\kappa_1-\varsigma_1}{2}\right)|\kappa_2|, \left(\frac{\varrho_1+\tau_1}{2}\right)\varrho_2 - \left(\frac{\tau_1-\varrho_1}{2}\right)|\tau_2|, \left(\frac{\varrho_1+c_1}{2}\right)\tau_2 + \left(\frac{c_1-\varrho_1}{2}\right)|\tau_2|, \left(\frac{\varsigma_1+\kappa_1}{2}\right)\kappa_2 + \left(\frac{\kappa_1-\varsigma_1}{2}\right)|\kappa_2|\right]$, if $|\varsigma_2| > |\kappa_2|$ and $|\varrho_2| > |\tau_2|$

7 Correct Total TrFTC

It can be easily verified that if the modified multiplication of TrFNs, defined in Section 4, is considered. Then, the total TrFTC of the TrFTP, represented by Table 1, corresponding to the FFS obtained by the following methods:

(i) Fuzzy north-west corner method is $[-411, 16, 251, 776]$
(ii) Fuzzy least cost method is $[-289, 42, 250.5, 709]$
(iii) Fuzzy Vogel's approximation method is $[-231, 35, 236.5, 606]$

8 Conclusion

This chapter has examined and corrected a fundamental limitation in the multiplication operation of TrFNs used in TrFTPs. The earlier formula proposed by Raj et al. [1] was restricted to non-negative TrFNs, resulting in computational inaccuracies for mixed-sign data. The generalized multiplication operation introduced here ensures validity across both positive and unrestricted TrFNs, maintaining their fuzzy properties.

The revised process utilized on a benchmark TrFTP validated the accuracy of total TrFTCs and de-fuzzified results of the process. This extension will advance the theoretical underpinnings of fuzzy arithmetic in general, and the methodology can be of great use in fuzzy optimization and decision-making problems.

Future work will be able to follow this methodology on forms of fuzzy arithmetic and multi-objective optimization problems with uncertainty structure that are more complex.

References

[1] Raj MEA, Sivaraman G, Vishnukumar P. A novel kind of arithmetic operations on trapezoidal fuzzy numbers and its applications to optimize the transportation cost. Int J Fuzzy Syst. 2023;25:1069–1076.

[2] Chen SJ, Chen SM. Fuzzy risk analysis based on the ranking of generalized trapezoidal fuzzy numbers. Appl Intell. 2007;26:1–11.

[3] Kaur A, Kumar A. A new method for solving fuzzy transportation problems using ranking function. Appl Math Model. 2011;35:5652 5661.

[4] Rezvani S, Molani M. Representation of trapezoidal fuzzy numbers with shape function. Ann Fuzzy Math Inform. 2014;8:89–112.

[5] Dubois D, Prade H. Fuzzy Sets and Systems: Theory and Applications. 1980. Academic Press.

Kirti

Modified Zero-Sum Matrix Game-Based to Solve MADM Problems Under Probabilistic Triangular Intuitionistic Hesitant Fuzzy Environment

Abstract: To the best of authors' knowledge, there exists only one zero-sum matrix game method to solve the problems of MADM with payoffs of probabilistic triangular intuitionistic hesitant fuzzy (PTIHF) environment. In this chapter, we first pointed out that this method is invalid because it assumed the ranking function – used to transform a PTIHF set into a real number – is linear over the used arithmetic operations of PTIHF sets. While, in actual case, the ranking function is not linear over the used arithmetic operations of the PTIHF sets. Then, for the validity of the existing method, new arithmetic operations of PTIHF sets are proposed. Furthermore, it is proved that the ranking function is linear over the proposed arithmetic operations. Finally, it is concluded that the existing method is valid only if the proposed arithmetic operations of PTIHF sets are considered instead of the existing arithmetic operations of PTIHF sets.

Keywords: fuzzy MADM, zero-sum matrix game, PTIHF set, ranking function.

1 Introduction

Yang and Xu [1], firstly, discussed the importance of multi-attribute decision-making (MADM) problems as well as the need of using fuzzy set [2] in MADM problems. Then, Yang and Xu [1] discussed a brief review of the existing game theory-based methods [3–6] to solve problems of MADM with fuzzy payoffs. Thereafter, Yang and Xu [1] discussed the importance of the intuitionistic fuzzy set [7], hesitant fuzzy set [8] and probabilistic sets [9]. Finally, using the triangular intuitionistic fuzzy number, the hesitant fuzzy set and the probabilistic set, Yang and Xu [1, Section 2, Definition 4] proposed the following definition of a probabilistic triangular intuitionistic hesitant fuzzy (PTIHF) set.

"A set $\tilde{A} = \left\| \begin{matrix} (\langle(\underline{a}_{11}, a_1, \bar{a}_{11}), (\underline{a}_{21}, a_1, \bar{a}_{21})\rangle, p_1); (\langle(\underline{a}_{12}, a_2, \bar{a}_{12}), (\underline{a}_{22}, a_2, \bar{a}_{22})\rangle, p_2); \ldots; \\ (\langle(\underline{a}_{1l}, a_l, \bar{a}_{1l}), (\underline{a}_{2l}, a_l, \bar{a}_{2l})\rangle, p_l); \ldots; (\langle(\underline{a}_{1L}, a_L, \bar{a}_{1L}), (\underline{a}_{2L}, a_L, \bar{a}_{2L})\rangle, p_L) \end{matrix} \right\|$ is said to be a PTIHF set,

Kirti, Department of Mathematics, National Institute of Technology, Kurukshetra 136119, Haryana, India

 | https://doi.org/10.1515/9783112222249-013

where

(i) $(\langle(\underline{a}_{1i}, a_i, \bar{a}_{1i});(\underline{a}_{2i}, a_i, \bar{a}_{2i})\rangle, p_i)$ represents the ith PTIHF element in which $\langle(\underline{a}_{1i}, a_i, \bar{a}_{1i});(\underline{a}_{2i}, a_i, \bar{a}_{2i})\rangle$ is a triangular intuitionistic fuzzy number such that $\underline{a}_{2i} \le \underline{a}_{1i} \le a_i \le \bar{a}_{1i} \le \bar{a}_{2i}$,

(ii) p_i represents the probability associated with the ith PTIHF element such that $0 \le p_i \le 1$ and $\sum_{i=1}^{L} p_i \le 1$, and

(iii) l represents the number of PTIHF elements in the PTIHF set $\tilde{A}$."

Yang and Xu [1] also proposed the following:

(i) The aggregation operator (1) [1, Section 4.2, eq. (31), p. 8] to aggregate n PTIHF sets

$$\tilde{A}_j = \left\| \begin{array}{l} \left(\left\langle\left(\underline{a}_{1j1}, a_{j1}, \bar{a}_{1j1}\right), \left(\underline{a}_{2j1}, a_{j1}, \bar{a}_{2j1}\right)\right\rangle, p_{j1}\right); \left(\left\langle\left(\underline{a}_{1j2}, a_{j2}, \bar{a}_{1j2}\right), \left(\underline{a}_{2j2}, a_{j2}, \bar{a}_{2j2}\right)\right\rangle, p_{j2}\right); \ldots; \\ \left(\left\langle\left(\underline{a}_{1jl}, a_{jl}, \bar{a}_{1jl}\right), \left(\underline{a}_{2jl}, a_{jl}, \bar{a}_{2jl}\right)\right\rangle, p_{jl}\right); \ldots; \left(\left\langle\left(\underline{a}_{1jL}, a_{jL}, \bar{a}_{1jL}\right), \left(\underline{a}_{2jL}, a_{jL}, \bar{a}_{2jL}\right)\right\rangle, p_{jL}\right) \end{array} \right\|,$$

$j = 1, 2, \ldots, n$

$$\sum_{j=1}^{n} \tilde{A}_j y_j = \left\| \begin{array}{l} \left(\left\langle \begin{array}{l} \left(\sum_{j=1}^{n} \underline{a}_{1j1}\, y_j, \sum_{j=1}^{n} a_{j1}\, y_j, \sum_{j=1}^{n} \bar{a}_{1j1}\, y_j\right), \\ \left(\sum_{j=1}^{n} \underline{a}_{2j1}\, y_j, \sum_{j=1}^{n} a_{j1}\, y_j, \sum_{j=1}^{n} \bar{a}_{2j1}\, y_j\right) \end{array} \right\rangle, \sum_{j=1}^{n} p_{j1}\, y_j \right); \\ \left(\left\langle \begin{array}{l} \left(\sum_{j=1}^{n} \underline{a}_{1j2}\, y_j, \sum_{j=1}^{n} a_{j2}\, y_j, \sum_{j=1}^{n} \bar{a}_{1j2}\, y_j\right), \\ \left(\sum_{j=1}^{n} \underline{a}_{2j2}\, y_j, \sum_{j=1}^{n} a_{j2}\, y_j, \sum_{j=1}^{n} \bar{a}_{2j2}\, y_j\right) \end{array} \right\rangle, \sum_{j=1}^{n} p_{j2}\, y_j \right); \ldots; \\ \left(\left\langle \begin{array}{l} \left(\sum_{j=1}^{n} \underline{a}_{1jl}\, y_j, \sum_{j=1}^{n} a_{jl}\, y_j, \sum_{j=1}^{n} \bar{a}_{1jl}\, y_j\right), \\ \left(\sum_{j=1}^{n} \underline{a}_{2jl}\, y_j, \sum_{j=1}^{n} a_{jl}\, y_j, \sum_{j=1}^{n} \bar{a}_{2jl}\, y_j\right) \end{array} \right\rangle, \sum_{j=1}^{n} p_{jl} y_j \right); \ldots; \\ \left(\left\langle \begin{array}{l} \left(\sum_{j=1}^{n} \underline{a}_{1jL}\, y_j, \sum_{j=1}^{n} a_{jL}\, y_j, \sum_{j=1}^{n} \bar{a}_{1jL}\, y_j\right), \\ \left(\sum_{j=1}^{n} \underline{a}_{2jL}\, y_j, \sum_{j=1}^{n} a_{jL}\, y_j, \sum_{j=1}^{n} \bar{a}_{2jL}\, y_j\right) \end{array} \right\rangle, \sum_{j=1}^{n} p_{jL}\, y_j \right) \end{array} \right\| \quad (1)$$

where $0 \le y_j \le 1$ and $\sum_{j=1}^{n} y_j = 1$.

(ii) The ranking function (eq. (2)) [1, Section 3.3, eq. (22), p. 6] to transform the PTIHF set $\tilde{A} = \left\| \begin{array}{l} (\langle(\underline{a}_{11}, a_1, \bar{a}_{11}), (\underline{a}_{21}, a_1, \bar{a}_{21})\rangle, p_1); (\langle(\underline{a}_{12}, a_2, \bar{a}_{12}), (\underline{a}_{22}, a_2, \bar{a}_{22})\rangle, p_2); \ldots; \\ (\langle(\underline{a}_{1l}, a_l, \bar{a}_{1l}), (\underline{a}_{2l}, a_l, \bar{a}_{2l})\rangle, p_l); \ldots; (\langle(\underline{a}_{1L}, a_L, \bar{a}_{1L}), (\underline{a}_{2L}, a_L, \bar{a}_{2L})\rangle, p_L) \end{array} \right\|$ into its equivalent real number $\eta_\lambda(\tilde{A})$.

$$\eta_\lambda(\tilde{A}) = V_\lambda(\tilde{A}) - W_\lambda(\tilde{A}),\ \lambda \in [0,1] \tag{2}$$

where

(i) $$V_\lambda(\tilde{A}) = \lambda V_\nu(\tilde{A}) + (1-\lambda)V_\mu(\tilde{A}) \tag{3}$$

(ii) $$W_\lambda(\tilde{A}) = \lambda W_\mu(\tilde{A}) + (1-\lambda)W_\nu(\tilde{A}) \tag{4}$$

(iii) $$V_\mu(\tilde{A}) = \frac{\sum_{i=1}^{L} p_i(\underline{a}_{1i} + 7a_i + \bar{a}_{1i})}{18}, \tag{5}$$

(iv) $$V_\nu(\tilde{A}) = \frac{\sum_{i=1}^{L} p_i(\underline{a}_{2i} + 7a_i + \bar{a}_{2i})}{18}, \tag{6}$$

(v) $$W_\mu(\tilde{A}) = \frac{\sum_{i=1}^{L} p_i(\bar{a}_{1i} - \underline{a}_{1i})}{6}, \tag{7}$$

(vi) $$W_\nu(\tilde{A}) = \frac{\sum_{i=1}^{L} p_i(\bar{a}_{2i} - \underline{a}_{2i})}{6}, \tag{8}$$

Thereafter, Yang and Xu [1] proved that the proposed ranking function (2) is linear, i.e.,

$$\eta_\lambda(\kappa\tilde{A}_1 + \tilde{A}_2) = \kappa\eta_\lambda(\tilde{A}_1) + \eta_\lambda(\tilde{A}_2) \tag{9}$$

Finally, using eq. (2), Yang and Xu [1] proposed a zero-sum matrix game based method to solve the problems of MADM with PTIHF payoffs.

In this chapter, it is shown that according to the arithmetic operations, considered by Yang and Xu [1] to obtain their proposed aggregation operator (eq. (1)), the linearity property is not satisfying for Yang and Xu's [1] ranking function (2) i.e., $\eta_\lambda(\kappa\tilde{A}_1 + \tilde{A}_2) \neq \kappa\eta_\lambda(\tilde{A}_1) + \eta_\lambda(\tilde{A}_2)$. Hence, according to Yang and Xu's [1] arithmetic operations of PTIHF sets, Yang and Xu's [1] zero-sum matrix game-based method is not valid. Also, we have proposed the new arithmetic operations of PTIHF sets for the validity of Yang and Xu's [1] method. Moreover, it is shown that according to the proposed arithmetic operations, the linearity property is satisfying for Yang and Xu's [1] ranking function (2). Hence, Yang and Xu's [1] zero-sum matrix game-based method is valid only if the proposed arithmetic operations of PTIHF sets are considered instead of Yang and Xu's [1] arithmetic operations of PTIHF sets.

2 Yang and Xu's Arithmetic Operations of PTIHF Sets

Yang and Xu [1] have not explicitly defined the arithmetic operations of PTIHF sets. However, Yang and Xu [1, Section 4.2, eq. (31), p. 8] have proposed the aggregation operator (eq. (1)) to aggregate PTIHF sets. Hence, the following arithmetic operations of PTIHF sets, used to obtain the aggregation operator (eq. (1)), may be considered as the arithmetic operations of PTIHF sets which are implicitly defined by Yang and Xu [1].

Let $\tilde{A}_1 = \left\| \begin{matrix} (\langle(\underline{a}_{111}, a_{11}, \bar{a}_{111}), (\underline{a}_{211}, a_{11}, \bar{a}_{211})\rangle, p_{11}); \\ (\langle(\underline{a}_{112}, a_{12}, \bar{a}_{112}), (\underline{a}_{212}, a_{12}, \bar{a}_{212})\rangle, p_{12}); \ldots; \\ (\langle(\underline{a}_{11L}, a_{1L}, \bar{a}_{11L}), (\underline{a}_{21L}, a_{1L}, \bar{a}_{21L})\rangle, p_{1L}) \end{matrix} \right\|$,

$\tilde{A}_2 = \left\| \begin{matrix} (\langle(\underline{a}_{121}, a_{21}, \bar{a}_{121}), (\underline{a}_{221}, a_{21}, \bar{a}_{221})\rangle, p_{21}); \\ (\langle(\underline{a}_{122}, a_{22}, \bar{a}_{122}), (\underline{a}_{222}, a_{22}, \bar{a}_{222})\rangle, p_{22}); \ldots; \\ (\langle(\underline{a}_{12L}, a_{2L}, \bar{a}_{12L}), (\underline{a}_{22L}, a_{2L}, \bar{a}_{22L})\rangle, p_{2L}) \end{matrix} \right\|$ be two PTIHF sets and $\lambda, \lambda_1, \lambda_2$ be real numbers. Then:

$$\text{(i)} \quad \tilde{A}_1 + \tilde{A}_2 = \left\| \begin{matrix} \left(\left\langle \begin{matrix} (\underline{a}_{111} + \underline{a}_{121}, a_{11} + a_{21}, \bar{a}_{111} + \bar{a}_{121}); \\ (\underline{a}_{211} + \underline{a}_{221}, a_{11} + a_{21}, \bar{a}_{211} + \bar{a}_{221}) \end{matrix} \right\rangle, p_{11} + p_{21} \right); \\ \left(\left\langle \begin{matrix} (\underline{a}_{112} + \underline{a}_{122}, a_{12} + a_{22}, \bar{a}_{112} + \bar{a}_{122}); \\ (\underline{a}_{212} + \underline{a}_{222}, a_{12} + a_{22}, \bar{a}_{212} + \bar{a}_{222}) \end{matrix} \right\rangle, p_{12} + p_{22} \right); \ldots; \\ \left(\left\langle \begin{matrix} (\underline{a}_{11L} + \underline{a}_{12L}, a_{1L} + a_{2L}, \bar{a}_{11L} + \bar{a}_{12L}); \\ (\underline{a}_{21L} + \underline{a}_{22L}, a_{1L} + a_{2L}, \bar{a}_{21L} + \bar{a}_{22L}) \end{matrix} \right\rangle, p_{1L} + p_{2L} \right) \end{matrix} \right\| \tag{10}$$

$$\text{(ii)} \quad \lambda\tilde{A}_1 = \left\| \begin{matrix} (\langle(\lambda\underline{a}_{111}, \lambda a_{11}, \lambda\bar{a}_{111}); (\lambda\underline{a}_{211}, \lambda a_{11}, \lambda\bar{a}_{211})\rangle, \lambda p_{11}); \\ (\langle(\lambda\underline{a}_{112}, \lambda a_{12}, \lambda\bar{a}_{112}); (\lambda\underline{a}_{212}, \lambda a_{12}, \lambda\bar{a}_{212})\rangle, \lambda p_{12}); \ldots; \\ (\langle(\lambda\underline{a}_{11L}, \lambda a_{1L}, \lambda\bar{a}_{11L}); (\lambda\underline{a}_{21L}, \lambda a_{1L}, \lambda\bar{a}_{21L})\rangle, \lambda p_{1L}) \end{matrix} \right\| \tag{11}$$

$$\text{(iii)} \quad \lambda_1\tilde{A}_1 + \lambda_2\tilde{A}_2 = \left\| \begin{matrix} \left(\left\langle \begin{matrix} (\lambda_1\underline{a}_{111} + \lambda_2\underline{a}_{121}, \lambda_1 a_{11} + \lambda_2 a_{21}, \lambda_1\bar{a}_{111} + \lambda_2\bar{a}_{121}), \\ (\lambda_1\underline{a}_{211} + \lambda_2\underline{a}_{221}, \lambda_1 a_{11} + \lambda_2 a_{21}, \lambda_1\bar{a}_{211} + \lambda_2\bar{a}_{221}) \end{matrix} \right\rangle, \lambda_1 p_{11} + \lambda_2 p_{21} \right); \\ \left(\left\langle \begin{matrix} (\lambda_1\underline{a}_{112} + \lambda_2\underline{a}_{122}, \lambda_1 a_{12} + \lambda_2 a_{22}, \lambda_1\bar{a}_{112} + \lambda_2\bar{a}_{122}), \\ (\lambda_1\underline{a}_{212} + \lambda_2\underline{a}_{222}, \lambda_1 a_{12} + \lambda_2 a_{22}, \lambda_1\bar{a}_{212} + \lambda_2\bar{a}_{222}) \end{matrix} \right\rangle, \lambda_1 p_{12} + \lambda_2 p_{22} \right); \ldots; \\ \left(\left\langle \begin{matrix} (\lambda_1\underline{a}_{11L} + \lambda_2\underline{a}_{12L}, \lambda_1 a_{1L} + \lambda_2 a_{2L}, \lambda_1\bar{a}_{11L} + \lambda_2\bar{a}_{12L}), \\ (\lambda_1\underline{a}_{21L} + \lambda_2\underline{a}_{22L}, \lambda_1 a_{1L} + \lambda_2 a_{2L}, \lambda_1\bar{a}_{21L} + \lambda_2\bar{a}_{22L}) \end{matrix} \right\rangle, \lambda_1 p_{1L} + \lambda_2 p_{2L} \right) \end{matrix} \right\| \tag{12}$$

3 Existing Results

Yang and Xu [1] stated and proved some results. Also, using the proposed results, Yang and Xu [1] proposed a method (zero-sum matrix game based method) to solve MADM problems under PTIHF environment. In this section, Yang and Xu's [1] results are discussed.

Yang and Xu [1, Section 3.2, Theorem 1, p. 6] stated and proved the following result.

If $\tilde{A}_1$ and $\tilde{A}_2$ are two PTIHF sets and κ is a real number. Then, the seven existing results of eqs. (13)– (19) are as follows:

$$V_\mu(\kappa\tilde{A}_1 + \tilde{A}_2) = \kappa V_\mu(\tilde{A}_1) + V_\mu(\tilde{A}_2) \tag{13}$$

$$V_\nu(\kappa\tilde{A}_1 + \tilde{A}_2) = \kappa V_\nu(\tilde{A}_1) + V_\nu(\tilde{A}_2) \tag{14}$$

$$W_\mu(\kappa\tilde{A}_1 + \tilde{A}_2) = \kappa W_\mu(\tilde{A}_1) + W_\mu(\tilde{A}_2) \tag{15}$$

$$W_\nu(\kappa\tilde{A}_1 + \tilde{A}_2) = \kappa W_\nu(\tilde{A}_1) + W_\nu(\tilde{A}_2) \tag{16}$$

$$V_\lambda(\kappa\tilde{A}_1 + \tilde{A}_2) = \kappa V_\lambda(\tilde{A}_1) + V_\lambda(\tilde{A}_2) \tag{17}$$

$$W_\lambda\left(\kappa\tilde{A}_1+\tilde{A}_2\right)=\kappa W_\lambda\left(\tilde{A}_1\right)+W_\lambda\left(\tilde{A}_2\right) \tag{18}$$

$$\eta_\lambda\left(\kappa\tilde{A}+\tilde{A}'\right)=\kappa\eta_\lambda\left(\tilde{A}\right)+\eta_\lambda\left(\tilde{A}'\right) \tag{19}$$

4 Invalidity of Existing Results

In the section, two PTIHF sets $\tilde{A}_1=\|(\langle(2,3,4),\ (1,3,5)\rangle,\ 0.8);(\langle(3,4,5),\ (2,4,6)\rangle,\ 0.2)\|$ and $\tilde{A}_2=\|(\langle(3,4,5),\ (2,4,6)\rangle,\ 0.7);(\langle(4,5,6),\ (3,5,7)\rangle,\ 0.3)\|$ are considered to show that Yang and Xu's [1] results, discussed in Section 3, are not valid.

4.1 Invalidity of the First Existing Result

According to Yang and Xu's [1] arithmetic operations, discussed in Section 2:

$$\kappa\tilde{A}_1+\tilde{A}_2=\left\|\begin{array}{l}(\langle(2\kappa+3,3\kappa+4,4\kappa+5),(1\kappa+2,3\kappa+4,5\kappa+6)\rangle,0.8\kappa+0.7);\\(\langle(3\kappa+4,4\kappa+5,5\kappa+6),(2\kappa+3,4\kappa+5,6\kappa+7)\rangle,0.2\kappa+0.3)\end{array}\right\|$$

Using expression (5), we get:

$$\begin{aligned}
&V_\mu\left(\kappa\tilde{A}_1+\tilde{A}_2\right)\\
&=\left(\frac{(0.8\kappa+0.7)(2\kappa+3+21\kappa+28+4\kappa+5)+(0.2\kappa+0.3)(3\kappa+4+28\kappa+35+5\kappa+6)}{18}\right)\\
&=\left(\frac{(0.8\kappa+0.7)(27\kappa+36)+(0.2\kappa+0.3)(36\kappa+45)}{18}\right)\\
&=\left(\frac{21.6\kappa^2+18.9\kappa+28.8\kappa+25.2+7.2\kappa^2+10.8\kappa+9\kappa+13.5}{18}\right)\\
&=\left(\frac{28.8\kappa^2+67.5\kappa+38.7}{18}\right)
\end{aligned} \tag{20}$$

$$V_\mu\left(\tilde{A}_1\right)=\frac{(0.8(2+21+4)+0.2(3+28+5))}{18}=\frac{21.6+7.2}{18}=1.6 \tag{21}$$

$$V_\mu\left(\tilde{A}_2\right)=\frac{(0.7(3+28+5)+0.3(4+35+6))}{18}=\frac{25.2+13.5}{18}=2.15 \tag{22}$$

Using eqs. (21) and (22), we get:

$$\kappa V_\mu\left(\tilde{A}_1\right)+V_\mu\left(\tilde{A}_2\right)=1.6\kappa+2.15 \tag{23}$$

It is obvious from eqs. (20) and (23) that $V_\mu\left(\kappa\tilde{A}_1+\tilde{A}_2\right)\neq\kappa V_\mu\left(\tilde{A}_1\right)+V_\mu\left(\tilde{A}_2\right)$.

4.2 Invalidity of the Second Existing Result

According to Yang and Xu's [1] arithmetic operations, discussed in Section 2:

$$\kappa\tilde{A}_1+\tilde{A}_2=\left\|\begin{matrix}(\langle(2\kappa+3,3\kappa+4,4\kappa+5),(1\kappa+2,3\kappa+4,5\kappa+6)\rangle,0.8\kappa+0.7);\\(\langle(3\kappa+4,4\kappa+5,5\kappa+6),(2\kappa+3,4\kappa+5,6\kappa+7)\rangle,0.2\kappa+0.3)\end{matrix}\right\|$$

Using expression (6), we get:

$$\begin{aligned}&V_\nu(\kappa\tilde{A}_1+\tilde{A}_2)\\&=\left(\frac{(0.8\kappa+0.7)(\kappa+2+21\kappa+28+5\kappa+6)+(0.2\kappa+0.3)(2\kappa+3+28\kappa+35+6\kappa+7)}{18}\right)\\&=\left(\frac{(0.8\kappa+0.7)(27\kappa+36)+(0.2\kappa+0.3)(36\kappa+45)}{18}\right)\\&=\left(\frac{21.6\kappa^2+18.9\kappa+28.8\kappa+25.2+7.2\kappa^2+10.8\kappa+9\kappa+13.5}{18}\right)\\&=\left(\frac{28.8\kappa^2+67.5\kappa+38.7}{18}\right)\end{aligned}\tag{24}$$

$$V_\nu(\tilde{A}_1)=\frac{(0.8(1+21+5)+0.2(2+28+6))}{18}=\frac{21.6+7.2}{18}=1.6\tag{25}$$

$$V_\nu(\tilde{A}_2)=\frac{(0.7(2+28+6)+0.3(3+35+7))}{18}=\frac{25.2+13.5}{18}=2.15\tag{26}$$

Using eqs. (25) and (26), we get:

$$\kappa V_\nu(\tilde{A}_1)+V_\nu(\tilde{A}_2)=1.6\kappa+2.15\tag{27}$$

It is obvious from eqs. (24) and (27) that $V_\nu(\kappa\tilde{A}_1+\tilde{A}_2)\neq\kappa V_\nu(\tilde{A}_1)+V_\nu(\tilde{A}_2)$.

4.3 Invalidity of the Third Existing Result

According to Yang and Xu's [1] arithmetic operations, discussed in Section 2:

$$\kappa\tilde{A}_1+\tilde{A}_2=\left\|\begin{matrix}(\langle(2\kappa+3,3\kappa+4,4\kappa+5),(1\kappa+2,3\kappa+4,5\kappa+6)\rangle,0.8\kappa+0.7);\\(\langle(3\kappa+4,4\kappa+5,5\kappa+6),(2\kappa+3,4\kappa+5,6\kappa+7)\rangle,0.2\kappa+0.3)\end{matrix}\right\|$$

Using expression (7), we get:

$$W_\mu(\kappa\tilde{A}_1+\tilde{A}_2)=\left(\frac{(0.8\kappa+0.7)(4\kappa+5-2\kappa-3)+(0.2\kappa+0.3)(5\kappa+6-3\kappa-4)}{6}\right)$$

$$=\left(\frac{(0.8\kappa+0.7)(2\kappa+2)+(0.2\kappa+0.3)(2\kappa+2)}{6}\right)$$

$$=\left(\frac{(2\kappa+2)(0.8\kappa+0.7+0.2\kappa+0.3)}{6}\right)$$

$$=\left(\frac{(2\kappa+2)(\kappa+1)}{6}\right)=\left(\frac{2\kappa^2+4\kappa+2}{6}\right) \tag{28}$$

$$W_\mu(\tilde{A}_1)=\frac{(0.8(4-2)+0.2(5-3))}{6}=0.3333 \tag{29}$$

$$W_\mu(\tilde{A}_2)=\frac{(0.7(5-3)+0.3(6-4))}{6}=0.3333 \tag{30}$$

Using eqs. (29) and (30), we get:

$$\kappa W_\mu(\tilde{A}_1)+W_\mu(\tilde{A}_2)=0.3333(\kappa+1) \tag{31}$$

It is obvious from eqs. (28) and (31) that $W_\mu(\kappa\tilde{A}_1+\tilde{A}_2)\neq\kappa W_\mu(\tilde{A}_1)+W_\mu(\tilde{A}_2)$.

4.4 Invalidity of the Fourth Existing Result

According to Yang and Xu's [1] arithmetic operations, discussed in Section 2:

$$\kappa\tilde{A}_1+\tilde{A}_2=\left\|\begin{matrix}(\langle(2\kappa+3,3\kappa+4,4\kappa+5),(1\kappa+2,3\kappa+4,5\kappa+6)\rangle,0.8\kappa+0.7);\\(\langle(3\kappa+4,4\kappa+5,5\kappa+6),(2\kappa+3,4\kappa+5,6\kappa+7)\rangle,0.2\kappa+0.3)\end{matrix}\right\|$$

Using expression (8), we get:

$$W_\nu(\kappa\tilde{A}_1+\tilde{A}_2)=\left(\frac{(0.8\kappa+0.7)(5\kappa+6-\kappa-2)+(0.2\kappa+0.3)(6\kappa+7-2\kappa-3)}{6}\right)$$

$$=\left(\frac{(0.8\kappa+0.7)(4\kappa+4)+(0.2\kappa+0.3)(4\kappa+4)}{6}\right)$$

$$=\left(\frac{(4\kappa+4)(0.8\kappa+0.7+0.2\kappa+0.3)}{6}\right)$$

$$=\left(\frac{(4\kappa+4)(\kappa+1)}{6}\right)=\left(\frac{4\kappa^2+8\kappa+4}{6}\right) \tag{32}$$

$$W_\nu(\tilde{A}_1) = \frac{(0.8(5-1)+0.2(6-2))}{6} = 0.6666 \tag{33}$$

$$W_\nu(\tilde{A}_2) = \frac{(0.7(6-2)+0.3(7-3))}{6} = 0.6666 \tag{34}$$

Using eqs. (33) and (34), we get:

$$\kappa W_\nu(\tilde{A}_1) + W_\nu(\tilde{A}_2) = 0.6666(\kappa+1) \tag{35}$$

It is obvious from eqs. (32) and (35) that $W_\nu(\kappa\tilde{A}_1+\tilde{A}_2) \neq \kappa W_\nu(\tilde{A}_1) + W_\nu(\tilde{A}_2)$.

4.5 Invalidity of the Fifth Existing Result

According to Yang and Xu's [1] arithmetic operations, discussed in Section 2:

$$\kappa\tilde{A}_1+\tilde{A}_2 = \left\| \begin{matrix} (\langle(2\kappa+3,3\kappa+4,4\kappa+5),(1\kappa+2,3\kappa+4,5\kappa+6)\rangle, 0.8\kappa+0.7); \\ (\langle(3\kappa+4,4\kappa+5,5\kappa+6),(2\kappa+3,4\kappa+5,6\kappa+7)\rangle, 0.2\kappa+0.3) \end{matrix} \right\|$$

Using expression (3), we get:

$$\begin{aligned} V_\lambda(\kappa\tilde{A}_1+\tilde{A}_2) &= \lambda\left(\frac{28.8\kappa^2+67.5\kappa+38.7}{18}\right) + (1-\lambda)\left(\frac{28.8\kappa^2+67.5\kappa+38.7}{18}\right) \\ &= \left(\frac{28.8\kappa^2+67.5\kappa+38.7}{18}\right) + \lambda\left(\frac{28.8\kappa^2+67.5\kappa+38.7-28.8\kappa^2-67.5\kappa-38.7}{18}\right) \\ &= \left(\frac{28.8\kappa^2+67.5\kappa+38.7}{18}\right) \end{aligned} \tag{36}$$

$$V_\lambda(\tilde{A}_1) = 1.6\lambda + (1-\lambda)1.6 = 1.6 \tag{37}$$

$$V_\lambda(\tilde{A}_2) = 2.15\lambda + (1-\lambda)2.15 = 2.15 \tag{38}$$

Using eqs. (37) and (38), we get:

$$\kappa V_\lambda(\tilde{A}_1) + V_\lambda(\tilde{A}_2) = 1.6\kappa + 2.15 \tag{39}$$

It is obvious from eqs. (36) and (39) that $V_\lambda(\kappa\tilde{A}_1+\tilde{A}_2) \neq \kappa V_\lambda(\tilde{A}_1) + V_\lambda(\tilde{A}_2)$.

4.6 Invalidity of the Sixth Existing Result

According to Yang and Xu's [1] arithmetic operations, discussed in Section 2:

$$\kappa\tilde{A}_1+\tilde{A}_2=\left\|\begin{matrix}((2\kappa+3,3\kappa+4,4\kappa+5),(1\kappa+2,3\kappa+4,5\kappa+6),0.8\kappa+0.7);\\((3\kappa+4,4\kappa+5,5\kappa+6),(2\kappa+3,4\kappa+5,6\kappa+7),0.2\kappa+0.3)\end{matrix}\right\|$$

Using expression (4), we get:

$$\begin{aligned}W_\lambda\left(\kappa\tilde{A}_1+\tilde{A}_2\right)&=\lambda\left(\frac{2\kappa^2+4\kappa+2}{6}\right)+(1-\lambda)\left(\frac{4\kappa^2+8\kappa+4}{6}\right)\\&=\left(\frac{4\kappa^2+8\kappa+4}{6}\right)+\lambda\left(\frac{2\kappa^2+4\kappa+2-4\kappa^2-8\kappa-4}{6}\right)\\&=\left(\frac{4\kappa^2+8\kappa+4}{6}\right)-\lambda\left(\frac{2\kappa^2+4\kappa+2}{6}\right)\end{aligned}\tag{40}$$

$$W_\lambda\left(\tilde{A}_1\right)=0.3333\lambda+(1-\lambda)0.6666=0.6666-0.3333\lambda\tag{41}$$

$$W_\lambda\left(\tilde{A}_2\right)=0.3333\lambda+(1-\lambda)0.6666=0.6666-0.3333\lambda\tag{42}$$

Using eqs. (41) and (42), we get:

$$\kappa W_\lambda\left(\tilde{A}_1\right)+W_\lambda\left(\tilde{A}_2\right)=(0.6666-0.3333\lambda)(\kappa+1)\tag{43}$$

It is obvious from eqs. (40) and (43) that $W_\lambda\left(\kappa\tilde{A}_1+\tilde{A}_2\right)\neq\kappa W_\lambda\left(\tilde{A}_1\right)+W_\lambda\left(\tilde{A}_2\right)$.

4.7 Invalidity of the Seventh Result

According to Yang and Xu's [1] arithmetic operations, discussed in Section 2:

$$\kappa\tilde{A}_1+\tilde{A}_2=\left\|\begin{matrix}(\langle(2\kappa+3,3\kappa+4,4\kappa+5),(1\kappa+2,3\kappa+4,5\kappa+6)\rangle,0.8\kappa+0.7);\\(\langle(3\kappa+4,4\kappa+5,5\kappa+6),(2\kappa+3,4\kappa+5,6\kappa+7)\rangle,0.2\kappa+0.3)\end{matrix}\right\|$$

Using expression (2), we get:

$$\begin{aligned}\eta_\lambda\left(\kappa\tilde{A}_1+\tilde{A}_2\right)&=\left(\frac{28.8\kappa^2+67.5\kappa+38.7}{18}\right)-\left(\frac{4\kappa^2+8\kappa+4}{6}\right)+\lambda\left(\frac{2\kappa^2+4\kappa+2}{6}\right)\\&=\left(\frac{16.8\kappa^2+43.5\kappa+26.7}{18}\right)+\lambda\left(\frac{2\kappa^2+4\kappa+2}{6}\right)\end{aligned}\tag{44}$$

$$\eta_\lambda\left(\tilde{A}_1\right)=V_\lambda\left(\tilde{A}_1\right)-W_\lambda\left(\tilde{A}_1\right)=1.6-0.6666+0.3333\lambda=0.9334+0.3333\lambda\tag{45}$$

$$\eta_\lambda\left(\tilde{A}_2\right)=V_\lambda\left(\tilde{A}_2\right)-W_\lambda\left(\tilde{A}_2\right)=2.15-0.6666+0.3333\lambda=1.4834+0.3333\lambda\tag{46}$$

Using eqs. (45) and (46), we get:

$$\kappa\eta_{\lambda}(\tilde{A}_1)+\eta_{\lambda}(\tilde{A}_2)=0.9334\kappa+1.4834+0.3333\lambda(\kappa+1) \tag{47}$$

It is obvious from eqs. (44) and (47) that $\eta_{\lambda}(\kappa\tilde{A}_1+\tilde{A}_2)\neq\kappa\eta_{\lambda}(\tilde{A}_1)+\eta_{\lambda}(\tilde{A}_2)$.

5 Invalidity of Yang and Xu's Method

Using the results (13)–(19), Yang and Xu [1] proposed a zero-sum matrix game-based method to solve such MADM problem in which the ranking value of each alternative over each attribute is represented by a PTIHF set. However, as discussed in Section 3, the results from eqs. (13) to (19) are not valid according to Yang and Xu's [1] arithmetic operations of PTIHF sets. Hence, Yang and Xu's [1] zero-sum matrix game-based method is also not valid according to Yang and Xu's [1] arithmetic operations of PTIHF sets.

6 Invalidity of Yang and Xu's Arithmetic Operations of PTIHF Sets

The following clearly indicates that Yang and Xu's [1] arithmetic operations of PTIHF sets, discussed in Section 2, are not valid.

Let $\tilde{A}_1=\|(\langle(2,3,4),(1,3,5)\rangle,0.8);(\langle(3,4,5),(2,4,6)\rangle,0.2)\|$ and $\tilde{A}_2=\|(\langle(3,4,5),(2,4,6)\rangle,0.7);(\langle(4,5,6),(3,5,7)\rangle,0.3)\|$ be two PTIHF sets. Then, according to Yang and Xu's [1] arithmetic operations of PTIHF sets, discussed in Section 2:

(i) $\tilde{A}_1+\tilde{A}_2=\|(\langle(5,7,9),(3,7,11)\rangle,1.5);(\langle(7,9,11),(5,9,13)\rangle,0.5)\|$. Since, $p_1=1.5, p_2=0.5$ and hence $p_1+p_2=2$. So, $\tilde{A}_1+\tilde{A}_2$ is not a PTIHFs set as none of the necessary conditions $0\le p_i\le 1$ and $\sum_{i=1}^{L}p_i\le 1$ is satisfying.

(ii) $100\tilde{A}_1=\|(\langle(200,300,400),(100,300,500)\rangle,80);(\langle(300,400,500),(200,400,600)\rangle,20)\|$. Since, $p_1=80,\ p_2=20$ and hence, $p_1+p_2=100$. So, $100\tilde{A}_1$ is not a PTIHFs set as none of the necessary conditions $0\le p_i\le 1$ and $\sum_{i=1}^{L}p_i\le 1$ is satisfying.

(iii) Since, the cardinality of $\tilde{A}_1$ and $\tilde{A}_2$ is 2. So, the cardinality of $\tilde{A}_1+\tilde{A}_2$ should be 4. While, it is obvious from (i) that the cardinality of $\tilde{A}_1+\tilde{A}_2$ is 2.

7 Proposed Arithmetic Operations of PTHIF Sets

In this section, the following arithmetic operations of PTIHF sets are proposed.

Let $\tilde{A}_1 = \mathrm{U}_{j=1,2,\ldots,L_1}, \{(\langle(\underline{a}_{11j}, a_{1j}, \bar{a}_{11j}); (\underline{a}_{21j}, a_{1j}, \bar{a}_{21j})\rangle, p_{1j})\}$, and $\tilde{A}_2 = \mathrm{U}_{k=1,2,\ldots,L_2}, \{(\langle(\underline{a}_{12k}, a_{2k}, \bar{a}_{12k}); (\underline{a}_{22k}, a_{2k}, \bar{a}_{22k})\rangle, p_{2k})\}$ be two PTIHF sets and $\lambda, \lambda_1, \lambda_2$ be real numbers. Then:

(i) $\tilde{A}_1 + \tilde{A}_2 = \mathrm{U}_{j=1,2,\ldots,L_1, k=1,2,\ldots,L_2}$

$$\left\{\left(\left\langle\left(\underline{a}_{11j}+\underline{a}_{12k}, a_{1j}+a_{2k}, \bar{a}_{11j}+\bar{a}_{12k}\right);\left(\underline{a}_{21j}+\underline{a}_{22k}, a_{1j}+a_{2k}, \bar{a}_{21j}+\bar{a}_{22k}\right)\right\rangle, p_{1j}p_{2k}\right)\right\}$$

$$\lambda\tilde{A}_1 = \begin{cases} \mathrm{U}_{j=1,2,\ldots,L_1}, \left\{\left(\left\langle\left(\lambda\underline{a}_{11j}, \lambda a_{1j}, \lambda\bar{a}_{11j}\right);\left(\lambda\underline{a}_{21j}, \lambda a_{1j}, \lambda\bar{a}_{21j}\right)\right\rangle, p_{1j}\right)\right\}, & \lambda \geq 0 \\ \mathrm{U}_{j=1,2,\ldots,L_1}, \left\{\left(\left\langle\left(\lambda\bar{a}_{11j}, \lambda a_{1j}, \lambda\underline{a}_{11j}\right);\left(\lambda\bar{a}_{21j}, \lambda a_{1j}, \lambda\underline{a}_{21j}\right)\right\rangle, p_{1j}\right)\right\}, & \lambda < 0 \end{cases}$$

8 Validity of the Proposed Arithmetic Operations

It is proved in this section that the proposed arithmetic operations are valid.

To prove that the first arithmetic operation is valid, there is need to prove that $0 \leq p_{1j}p_{2k} \leq 1$ and $\sum_{j=1}^{L_1}\sum_{k=1}^{L_2} p_{1j}p_{2k} = 1$

Since $0 \leq p_{1j} \leq 1$ and $0 \leq p_{2k} \leq 1$ so, $0 \leq p_{1j}p_{2k} \leq 1$.

Since $\sum_{j=1}^{L_1} p_{1j} = 1$ and $\sum_{k=1}^{L_2} p_{2k} = 1$ so, $\sum_{j=1}^{L_1}\sum_{k=1}^{L_2} p_{1j}p_{2k} = 1$.

It is pertinent to mention that as in the second arithmetic operation p_{1j} is independent from λ so the condition $0 \leq p_{1j} \leq 1$ and $\sum_{j=1}^{L_1} p_{1j} = 1$ will be automatically satisfied.

9 Validity of the Existing Results

In Section 3, two PTIHF sets are considered to show that if $\tilde{A}_1$ and $\tilde{A}_2$ are two PTIHF sets and κ is a real number. Then, according to Yang and Xu's [1] arithmetic operations of PTIHF sets, discussed in Section 2:

$$V_\mu(\kappa\tilde{A}_1 + \tilde{A}_2) \neq \kappa V_\mu(\tilde{A}_1) + V_\mu(\tilde{A}_2)$$

$$V_\nu(\kappa\tilde{A}_1 + \tilde{A}_2) \neq \kappa V_\nu(\tilde{A}_1) + V_\nu(\tilde{A}_2)$$

$$W_\mu(\kappa\tilde{A}_1 + \tilde{A}_2) \neq \kappa W_\mu(\tilde{A}_1) + W_\mu(\tilde{A}_2)$$

$$W_\nu(\kappa\tilde{A}_1 + \tilde{A}_2) \neq \kappa W_\nu(\tilde{A}_1) + W_\nu(\tilde{A}_2)$$

$$V_\lambda(\kappa\tilde{A}_1 + \tilde{A}_2) \neq \kappa V_\lambda(\tilde{A}_1) + V_\lambda(\tilde{A}_2)$$

$$W_\lambda(\kappa\tilde{A}_1+\tilde{A}_2)\neq\kappa W_\lambda(\tilde{A}_1)+W_\lambda(\tilde{A}_2)$$

$$\eta_\lambda(\kappa\tilde{A}_1+\tilde{A}_2)\neq\kappa\eta_\lambda(\tilde{A}_1)+\eta_\lambda(\tilde{A}_2)$$

Here, it is proved that if the proposed arithmetic operations are considered instead of Yang and Xu's [1] arithmetic operations. Then, all results stated below will be valid:

$$V_\mu(\kappa\tilde{A}_1+\tilde{A}_2)=\kappa V_\mu(\tilde{A}_1)+V_\mu(\tilde{A}_2)$$

$$V_\nu(\kappa\tilde{A}_1+\tilde{A}_2)=\kappa V_\nu(\tilde{A}_1)+V_\nu(\tilde{A}_2)$$

$$W_\mu(\kappa\tilde{A}_1+\tilde{A}_2)=\kappa W_\mu(\tilde{A}_1)+W_\mu(\tilde{A}_2)$$

$$W_\nu(\kappa\tilde{A}_1+\tilde{A}_2)=\kappa W_\nu(\tilde{A}_1)+W_\nu(\tilde{A}_2)$$

$$V_\lambda(\kappa\tilde{A}_1+\tilde{A}_2)=\kappa V_\lambda(\tilde{A}_1)+V_\lambda(\tilde{A}_2)$$

$$W_\lambda(\kappa\tilde{A}_1+\tilde{A}_2)=\kappa W_\lambda(\tilde{A}_1)+W_\lambda(\tilde{A}_2)$$

$$\eta_\lambda(\kappa\tilde{A}_1+\tilde{A}_2)=\kappa\eta_\lambda(\tilde{A}_1)+\eta_\lambda(\tilde{A}_2).$$

9.1 Validity of the First Existing Result

$$V_\mu(\tilde{A}_1+\tilde{A}_2)=V_\mu\Big(\cup_{j=1,2,\ldots,L_1,k=1,2,\ldots,L_2}\Big\{\Big(\Big\langle\Big(\underline{a}_{11j}+\underline{a}_{12k},a_{1j}+a_{2k},\bar{a}_{11j}+\bar{a}_{12k}\Big);\Big(\underline{a}_{21j}+\underline{a}_{22k},a_{1j}+a_{2k},\bar{a}_{21j}+\bar{a}_{22k}\Big)\Big\rangle,p_{1j}p_{2k}\Big)\Big\}\Big)$$

$$=\frac{\sum_{j=1}^{L_1}\sum_{k=1}^{L_2}p_{1j}p_{2k}\left(\underline{a}_{11j}+\underline{a}_{12k}+7\left(a_{1j}+a_{2k}\right)+\bar{a}_{11j}+\bar{a}_{12k}\right)}{18}$$

$$=\frac{\sum_{j=1}^{L_1}\sum_{k=1}^{L_2}p_{1j}p_{2k}\left(\underline{a}_{11j}+7a_{1j}+\bar{a}_{11j}\right)}{18}+\frac{\sum_{j=1}^{L_1}\sum_{k=1}^{L_2}p_{1j}p_{2k}(\underline{a}_{12k}+7a_{2k}+\bar{a}_{12k})}{18}$$

$$=\frac{\sum_{j=1}^{L_1}p_{1j}\left(\underline{a}_{11j}+7a_{1j}+\bar{a}_{11j}\right)\sum_{k=1}^{L_2}p_{2k}}{18}+\frac{\sum_{k=1}^{L_2}p_{2k}(\underline{a}_{12k}+7a_{2k}+\bar{a}_{12k})\sum_{j=1}^{L_1}p_{1j}}{18}\cdot$$

$$=\frac{\sum_{j=1}^{L1}p_{1j}\left(\underline{a}_{11j}+7a_{1j}+\bar{a}_{11j}\right)}{18}+\frac{\sum_{k=1}^{L2}p_{2k}(\underline{a}_{12k}+7a_{2k}+\bar{a}_{12k})}{18}$$

$$=V_\mu(\tilde{A}_1)+V_\mu(\tilde{A}_2)$$

9.2 Validity of the Second Existing Result

$$V_\nu\left(\tilde{A}_1+\tilde{A}_2\right)=V_\nu\left(\mathrm{U}_{j=1,2,..,L_1,\,k=1,2,..,L_2}\left\{\left(\left\langle\left(\underline{a}_{11j}+\underline{a}_{12k},a_{1j}+a_{2k},\bar{a}_{11j}+\bar{a}_{12k}\right);\left(\underline{a}_{21j}+\underline{a}_{22k},a_{1j}+a_{2k},\bar{a}_{21j}+\bar{a}_{22k}\right)\right\rangle,p_{1j}p_{2k}\right)\right\}\right)$$

$$=\frac{\sum_{j=1}^{L_1}\sum_{k=1}^{L_2}p_{1j}p_{2k}\left(\underline{a}_{21j}+\underline{a}_{22k}+7\left(a_{1j}+a_{2k}\right)+\bar{a}_{21j}+\bar{a}_{22k}\right)}{18}$$

$$=\frac{\sum_{j=1}^{L_1}\sum_{k=1}^{L_2}p_{1j}p_{2k}\left(\underline{a}_{21j}+7a_{1j}+\bar{a}_{21j}\right)}{18}+\frac{\sum_{j=1}^{L_1}\sum_{k=1}^{L_2}p_{1j}p_{2k}(\underline{a}_{22k}+7a_{2k}+\bar{a}_{22k})}{18}$$

$$=\frac{\sum_{j=1}^{L_1}p_{1j}\left(\underline{a}_{21j}+7a_{1j}+\bar{a}_{21j}\right)\sum_{k=1}^{L_2}p_{2k}}{18}+\frac{\sum_{k=1}^{L_2}p_{2k}(\underline{a}_{22k}+7a_{2k}+\bar{a}_{22k})\sum_{j=1}^{L_1}p_{1j}}{18}$$

$$=\frac{\sum_{j=1}^{L1}p_{1j}\left(\underline{a}_{21j}+7a_{1j}+\bar{a}_{21j}\right)}{18}+\frac{\sum_{k=1}^{L2}p_{2k}(\underline{a}_{22k}+7a_{2k}+\bar{a}_{22k})}{18}$$

$$=V_\nu\left(\tilde{A}_1\right)+V_\nu\left(\tilde{A}_2\right)$$

9.3 Validity of the Third Existing Result

$$W_\mu\left(\tilde{A}_1+\tilde{A}_2\right)=W_\mu\left(\mathrm{U}_{j=1,2,..,L_1,\,k=1,2,..,L_2}\left\{\left(\left\langle\left(\underline{a}_{11j}+\underline{a}_{12k},a_{1j}+a_{2k},\bar{a}_{11j}+\bar{a}_{12k}\right);\left(\underline{a}_{21j}+\underline{a}_{22k},a_{1j}+a_{2k},\bar{a}_{21j}+\bar{a}_{22k}\right)\right\rangle,p_{1j}p_{2k}\right)\right\}\right)$$

$$=\frac{\sum_{j=1}^{L_1}\sum_{k=1}^{L_2}p_{1j}p_{2k}\left(\left(\bar{a}_{11j}+\bar{a}_{12k}\right)-\left(\underline{a}_{11j}+\underline{a}_{12k}\right)\right)}{6}$$

$$=\frac{\sum_{j=1}^{L_1}\sum_{k=1}^{L_2}p_{1j}p_{2k}\left(\bar{a}_{11j}-\underline{a}_{11j}\right)}{6}+\frac{\sum_{j=1}^{L_1}\sum_{k=1}^{L_2}p_{1j}p_{2k}(\bar{a}_{12k}-\underline{a}_{12k})}{6}$$

$$=\frac{\sum_{j=1}^{L_1}p_{1j}\left(\bar{a}_{11j}-\underline{a}_{11j}\right)\sum_{k=1}^{L_2}p_{2k}}{6}+\frac{\sum_{k=1}^{L_2}p_{1j}(\bar{a}_{12k}-\underline{a}_{12k})\sum_{l=1}^{L_1}p_{1j}}{6}$$

$$=\frac{\sum_{j=1}^{L_1}p_{1j}\left(\bar{a}_{11j}-\underline{a}_{11j}\right)}{6}+\frac{\sum_{k=1}^{L_2}p_{1j}(\bar{a}_{12k}-\underline{a}_{12k})}{6}$$

$$=W_\mu\left(\tilde{A}_1\right)+W_\mu\left(\tilde{A}_2\right)$$

9.4 Validity of the Fourth Existing Result

$$W_\nu\left(\tilde{A}_1+\tilde{A}_2\right)=W_\nu\left(U_{j=1,2,..,L_1,\,k=1,2,..,L_2}\left\{\left(\left\langle\left(\underline{a}_{11j}+\underline{a}_{12k},a_{1j}+a_{2k},\bar{a}_{11j}+\bar{a}_{12k}\right);\left(\underline{a}_{21j}+\underline{a}_{22k},a_{1j}+a_{2k},\bar{a}_{21j}+\bar{a}_{22k}\right)\right\rangle,p_{1j}p_{2k}\right)\right\}\right)$$

$$=\frac{\sum_{j=1}^{L_1}\sum_{k=1}^{L_2}p_{1j}p_{2k}\left(\left(\bar{a}_{21j}+\bar{a}_{22k}\right)-\left(\underline{a}_{21j}+\underline{a}_{22k}\right)\right)}{6}$$

$$=\frac{\sum_{j=1}^{L_1}\sum_{k=1}^{L_2}p_{1j}p_{2k}\left(\bar{a}_{21j}-\underline{a}_{21j}\right)}{6}+\frac{\sum_{j=1}^{L_1}\sum_{k=1}^{L_2}p_{1j}p_{2k}(\bar{a}_{22k}-\underline{a}_{22k})}{6}$$

$$=\frac{\sum_{j=1}^{L_1}p_{1j}\left(\bar{a}_{21j}-\underline{a}_{21j}\right)\sum_{k=1}^{L_2}p_{2k}}{6}+\frac{\sum_{k=1}^{L_2}p_{1j}(\bar{a}_{22k}-\underline{a}_{22k})\sum_{l=1}^{L_1}p_{1j}}{6}$$

$$=\frac{\sum_{j=1}^{L_1}p_{1j}\left(\bar{a}_{21j}-\underline{a}_{21j}\right)}{6}+\frac{\sum_{k=1}^{L_2}p_{1j}(\bar{a}_{22k}-\underline{a}_{22k})}{6}$$

$$=W_\nu\left(\tilde{A}_1\right)+W_\nu\left(\tilde{A}_1\right)$$

9.5 Validity of the Fifth Existing Result

$$V_\lambda\left(\tilde{A}_1+\tilde{A}_2\right)=V_\lambda\left(\text{U}_{j=1,2,..,\,L_1,\,k=1,2,..,L_2}\left\{\left(\left\langle\left(\underline{a}_{11j}+\underline{a}_{12k},a_{1j}+a_{2k},\bar{a}_{11j}+\bar{a}_{12k}\right);\left(\underline{a}_{21j}+\underline{a}_{22k},a_{1j}+a_{2k},\bar{a}_{21j}+\bar{a}_{22k}\right)\right\rangle,p_{1j}p_{2k}\right)\right\}\right)$$

$$=\lambda\frac{\sum_{j=1}^{L_1}\sum_{k=1}^{L_2}p_{1j}p_{2k}\left(\underline{a}_{21j}+\underline{a}_{22k}+7\left(a_{1j}+a_{2k}\right)+\bar{a}_{21j}+\bar{a}_{22k}\right)}{18}$$

$$+(1-\lambda)\frac{\sum_{j=1}^{L_1}\sum_{k=1}^{L_2}p_{1j}p_{2k}\left(\underline{a}_{11j}+\underline{a}_{12k}+7\left(a_{1j}+a_{2k}\right)+\bar{a}_{11j}+\bar{a}_{12k}\right)}{18}$$

$$=\lambda\frac{\sum_{j=1}^{L_1}\sum_{k=1}^{L_2}p_{1j}p_{2k}\left(\underline{a}_{21j}+7a_{1j}+\bar{a}_{21j}\right)}{18}+(1-\lambda)\frac{\sum_{j=1}^{L_1}\sum_{k=1}^{L_2}p_{1j}p_{2k}\left(\underline{a}_{11j}+7a_{1j}+\bar{a}_{11j}\right)}{18}$$

$$+\lambda\frac{\sum_{j=1}^{L_1}\sum_{k=1}^{L_2}p_{1j}p_{2k}(\underline{a}_{22k}+7a_{2k}+\bar{a}_{22k})}{18}+(1-\lambda)\frac{\sum_{j=1}^{L_1}\sum_{k=1}^{L_2}p_{1j}p_{2k}(\underline{a}_{12k}+7a_{2k}+\bar{a}_{12k})}{18}$$

$$=\lambda\frac{\sum_{j=1}^{L_1}p_{1j}\left(\underline{a}_{21j}+7a_{1j}+\bar{a}_{21j}\right)\sum_{k=1}^{L_2}p_{2k}}{18}+(1-\lambda)\frac{\sum_{j=1}^{L_1}p_{1j}\left(\underline{a}_{11j}+7a_{1j}+\bar{a}_{11j}\right)\sum_{k=1}^{L_2}p_{2k}}{18}$$

$$+\lambda\frac{\sum_{k=1}^{L_2}p_{2k}(\underline{a}_{22k}+7a_{2k}+\bar{a}_{22k})\sum_{j=1}^{L_1}p_{1j}}{18}+(1-\lambda)\frac{\sum_{k=1}^{L_2}p_{2k}(\underline{a}_{12k}+7a_{2k}+\bar{a}_{12k})\sum_{j=1}^{L_1}p_{1j}}{18}$$

$$=\lambda\frac{\sum_{j=1}^{L_1}p_{1j}\left(\underline{a}_{21j}+7a_{1j}+\bar{a}_{21j}\right)}{18}+(1-\lambda)\frac{\sum_{j=1}^{L_1}p_{1j}\left(\underline{a}_{11j}+7a_{1j}+\bar{a}_{11j}\right)}{18}$$

$$+\lambda\frac{\sum_{k=1}^{L_2}p_{2k}(\underline{a}_{22k}+7a_{2k}+\bar{a}_{22k})}{18}+(1-\lambda)\frac{\sum_{k=1}^{L_2}p_{2k}(\underline{a}_{12k}+7a_{2k}+\bar{a}_{12k})}{18}$$

$$=\lambda V_\nu\left(\tilde{A}_1\right)+(1-\lambda)V_\mu\left(\tilde{A}_1\right)+\lambda V_\nu\left(\tilde{A}_2\right)+(1-\lambda)V_\mu\left(\tilde{A}_2\right)$$

$$=V_\lambda\left(\tilde{A}_1\right)+V_\lambda\left(\tilde{A}_2\right)$$

9.6 Validity of the Sixth Existing Result

$$W_\lambda\left(\tilde{A}_1+\tilde{A}_2\right)=W_\lambda\left(\mathrm{U}_{j=1,2,\ldots,L_1,k=1,2,\ldots,L_2}\left\{\left(\left\langle\left(\underline{a}_{11j}+\underline{a}_{12k},a_{1j}+a_{2k},\bar{a}_{11}+\bar{a}_{12k}\right);\left(\underline{a}_{21j}+\underline{a}_{22k},a_{1j}+a_{2k},\bar{a}_{21j}+\bar{a}_{22k}\right)\right\rangle,p_{1j}p_{2k}\right)\right\}\right)$$

$$=\lambda\frac{\sum_{j=1}^{L_1}\sum_{k=1}^{L_2}p_{1j}p_{2k}\left(\left(\bar{a}_{11j}+\bar{a}_{12k}\right)-\left(\underline{a}_{11j}+\underline{a}_{12k}\right)\right)}{6}$$

$$+(1-\lambda)\frac{\sum_{j=1}^{L_1}\sum_{k=1}^{L_2}p_{1j}p_{2k}\left(\left(\bar{a}_{21j}+\bar{a}_{22k}\right)-\left(\underline{a}_{21j}+\underline{a}_{22k}\right)\right)}{6}$$

$$=\lambda\frac{\sum_{j=1}^{L_1}p_{1j}\left(\bar{a}_{11j}-\underline{a}_{11j}\right)\sum_{k=1}^{L_2}p_{2k}}{6}+(1-\lambda)\frac{\sum_{j=1}^{L_1}p_{1j}\left(\bar{a}_{21j}-\underline{a}_{21j}\right)\sum_{l=1}^{L_2}p_{2k}}{6}$$

$$+\lambda\frac{\sum_{k=1}^{L_2}p_{2k}(\bar{a}_{12k}-\underline{a}_{12k})\sum_{j=1}^{L_1}p_{1j}}{6}+(1-\lambda)\frac{\sum_{k=1}^{L_2}p_{2k}(\bar{a}_{22k}-\underline{a}_{22k})\sum_{j=1}^{L_1}p_{1j}}{6}$$

$$=\lambda\frac{\sum_{j=1}^{L_1}p_{1j}\left(\bar{a}_{11j}-\underline{a}_{11j}\right)}{6}+(1-\lambda)\frac{\sum_{j=1}^{L_1}p_{1j}\left(\bar{a}_{21j}-\underline{a}_{21j}\right)}{6}$$

$$+\lambda\frac{\sum_{k=1}^{L_2}p_{2k}(\bar{a}_{12k}-\underline{a}_{12k})}{6}+(1-\lambda)\frac{\sum_{k=1}^{L_2}p_{2k}(\bar{a}_{22k}-\underline{a}_{22k})}{6}$$

$$= \lambda W_\mu(\tilde{A}_1) + (1-\lambda) W_\nu(\tilde{A}_1) + \lambda W_\mu(\tilde{A}_2) + (1-\lambda) W_\nu(\tilde{A}_2)$$
$$= W_\lambda(\tilde{A}_1) + W_\lambda(\tilde{A}_2)$$

9.7 Validity of the Seventh Existing Result

$$\eta_\lambda(\tilde{A}_1 + \tilde{A}_2) = \eta_\lambda\Big(\mathrm{U}_{j=1,2,..,L_1,\,k=1,2,..,L_2}\Big\{\Big(\big\langle\big(\underline{a}_{11j} + \underline{a}_{12k}, a_{1j} + a_{2k}, \bar{a}_{11j} + \bar{a}_{12k}\big);$$
$$\big(\underline{a}_{21j} + \underline{a}_{22k}, a_{1j} + a_{2k}, \bar{a}_{21j} + \bar{a}_{22k}\big)\big\rangle, p_{1j}p_{2k}\Big)\Big\}\Big)$$
$$= \left(\lambda \frac{\sum_{j=1}^{L_1}\sum_{k=1}^{L_2} p_{1j}p_{2k}\left(\underline{a}_{21j} + \underline{a}_{22k} + 7(a_{1j} + a_{2k}) + \bar{a}_{21j} + \bar{a}_{22k}\right)}{18}\right.$$
$$\left. + (1-\lambda)\frac{\sum_{j=1}^{L_1}\sum_{k=1}^{L_2} p_{1j}p_{2k}\left(\underline{a}_{11j} + \underline{a}_{12k} + 7(a_{1j} + a_{2k}) + \bar{a}_{11j} + \bar{a}_{12k}\right)}{18}\right)$$
$$- \left(\lambda \frac{\sum_{j=1}^{L_1}\sum_{k=1}^{L_2} p_{1j}p_{2k}\left((\bar{a}_{11j} + \bar{a}_{12k}) - (\underline{a}_{11j} + \underline{a}_{12k})\right)}{6}\right.$$
$$\left. + (1-\lambda)\frac{\sum_{j=1}^{L_1}\sum_{k=1}^{L_2} p_{1j}p_{2k}\left((\bar{a}_{21j} + \bar{a}_{22k}) - (\underline{a}_{21j} + \underline{a}_{22k})\right)}{6}\right)$$
$$= \left(\lambda \frac{\sum_{j=1}^{L_1}\sum_{k=1}^{L_2} p_{1j}p_{2k}\left(\underline{a}_{21j} + 7a_{1j} + \bar{a}_{21j}\right)}{18} + (1-\lambda)\frac{\sum_{j=1}^{L_1}\sum_{k=1}^{L_2} p_{1j}p_{2k}\left(\underline{a}_{11j} + 7a_{1j} + \bar{a}_{11j}\right)}{18}\right.$$
$$\left. + \lambda \frac{\sum_{j=1}^{L_1}\sum_{k=1}^{L_2} p_{1j}p_{2k}(\underline{a}_{22k} + 7a_{2k} + \bar{a}_{22k})}{18} + (1-\lambda)\frac{\sum_{j=1}^{L_1}\sum_{k=1}^{L_2} p_{1j}p_{2k}(\underline{a}_{12k} + 7a_{2k} + \bar{a}_{12k})}{18}\right)$$
$$- \left(\lambda \frac{\sum_{j=1}^{L_1}\sum_{k=1}^{L_2} p_{1j}p_{2k}\left(\bar{a}_{11j} - \underline{a}_{11j}\right)}{6} + (1-\lambda)\frac{\sum_{j=1}^{L_1}\sum_{k=1}^{L_2} p_{1j}p_{2k}\left(\bar{a}_{21j} - \underline{a}_{21j}\right)}{6}\right.$$

$$
\left. + \lambda \frac{\sum_{j=1}^{L_1}\sum_{k=1}^{L_2} p_{1j}p_{2k}(\bar{a}_{12k} - \underline{a}_{12k})}{6} + (1-\lambda)\frac{\sum_{j=1}^{L_1}\sum_{k=1}^{L_2} p_{1j}p_{2k}(\bar{a}_{22k} - \underline{a}_{22k})}{6} \right)
$$

$$
= \left(\lambda \frac{\sum_{j=1}^{L_1} p_{1j}\left(\underline{a}_{21j} + 7a_{1j} + \bar{a}_{21j}\right)\sum_{k=1}^{L_2} p_{2k}}{18} + (1-\lambda)\frac{\sum_{j=1}^{L_1} p_{1j}\left(\underline{a}_{11j} + 7a_{1j} + \bar{a}_{11j}\right)\sum_{k=1}^{L_2} p_{2k}}{18} \right.
$$

$$
\left. + \lambda \frac{\sum_{k=1}^{L_2} p_{2k}(\underline{a}_{22k} + 7a_{2k} + \bar{a}_{22k})\sum_{j=1}^{L_1} p_{1j}}{18} + (1-\lambda)\frac{\sum_{k=1}^{L_2} p_{2k}(\underline{a}_{12k} + 7a_{2k} + \bar{a}_{12k})\sum_{j=1}^{L_1} p_{1j}}{18} \right)
$$

$$
- \left(\lambda \frac{\sum_{j=1}^{L_1} p_{1j}\left(\bar{a}_{11j} - \underline{a}_{11j}\right)\sum_{k=1}^{L_2} p_{2k}}{6} + (1-\lambda)\frac{\sum_{j=1}^{L_1} p_{1j}\left(\bar{a}_{21j} - \underline{a}_{21j}\right)\sum_{k=1}^{L_2} p_{2k}}{6} \right.
$$

$$
\left. + \lambda \frac{\sum_{k=1}^{L_2} p_{2k}(\bar{a}_{12k} - \underline{a}_{12k})\sum_{j=1}^{L_1} p_{1j}}{6} + (1-\lambda)\frac{\sum_{k=1}^{L_2} p_{2k}(\bar{a}_{22k} - \underline{a}_{22k})\sum_{j=1}^{L_1} p_{1j}}{6} \right)
$$

$$
= \left(\lambda \frac{\sum_{j=1}^{L_1} p_{1j}\left(\underline{a}_{21j} + 7a_{1j} + \bar{a}_{21j}\right)}{18} + (1-\lambda)\frac{\sum_{j=1}^{L_1} p_{1j}\left(\underline{a}_{11j} + 7a_{1j} + \bar{a}_{11j}\right)}{18} \right)
$$

$$
- \left(\lambda \frac{\sum_{j=1}^{L_1} p_{1j}\left(\bar{a}_{11j} - \underline{a}_{11j}\right)}{6} + (1-\lambda)\frac{\sum_{j=1}^{L_1} p_{1j}\left(\bar{a}_{21j} - \underline{a}_{21j}\right)}{6} \right)
$$

$$
+ \left(\lambda \frac{\sum_{k=1}^{L_2} p_{2k}(\underline{a}_{22k} + 7a_{2k} + \bar{a}_{22k})}{18} + (1-\lambda)\frac{\sum_{k=1}^{L_2} p_{2k}(\underline{a}_{12k} + 7a_{2k} + \bar{a}_{12k})}{18} \right)
$$

$$
- \left(\lambda \frac{\sum_{k=1}^{L_2} p_{2k}(\bar{a}_{12k} - \underline{a}_{12k})}{6} + (1-\lambda)\frac{\sum_{k=1}^{L_2} p_{2k}(\bar{a}_{22k} - \underline{a}_{22k})}{6} \right)
$$

$$
= V_\lambda(\tilde{A}_1) - W_\lambda(\tilde{A}_1) + V_\lambda(\tilde{A}_2) - W_\lambda(\tilde{A}_2)
$$

$$
= \eta_\lambda(\tilde{A}_1) + \eta_\lambda(\tilde{A}_2)
$$

10 Validity of Yang and Xu's Method

Using the results (13)–(19), Yang and Xu [1] proposed a method (zero-sum matrix game-based method) to solve such MADM problem in which the ranking value of each alternative over each attribute is represented by a PTIHF set. However, as discussed in Section 3, the results (13)–(19) are not valid according to Yang and Xu's [1] arithmetic operations of PTIHF sets. While, as discussed in Section 9, the results of eqs. (13)–(19) are valid according to the proposed arithmetic operations of PTIHF sets. Hence, Yang and Xu's [1] zero-sum matrix game-based method is valid only if the proposed arithmetic operations of PTIHF sets are considered instead of Yang and Xu's [1] arithmetic operations.

11 Conclusions

It has been demonstrated that the arithmetic operations of PTIHF sets introduced by Yang and Xu [1] are mathematically incorrect, thereby making their zero-sum matrix game-based approach inconsistent. In response, the correct arithmetic operations for PTIHF sets have been introduced to ensure the zero-sum matrix game framework remains consistent and applicable. These new operations not only rectify the errors at the base level but also offer a stronger mathematical foundation for uncertainty-based decision-making models.

Future studies could be aimed at applying these operations to multi-player matrix games and investigating their uses in solving realistic complex optimization and decision-making problems.

References

[1] Yang J, Xu Z. Matrix game-based approach for MADM with probabilistic triangular intuitionistic hesitant fuzzy information and its application. Comput Ind Eng. 2022;163:107787.
[2] Zadeh LA. Fuzzy sets. Inf Control. 1965;8:338–353.
[3] Chen YW, Larbani M. Two-person zero-sum game approach for fuzzy multiple attribute decision making problems. Fuzzy Sets Syst. 2006;157:34–51.
[4] Oderanti FO, De Wilde P. Dynamics of business games with management of fuzzy rules for decision making. Int J Prod Econ. 2010;128:96–109.
[5] Wu J, Liang L. A multiple criteria ranking method based on game cross- evaluation approach. Ann Oper Res. 2012;197:191–200.
[6] Yang WE, Wang JQ. Vague linguistic matrix game approach for multi- criteria decision making with uncertain weights. J Intell Fuzzy Syst. 2013;25:315–324.
[7] Atanassov KT. Intuitionistic fuzzy sets. Fuzzy Sets Syst. 1986;20:87–96.
[8] Torra V. Hesitant fuzzy sets. Int J Intell Syst. 2010;25:529–539.
[9] Lin M, Chen Z, Liao H, Xu Z. ELECTRE II method to deal with probabilistic linguistic term sets and its application to edge computing. Nonlinear Dyn. 2019;96:2125–2143.

Chirag Dhankhar, Anju Kumari, Amit Sharma*, Reeta Bhardwaj

A *q*-ROFS MARICA Method with Genetic Algorithm for Objective Weight Optimization: A Hybrid Multicriteria Decision-Making Approach

Abstract: Decision-making in real-world problems is often blurred by uncertainty, vagueness, and incomplete information. In this chapter, to tackle such challenges, a novel hybrid multicriteria decision-making approach using q-rung orthopair fuzzy sets (q-ROFS) is introduced. This approach integrates Multiattributive Real-Ideal Comparative Analysis (MARICA) method along with the genetic algorithm (GA) for weight optimization. One of the main features of q-ROFS is its ability to model the opinions of reluctant experts through paired membership and nonmembership grades, while MARICA ensures the stability of ranking by comparing the alternatives against the ideal and anti-ideal solutions. GA is incorporated to optimize the weights by maximizing the variance of the closeness coefficient, ensuring that the decision model does not rely on subjective judgments. A numerical example is incorporated for supplier selection against five evaluated criteria, which illustrates how the newly proposed method guarantees robustness and transparency. The comparison of results with previous works based on q-ROFS asserts the significance of the proposed GA-MARICA approach.

Keywords: MARICA method, genetic algorithm, multicriteria decision making, q-rung orthopair fuzzy sets

1 Introduction

In complex environments, decision-making often involves multiple criteria, uncertain data, and subjective judgments. To solve real-world problems, we must develop powerful tools to deal with such vagueness in an efficient and effective manner. Traditional crisp models are inadequate for such ambiguity, thus necessitating the use of

---

***Corresponding author: Amit Sharma,** Department of Mathematics, Amity School of Applied Sciences, Amity University, Gurugram, Haryana, India, e-mail: dba.amitsharma@gmail.com
Chirag Dhankhar, Amity College of Commerce, Amity University, Gurugram, Haryana, India, e-mail: dhankharchirag@gmail.com
Anju Kumari, Amity College of Commerce, Amity University, Gurugram, Haryana, India, e-mail: anju3131998@gmail.com
Reeta Bhardwaj, Department of Mathematics, Amity School of Applied Sciences, Amity University, Gurugram, Haryana, India, e-mail: bhardwajreeta84@gmail.com

 | https://doi.org/10.1515/9783112222249-014

modern fuzzy frameworks. This study addresses these challenges by integrating q-rung orthopair fuzzy sets (q-ROFS) with a new weight optimization-based multicriteria decision-making (MCDM) approach. Fuzzy set theory by Zadeh [1] and its extensions, including intuitionistic fuzzy sets by Atanassov [2] and Pythagorean fuzzy sets by Yager [3], offer richer semantics; among them, the q-ROFS framework provides enhanced flexibility by relaxing the orthogonality bound to $(\text{membership})^q + (\text{nonmembership})^q \leq 1$, enabling greater expressiveness for human judgments.

A main, persistent concerns is criterion weighting. Subjective weights are more risk bias while purely statistical weights sometimes do not capture decision performance. We address this gap by coupling q-ROFS Multiattributive Real-Ideal Comparative Analysis (MARICA) a genetic algorithm (GA) that objectively learns weights on the simplex to maximize the variance of MARICA-based closeness scores, thereby enhancing discrimination across alternatives without decision-maker-imposed weights.

Evolutionary algorithms, a "generic population-based" metaheuristic optimization technique, are the ancestors of the GA. Similar to evolutionary algorithm, GA is predicated on biologically inspired processes including selection, recombination, mutation, and reproduction (Immanuel and Chakraborty [4]). The method is based on the idea that the fittest offspring would either create the best group of offspring or provide the best solution. The initial population, which is a population of randomly produced solutions or people, is often where the evolution begins. Each population in this repeated process is referred to as a generation. Both the presently existing chromosomes from the previous generations and the entire newly created population are evaluated for fitness in each generation. Fitness refers to the value that an objective function in the optimization issue under study achieves. A new generation is created by randomly recombining or mutating each person's genome after a stochastic selection of the more fit people from the present population. The following generation (or iteration) of the GA makes use of the freshly produced candidate solutions. Usually, the method stops after the user-specified maximum number of generations has been achieved or the population reaches a fitness level that is deemed optimal or nearly optimal (satisfying).

Kamalinia et al. [5] integrate MADM with GA to enhance search efficiency and provide optimal DG placement solutions under multicriteria frameworks. Nestic et al. [6] show the integration of fuzzy modeling with GA to improve process quality optimization by enabling efficient parameter tuning and decision-making under imprecision. Choudhary et al. [7] propose a hybrid fuzzy-GAs to enhance WBAN performance by combining fuzzy logic's uncertainty handling with GA's global search capabilities. Roshandel et al. [8] integrate weighted fuzzy inference systems with data envelopment analysis and GAs to enhance robustness, optimize weighting schemes, and improve decision accuracy in alliance management. Raut et al. [9] extend TSP solutions into Fermatean neutrosophic environments, where GA is combined with higher-order fuzzy frameworks to model uncertainty and improve decision accuracy.

The MARICA method, developed by Pamučar and Ćirović [10], is a MCDM technique designed to overcome rank-reversal issues of classical methods. It compares alternatives against an ideal solution to ensure more stable and consistent decision outcomes. Cheng et al. [11] propose a decision-making model using R-numbers to handle uncertainty in multiexpert, multicriteria situations. Experts are grouped using k-means clustering, with group importance determined by the size and variety of opinions. LOPCOW is used for weighing, and CRADIS finds the best choice. The model is used in selecting hydrogen fuel supply routes. Özta¸s and Özta¸s [12] evaluate G20 innovation performance (2018–2022) using LOPCOW and MAIRCA methodologies. Human capital and research are important factors, with the United States showing the highest performance. The study also looks at the impact of COVID-19 and confirms the reliability of the method. Sedighi Khavidak et al. [13] use ENTROPY for attribute weighting and apply COBRA, MARCOS, and MAIRCA to assess six HDDs. The ENTROPY-COBRA combination is introduced, and Western Digital-My Passport is recognized as the best HDD. Teši´c et al. [14] suggest a FUCOM-FMAIRCA model to select military academy instructors. Criteria are prioritized by FUCOM, and candidates are evaluated with fuzzy MAIRCA. Sensitivity analysis confirms the reliability. Seikh and Mandal [15] introduce novel q-rung orthopair fuzzy Archimedean aggregation operators to handle uncertainty in decision-making. The proposed operators are applied to a site selection problem for software operating units, demonstrating their effectiveness and flexibility in MCDM under fuzzy environments.

1.1 Motivation of the Study

In real-world decision-making environments, uncertainty and vagueness are inevitable due to imprecise data, subjective expert opinions, and incomplete information. Traditional decision-making models often fail to adequately represent such uncertainty, leading to unreliable or biased outcomes. Though many fuzzy and MCDM strategies have been advanced, nonetheless, they always come with greater dependency on subjective weight decision-making and may lack ranking robustness. This poses a real challenge for an integrated data-driven framework that can determine weights objectively and handle complex uncertainty while delivering consistently reliable decision results. In this regard, it is necessary to develop an utmost decision alternative ranking system under uncertainty for better decisions. Encouraged by these situations, a hybrid approach is proposed that combines the uncertainty modeling flexibility of q-ROFS, the ranking stability of the MARICA method, and the output optimization ability of the GA for weight determination. This fusion aims to significantly improve MCDM in terms of both data reliability and discriminatory power under uncertain contexts.

1.2 Contribution of the Study

Grounded in this motivation, this study makes the following key contributions:

1. A distance-based MARICA formulation in the q-ROFS environment.
2. A GA-based objective weighting scheme (no subjective inputs) that maximizes ranking discrimination.
3. A reproducible numerical example aligning step-by-step with the proposed pipeline.

1.3 Structure of This Chapter

The remaining chapter is organized as follows: Section 2 provides the preliminaries of fuzzy set theory. Section 3 describes the proposed q-ROFS-MARICA-GA methodology. Section 4 presents a numerical example. Section 5 provides a comparative analysis with existing methods. Finally, Section 6 concludes the study and outlines future research directions.

2 Preliminaries

Definition 2.1 ([2]). An intuitionistic fuzzy set in a universal discourse A is defined as:

$$S = \{\langle a, \eta(a), \zeta(a)\rangle | a \in A\} \tag{1}$$

where $\forall$ a and $\eta(a), \zeta(a) \in [0,1]$ shows membership and non-membership degrees of a to S, respectively, for $0 \le \eta(a) + \zeta(a) \le 1$, and the hesitancy degree of a to S is defined as $\pi_H = 1 - \eta(a)\zeta(a)$. Usually the pair $\langle \eta, \zeta \rangle$ is called an IFN.

Definition 2.2 ([16]). A q-ROFS in a universe of discourse A is defined as:

$$S = \{\langle a, \eta(a), \zeta(a)\rangle | a \in A\} \tag{2}$$

where $\eta(a)$ and $\zeta(a)$ represent membership and non-membership degrees of a belonging to the q-ROFS S, such that $\forall$ a, $0 \le \eta(a), \zeta(a) \le 1$, and $0 \le \eta(a)^q + \zeta(a)^q \le 1$ holds, and the hesitancy degree of a to S is defined as $\pi_H = (1 - \eta(a)^q - \zeta(a)^q)^{1/q}$ where $q \ge 1$. Usually, the pair $\langle \eta, \zeta \rangle$ is called a q-ROFN.

Definition 2.3 ([17]). The Euclidean distance measure between two q-ROFNs $A = \langle \eta_A, \zeta_A \rangle$ and $B = \langle \eta_B, \zeta_B \rangle$ is defined as:

$$D(A,B) = \sqrt{\frac{1}{2}\left[(\eta_A^q - \eta_B^q)^2 + \left(\zeta_A^q - \zeta_B^q\right)^2\right]} \tag{3}$$

where η_A, η_B, and ζ_A, ζ_B represent the membership and non-membership degrees of A and B, respectively, and $q \geq 1$ controls the flexibility of the orthopair structure.

3 Methodology

In this study, we integrate the MARICA method with a GA to optimize the criteria weights under the q-ROFS framework. The methodology is divided into two main phases.

3.1 Phase I: MARICA Under q-ROFS

Step 1. **Decision matrix.** Let $U = \{U_1, U_2, \ldots, U_m\}$ be the set of m alternatives and $C = \{C_1, C_2, \ldots, C_n\}$ be the set of n criteria. By using the q-ROFS $\chi_{ij} = \langle \eta_{ij}, \zeta_{ij} \rangle$, the expert evaluates the alternative U_i with respect to the attribute C_j to construct the DMx $D = (\chi_{ij})_{m \times n}$, shown as follows:

$$D = \begin{array}{c} \\ U_1 \\ U_2 \\ \vdots \\ U_m \end{array} \begin{array}{c} \begin{array}{cccc} C_1 & C_1 & \cdots & C_n \end{array} \\ \begin{bmatrix} \chi_{11} & \chi_{12} & \cdots & \chi_{1n} \\ \chi_{21} & \chi_{22} & \cdots & \chi_{1n} \\ \vdots & \vdots & \ddots & \vdots \\ \chi_{m1} & \chi_{m2} & \cdots & \chi_{mn} \end{bmatrix} \end{array}$$

where $i = 1, 2, \ldots, m$ and $j = 1, 2, \ldots, n$.

Step 2. **Normalization.** For benefit-type criteria, normalize as:

$$D' = \begin{cases} \langle \eta_{ij}, \zeta_{ij} \rangle: \text{ if } C_j \text{ is a criteria of benefit type} \\ \langle \zeta_{ij}, \eta_{ij} \rangle: \text{ if } C_j \text{ is a criteria of cost type} \end{cases} \tag{4}$$

Step 3. **Ideal and anti-ideal solutions.** For each criterion C_j:

$$\eta_j^+ = \max_i \eta_{ij}, \quad \zeta_j^+ = \min_i \zeta_{ij}, \quad \eta_j^- = \min_i \eta_{ij}, \quad \zeta_j^- = \max_i \zeta_{ij}. \tag{5}$$

Step 4. **Distances.** Compute the q-ROFS Euclidean-type distances:

$$d_{ij}^+ = \sqrt{\frac{1}{2}[((\eta'_{ij})^q - (\eta_j^+)^q)^2 + ((\zeta'_{ij})^q - \left(\zeta_j^+\right)^q)^2]},$$

$$d_{ij}^- = \sqrt{\frac{1}{2}[((\eta'_{ij})^q - (\eta_j^-)^q)^2 + ((\zeta'_{ij})^q - \left(\zeta_j^-\right)^q)^2]}. \tag{6}$$

Step 5. Closeness coefficient along with weighted distances. With weight vector $W = (w_1, \ldots, w_n)$:

$$D_i^+ = \sum_{j=1}^{n} w_j d_{ij}^+, \qquad D_i^- = \sum_{j=1}^{n} w_j d_{ij}^-,$$
$$C_i = \frac{D_i^-}{D_i^+ + D_i^-}, \qquad i = 1, 2, \ldots, m. \tag{7}$$

Alternatives are ranked in descending order of C_i.

3.2 Phase II: GA for Weight Optimization

Step 1. **Encoding.** Each chromosome represents a weight vector $W = (w_1, \ldots, w_n)$ with $w_j \geq 0, \sum_j w_j = 1$.
Step 2. **Initialization.** Generate an initial population of random weight vectors.
Step 3. **Fitness function.** For each chromosome, apply MARICA and compute closeness coefficients C_i. The fitness is defined as:

$$f(w) = \mathrm{Var}(C) = \frac{1}{m} \sum_{i=1}^{m} (C_i - \overline{C})^2,$$

where higher variance indicates stronger discrimination among alternatives.
Step 4. **Genetic operators.** Use tournament selection, blend crossover (BLX-α), and Gaussian mutation to evolve new populations. Weights are renormalized after each operation.
Step 5. **Termination.** Stop when the maximum generation is reached or convergence occurs. The best-performing weight vector is selected.

The above steps are also exhibited in Figure 1.

4 Numerical Example

A green supplier selection problem is considered with four alternatives (U_1, U_2, U_3, and U_4) and five criteria: C_1 = Environmental Management, C_2 = Cost, C_3 = Quality, C_4 = Delivery, and C_5 = Emission Control.

Here, C_2 and C_5 are cost-type criteria, while C_1, C_3, and C_4 are benefit-type. The evaluation is carried out in a q-ROF environment with $q = 3$. The corresponding q-ROFS decision matrix is provided in Table 1.

Starting with applying the Phase I: MARICA under q-ROFS.

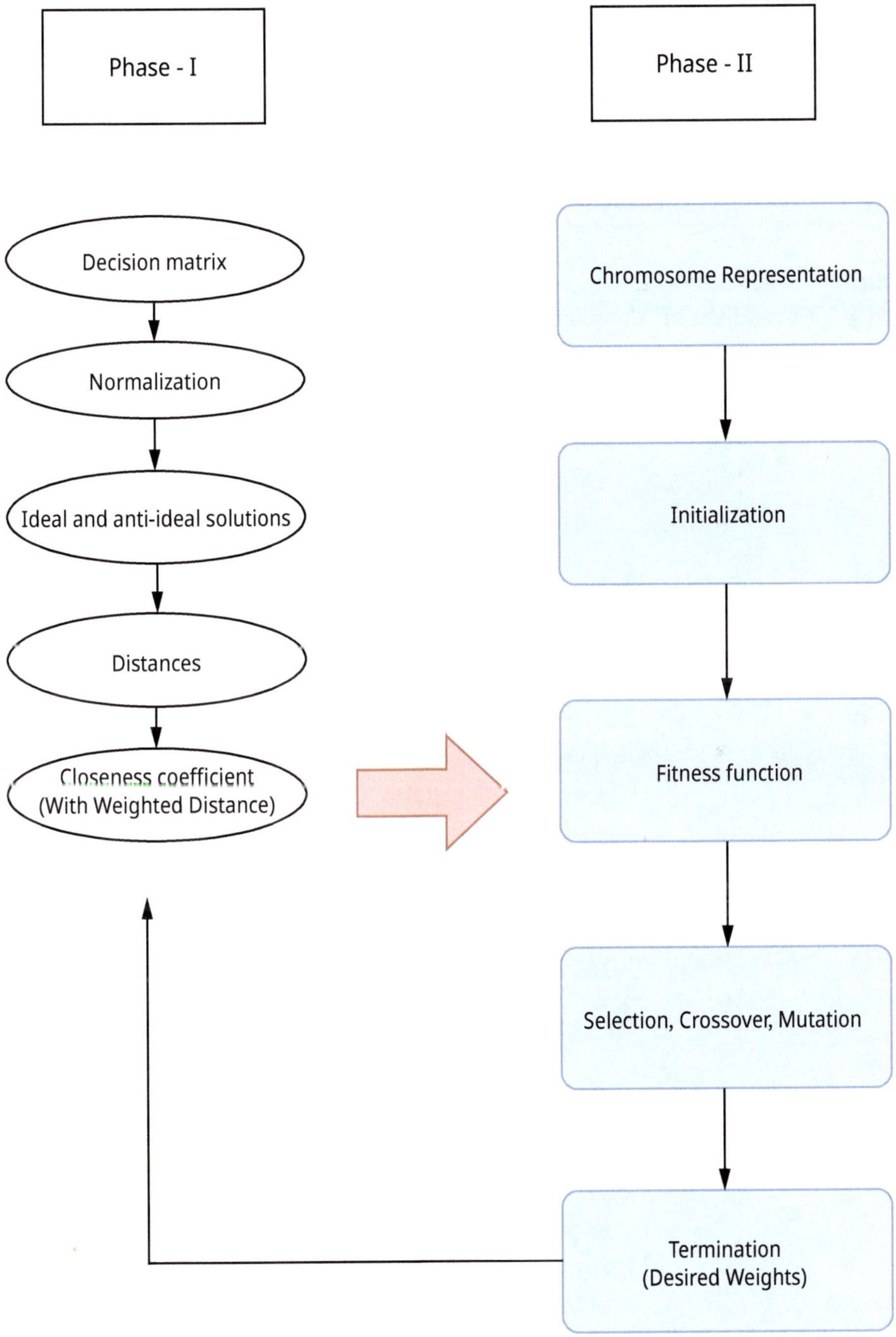

Figure 1: Flowchart of the proposed method.

Table 1: q-ROF decision matrix.

	C_1	C_2	C_3	C_4	C_5
U_1	(0.80,0.50)	(0.90,0.60)	(0.70,0.21)	(0.80,0.70)	(0.70,0.66)
U_2	(0.90,0.50)	(0.20,0.80)	(0.60,0.40)	(0.66,0.90)	(0.41,0.70)
U_3	(0.50,0.50)	(0.70,0.60)	(0.80,0.80)	(0.77,0.32)	(0.70,0.20)
U_4	(0.35,0.81)	(0.60,0.60)	(0.72,0.50)	(0.79,0.31)	(0.65,0.72)

Step 1. **Decision matrix.** Considering Table 1 as the decision matrix.

Step 2. **Normalization.** Considering C_1, C_3, C_4 as benefit-type criteria and C_2, C_5 as cost-type criteria, normalizing the matrix using eq. (4) as:

$$D' = \begin{bmatrix} (0.8, 0.5) & (0.6, 0.9) & (0.7, 0.21) & (0.8, 0.7) & (0.66, 0.7) \\ (0.9, 0.5) & (0.8, 0.2) & (0.6, 0.4) & (0.66, 0.9) & (0.70, 0.41) \\ (0.5, 0.5) & (0.6, 0.7) & (0.8, 0.8) & (0.77, 0.32) & (0.2, 0.7) \\ (0.35, 0.81) & (0.6, 0.6) & (0.72, 0.5) & (0.79, 0.31) & (0.72, 0.65) \end{bmatrix}$$

Step 3. **Ideal and anti-ideal solutions.** Using eq. (5) for finding the ideal and anti-ideal as:

$$\text{Ideal solution}\left(\eta_j^+, \zeta_j^+\right) = \begin{bmatrix} 0.9000 & 0.5000 \\ 0.8000 & 0.2000 \\ 0.8000 & 0.2100 \\ 0.8000 & 0.3100 \\ 0.7200 & 0.4100 \end{bmatrix}$$

$$\text{Anti - ideal solution}\left(\eta_j^-, \zeta_j^-\right) = \begin{bmatrix} 0.3500 & 0.8100 \\ 0.6000 & 0.9000 \\ 0.6000 & 0.8000 \\ 0.6600 & 0.9000 \\ 0.2000 & 0.7000 \end{bmatrix}$$

Step 4. **Distances.** Compute the q-ROFS Euclidean-type distances using eq. (6):

$$d^+ = \begin{bmatrix} 0.1202 & 0.5794 & 0.1061 & 0.2785 & 0.2350 \\ 0 & 0 & 0.2143 & 0.5251 & 0.0201 \\ 0.3960 & 0.3748 & 0.4214 & 0.0336 & 0.4077 \\ 0.5646 & 0.3007 & 0.1691 & 0.0112 & 0.1799 \end{bmatrix}$$

$$d^- = \begin{bmatrix} 0.4651 & 0 & 0.4313 & 0.2685 & 0.2797 \\ 0.5646 & 0.5794 & 0.3394 & 0 & 0.3912 \\ 0.3010 & 0.2263 & 0.1980 & 0.5126 & 0 \\ 0 & 0.3182 & 0.2976 & 0.5221 & 0.3416 \end{bmatrix}$$

Step 5. **Closeness coefficient along with weighted distances.** Using Phase II: GA for weight optimization, the optimized weight vector obtained is $W = (0.2363, 0.0555, 0.1547, 0.1469, 0.4065)$. Then using eq. (7), we get:

$$D^+ = \begin{bmatrix} 0.2134 \\ 0.1184 \\ 0.3502 \\ 0.2510 \end{bmatrix}, \quad D^- = \begin{bmatrix} 0.3298 \\ 0.3771 \\ 0.1896 \\ 0.2793 \end{bmatrix}$$

The closeness coefficients for each alternative are then obtained as:

$$C_i = \begin{bmatrix} 0.6071 \\ 0.7610 \\ 0.3512 \\ 0.5266 \end{bmatrix}, \quad i = 1, 2, 3, 4$$

Based on the closeness coefficients, the alternatives are ranked in descending order:

$$\text{Ranking: } U_2 \succ U_1 \succ U_4 \succ U_3$$

5 Comparative Analysis with Existing Methods

To further validate the effectiveness of the proposed method, its ranking results are compared with several well-established q-ROF-based approaches from the literature.

Table 2: Comparison of rankings between the proposed method and existing q-ROF-based methods.

Method	Ranking order	Best alternative
Proposed method (this study)	$U_2 \succ U_1 \succ U_4 \succ U_3$	U_2
q-ROFWA [18]	$U_2 \succ U_1 \succ U_4 \succ U_3$	U_2
q-ROF TOPSIS [19]	$U_2 \succ U_4 \succ U_1 \succ U_3$	U_2
q-ROF MULTIMOORA (RS) [20]	$U_2 \succ U_1 \succ U_4 \succ U_3$	U_2
Yq-ROFWG [21]	$U_4 \succ U_3 \succ U_1 \succ U_2$	U_4

The comparison of the proposed method with several existing q-ROF-based approaches, as presented in Table 2, clearly demonstrates the validity of the obtained results. It can be observed that the proposed method provides the same ranking order as q-ROFWA and q-ROF MULTIMOORA, both of which have been widely employed in earlier studies. The q-ROF TOPSIS method yields a slightly different intermediate ordering by placing U_4 ahead of U_1, but the best alternative remains consistent (U_2). Interestingly, the Yq-ROFWG approach identifies U_4 as the best alternative, which highlights the sensitivity of different operators in handling q-ROF information. Despite these variations, U_2 emerges as the optimal alternative across most methods, including the proposed approach. This consistency with earlier established methods proves the credibility of the proposed method and confirms that it is a reliable decision-making tool under q-rung orthopair fuzzy environments. Based on these and subsequent findings, this chapter highlights the advantages and novelty of the proposed GA-operator-applied multicriteria analysis procedure specific to q-ROFS. In contrast to traditional fuzzy MCDM techniques, which are heavily dependent on subjective judgments when determining criteria weights, the incorporation of a GA enables objective and data-driven optimization by maximizing the closeness coefficient variance. Similarly, the MARICA approach enhances ranking stability by comparing the alternatives with both positive and negative ideal solutions. The q-ROFS framework adds the capability to model uncertainty and hesitancy in expert judgments, thus providing greater flexibility for our purposes. This amalgamation offers a solid and robust way of making decisions with a high degree of discriminatory power, transparency, and reliability. Overall, comparative study supports the strength and relevance of the proposed method while emphasizing its originality through the integration of evolutionary optimization, comparative analysis, and advanced fuzzy representations within a single MCDM hybrid framework.

6 Conclusion

A hybrid MCDM framework is presented and is based on integrating the q-ROFS environment with the MARICA method and a GA for objective weight optimization. Our method succeeds well in dealing with two challenges caused by decision-making under uncertain situations. The first one is presenting imprecise, lenient, and unclear expert knowledge when needed, and the second one is avoiding the sense of vagueness in the weighting process of each attribute. The application of q-ROFS permits a more comprehensive encoding of information, providing richer semantic data than those of classical and lower-order fuzzy extensions to reflect human judgments more effectively. The MARICA method is a consistent and rank-preserving preference allocation means for alternatives. Additionally, GA is utilized to explore the space of weights to maximize the distances of closeness coefficients, resulting in more distinctive discrimination between alternatives. The transparency, strength, and practical nature of our formulation in decision-making are demonstrated by numerically working through a problem involving four alternatives and five criteria. The enhanced separation among the closeness coefficients computed by GA-aware weights protects the model from potentially producing a mild judgment on the final outcome. Comprehensive comparison with q-ROF-based validators such as q-ROFWA, q-ROF TOPSIS, q-ROF MULTIMOORA, and Yq-ROFWG supports the viability of our approach. The majority of the cases are consistent with literature, as the recommended alternative is U_2, thus giving confidence in the proposed technique. The minor differences with some operators demonstrate the model's flexibility and the importance of choosing a structure depending on the issue under study. As a result of the study, three new suggestions are made. The first is given by the q-ROFS-based MARICA distance model, followed by the second suggestion, which provides a GA-directed nonsubjective weight scheme, thus avoiding subjective bias, and then an experimental case study validating the proposed framework. Future work may extend the proposed framework to larger problems of decision-making in dynamic environments or integrate it with other optimization techniques like metaheuristics. The study outcome is that the integrated approach is a fair and critical decision-assisting instrument for complex setups in actual environments.

References

[1] Zadeh LA. Fuzzy sets. Inf Control. 1965;8(3):338–353.

[2] Atanassov KT. Intuitionistic fuzzy sets. Fuzzy Sets Syst. 1986;20(1):87–96.

[3] Yager RR. Pythagorean membership grades in multicriteria decision making. IEEE Trans Fuzzy Syst. 2013;22(4):958–965.

[4] Immanuel SD, Chakraborty UK, Genetic algorithm: An approach on optimization. In: 2019 international conference on communication and electronics systems (ICCES), IEEE, 2019, pp. 701–708.

[5] Kamalinia S, Afsharnia S, Khodayar M, Rahimikian A, Sharbafi M, A combination of MADM and genetic algorithm for optimal dg allocation in power systems. In: 2007 42nd International Universities Power Engineering Conference. IEEE, 2007, pp. 1031–1035.

[6] Nestic S, Djordjevic A, Puskaric H, Djordjevic MZ, Tadic D, Stefanovic M. The evaluation and improvement of process quality by using the fuzzy sets theory and genetic algorithm approach. J Intell Fuzzy Syst. 2015;29(5):2017–2028.

[7] Choudhary A, Nizamuddin M, Sachan VK. A hybrid fuzzy-genetic algorithm for performance optimization of cyber physical wireless body area networks. Int J Fuzzy Syst. 2020;22(2):548–569.

[8] Roshandel S, Karimi Govareshaki MH, Abbasi M. Assessing strategic alliance maturity through a weighted fuzzy inference system: Perspectives from data envelopment analysis and genetic algorithm. Ind Manag J. 2024;16(4):666–699.

[9] Raut PK, Pramanik S, Mohapatra DK, Sahoo SK. Solving the shortest path based on the traveling salesman problem with a genetic algorithm in a Fermatean neutrosophic environment. Neutrosophic Sets Syst. 2025;78:353–366.

[10] Pamučar D, Ćirović G. The selection of transport and handling resources in logistics centers using multiattributive border approximation area comparison (MABAC). Expert Syst Appl. 2015;42(6):3016–3028.

[11] Cheng R, Fan J, Wu M, Seiti H. A large-scale multi-attribute group decision-making method with rnumbers and its application to hydrogen fuel cell logistics path selection. Complex Intell Syst. 2024;10(4):5213–5260.

[12] Özta¸s T, Özta¸s GZ. Innovation performance analysis of g20 countries: A novel integrated LOPCOW-MAIRCA MCDM approach including the COVID-19 period. Verimlilik Dergisi. 2024;1–20.

[13] Sedighi Khavidak S, Hashemkhani Zolfani S, Nemati A. Selection of portable hard disk drive using ENTROPY-COBRA-MARCOS-MAIRCA: A case of the iranian market. Decis Mak Artif Intell Trends. 2024;1(1):52–76.

[14] Teši´c D, Delibaši´c B, Božani´c D, Loji´c R, Pamuˇcar D, Balassa BE. Application of the FUCOM-FUZZY MAIRCA model in human resource management. Acta Polytech Hung. 2023;20(3):231–249.

[15] Seikh MR, Mandal U. q-Rung orthopair fuzzy Archimedean aggregation operators: application in the site selection for software operating units. Symmetry. 2023;15(9):1680.

[16] Yager RR. Generalized orthopair fuzzy sets. IEEE Trans Fuzzy Syst. 2016;25(5):1222–1230.

[17] Liu D, Chen X, Peng D. Some cosine similarity measures and distance measures between q-rung orthopair fuzzy sets. Int J Intell Syst. 2019;34(7):1572–1587.

[18] Yager RR. Generalized orthopair fuzzy sets. IEEE Trans Fuzzy Syst. 2017;25(5):1222–1230.

[19] Alkan N, Kahraman C. Evaluation of government strategies against COVID-19 pandemic using q-rung orthopair fuzzy topsis method. Appl Soft Comput. 2021;110:107653.

[20] Liu H-C, Zhang L-J, Ping Y-J, Wang L. Failure mode and effects analysis for proactive healthcare risk evaluation: A systematic literature review. J Eval Clin Pract. 2020;26(4):1320–1337.

[21] Liu J, Mai J, Li H, Huang B, Liu Y. On three perspectives for deriving three-way decision with linguistic intuitionistic fuzzy information. Inf Sci. 2022;588:350–380.

About the Editors

Dr Kamal Kumar is listed in the top 2% of cited scientists list of the Stanford University and Elsevier B.V. He is an assistant professor at Amity University Haryana, Gurugram, India. He received his doctoral degree from Thapar Institute of Engineering & Technology Patiala, Punjab, India, in 2020. He has more than 9 years of teaching experience in different subjects of engineering mathematics and statistics. His current research interests are in fuzzy decision-making, aggregation operators, soft computing, and uncertainty theory. He has published more than 65 international articles in different reputed SCI journals such as *The Lancet*, *IEEE Transactions on Fuzzy Systems*, etc. He has published an edited book in CRC Press. He has also published five book chapters and three patents. His Google citation is over 2,700 with an h-index of 25 and i10 index of 30. He is an associate editor in the journals *PLOS One*, *Journal of Applied Mathematics*, and *International Journal of Mathematical Engineering and Management Sciences*.

Dr Shahid Ahmad Bhat is a postdoctoral researcher at LUT Business School, LUT University, Finland, focusing on research and teaching. He earned his Ph.D. in applied mathematics (operations research) from Thapar Institute of Engineering and Technology (TIET), Punjab, India, in collaboration with the Indian Institute of Technology (IIT) Roorkee. His research specializes in developing advanced multiattribute decision-making (MADM) methods in fuzzy environments. Previously, he worked as a postdoctoral fellow at the United Arab Emirates University, contributing to sustainable and resilient global supply chain models during the COVID-19 pandemic. His research interests include supply chain sustainability using AI, optimization techniques, fuzzy mathematical modeling, and MCMD. He has published in leading journals like *Engineering Applications of Artificial Intelligence and Applied Soft Computing*, *Expert Systems with Applications*, *Resources Policy*, *Knowledge-Based System*, and *IEEE Access*, with an h-index of 11.

Dr Paul Augustine Ejegwa is an experienced mathematics educator and researcher with specialization in fuzzy mathematics, decision-making, and computational intelligence. He is an assistant professor of Mathematics at Joseph Sarwuan Tarka University, Makurdi, Nigeria. He has published over 100 scholarly works, including journal articles, book chapters, and conference proceedings indexed in ISI and Scopus. He is listed among the top 2% of most-cited scientists by Stanford University and Elsevier B.V. He serves as an editorial board member and reviewer for several reputable journals, including *Heliyon*, *Mathematics*, and *AIMS Mathematics*. He is a member of the Asian Council of Science Editors, the International Society of Fuzzy Sets Extensions and Applications, the Teachers' Registration Council of Nigeria, the Nigerian Mathematical Society, and the MathTech Thinking Foundation.

Dr Reeta Bhardwaj is an assistant professor in the Department of Mathematics at the Amity School of Applied Sciences, Amity University Haryana, India. With a PhD in mathematics, specializing in stochastic and fuzzy modeling from Amity University Haryana, she has developed expertise in decision-making algorithms and techniques using fuzzy sets and logic. With over 18 years of experience in teaching and research, she has significantly contributed to the academic community. She has published more than 50 research articles/book chapters/conference proceedings in prestigious SCOPUS/SCI journals, and she has made significant contributions to the field of mathematics, particularly in the areas of queuing theory, fuzzy decision-making and stochastic modeling. In addition to research, she actively contributes to the academic community by reviewing articles for reputed journals.

 | https://doi.org/10.1515/9783112222249-015

Index

 | https://doi.org/10.1515/9783112222249-016

De Gruyter Series on the Applications of Mathematics in Engineering and Information Sciences

Already published in the series

Volume 25
Artificial Intelligence in Cancer: Detection, Diagnosis, and Treatment
Rishabha Malviya, Shivam Rajput, Mukesh Roy, Sathvik Belagodu Sridhar, Sabin Thomas
ISBN 978-3-11-167037-9, e-ISBN (PDF) 978-3-11-167074-4,
e-ISBN (EPUB) 978-3-11-167151-2

Volume 24
Complex Engineering Systems – Modeling and Optimization: Multi-Objective Analysis and Smart Manufacturing Applications
Satyvir Singh, Mukesh Kumar Awasthi (Eds.)
ISBN 978-3-11-172340-2, e-ISBN (PDF) 978-3-11-172346-4,
e-ISBN (EPUB) 978-3-11-172359-4

Volume 23
Fuzzy Clustering Algorithms: Symbiotic Relationship Between Mathematics and Similarity Measures
Rakesh Kumar, Geeta Arora
ISBN 978-3-11-149238-4, e-ISBN (PDF) 978-3-11-149427-2,
e-ISBN (EPUB) 978-3-11-149576-7

Volume 22
Metaverse and Digital Twins: Blockchain and Healthcare Technology
Bhupinder Singh, Rishabha Malviya, Christian Kaunert (Eds.)
ISBN 978-3-11-163629-0, e-ISBN (PDF) 978-3-11-164689-3,
e-ISBN (EPUB) 978-3-11-164464-6

Volume 21
Reengineering Cyber Security Process: Quality Management
Mehrdad Sepehri Sharbaf
ISBN 978-3-11-144382-9, e-ISBN (PDF) 978-3-11-144463-5,
e-ISBN (EPUB) 978-3-11-144480-2

Volume 20
Digital Blockchain: Big Data, Artificial Intelligence, and Virtual Reality in Healthcare Ecosystem
Rishabha Malviya, Shristy Verma, Sonali Sundram, Harshil Shah
ISBN 978-3-11-157386-1, e-ISBN (PDF) 978-3-11-157428-8,
e-ISBN (EPUB) 978-3-11-157485-1

www.degruyterbrill.com

Volume 19
Artificial Intelligence for Healthcare: Machine Learning and Diagnostics
Rishabha Malviya, Shivam Rajput, Mukesh Roy, Irfan Ahmad, Saurabh Srivastava
ISBN 978-3-11-154758-9, e-ISBN (PDF) 978-3-11-154972-9,
e-ISBN (EPUB) 978-3-11-154984-2

Volume 18
Operational Perspective of Modeling System Reliability: Research Tools for System Dynamics
Deepti Aggrawal, Adarsh Anand, Yoshinobu Tamura, Mohini Agarwal (Eds.)
ISBN 978-3-11-147540-0, e-ISBN (PDF) 978-3-11-147610-0,
e-ISBN (EPUB) 978-3-11-147713-8

Volume 17
Algorithms: Big Data, Optimization Techniques, Cyber Security
Sushil C. Dimri, Abhay Saxena, Bhuvan Unhelkar and Akshay Kumar (Eds.)
ISBN 978-3-11-122800-6, e-ISBN (PDF) 978-3-11-122915-7,
e-ISBN (EPUB) 978-3-11-122963-8

Volume 16
Distributed Transfer Function Method: One-Dimensional Problems in Engineering
Bingen Yang, Kyoungrae Noh
ISBN 978-3-11-075854-2, e-ISBN (PDF) 978-3-11-075893-1,
e-ISBN (EPUB) 978-3-11-075900-6

Volume 15
Machine Learning for Cyber Security
Malik Preeti, Nautiyal Lata, Ram Mangey (Eds.)
ISBN 978-3-11-076673-8, e-ISBN (PDF) 978-3-11-076674-5,
e-ISBN (EPUB) 978-3-11-076676-9

Volume 14
Multiple Criteria Decision-Making Methods: Applications for Managerial Discretion
Mohini Agarwal, Adarsh Anand, Deepti Aggrawal
ISBN 978-3-11-074356-2, e-ISBN (PDF) 978-3-11-074363-0,
e-ISBN (EPUB) 978-3-11-074371-6

Volume 13
Integral Transforms and Applications
Nita H. Shah, Monika K. Naik
ISBN 978-3-11-079282-9, e-ISBN (PDF) 978-3-11-079285-0,
e-ISBN (EPUB) 978-3-11-079292-8

Volume 12
Noise Filtering for Big Data Analytics
Souvik Bhattacharyya, Koushik Ghosh (Eds.)
ISBN 978-3-11-069709-4, e-ISBN (PDF) 978-3-11-069721-6,
e-ISBN (EPUB) 978-3-11-069726-1

Volume 11
Artificial Intelligence for Signal Processing and Wireless Communication
Abhinav Sharma, Arpit Jain, Ashwini Kumar Arya, Mangey Ram (Eds.)
ISBN 978-3-11-073882-7, e-ISBN (PDF) 978-3-11-073465-2,
e-ISBN (EPUB) 978-3-11-073472-0

Volume 10
Meta-heuristic Optimization Techniques: Applications in Engineering
Anuj Kumar, Sangeeta Pant, Mangey Ram, Om Yadav (Eds.)
ISBN 978-3-11-071617-7, e-ISBN (PDF) 978-3-11-071621-4,
e-ISBN (EPUB) 978-3-11-071625-2

Volume 9
Linear Integer Programming: Theory, Applications, Recent Developments
Elias Munapo, Santosh Kumar
ISBN 978-3-11-070292-7, e-ISBN (PDF) 978-3-11-070302-3
e-ISBN (EPUB) 978-3-11-070311-5

Volume 8
Mathematics for Reliability Engineering: Modern Concepts and Applications
Mangey Ram, Liudong Xing (Eds.)
ISBN 978-3-11-072556-8, e-ISBN (PDF) 978-3-11-072563-6
e-ISBN (EPUB) 978-3-11-072559-9

Volume 7
Mathematical Fluid Mechanics: Advances on Convection Instabilities and Incompressible Fluid Flow
B. Mahanthesh (Ed.)
ISBN 978-3-11-069603-5, e-ISBN (PDF) 978-3-11-069608-0
e-ISBN (EPUB) 978-3-11-069612-7

Volume 6
Distributed Denial of Service Attacks: Concepts, Mathematical and Cryptographic Solutions
Rajeev Singh, Mangey Ram (Eds.)
ISBN 978-3-11-061675-0, e-ISBN (PDF) 978-3-11-061975-1
e-ISBN (EPUB) 978-3-11-061985-0

Volume 5
Systems Reliability Engineering: Modeling and Performance Improvement
Amit Kumar, Mangey Ram (Eds.)
ISBN 978-3-11-060454-2, e-ISBN (PDF) 978-3-11-061737-5
e-ISBN (EPUB) 978-3-11-061754-2

Volume 4
Systems Performance Modeling
Adarsh Anand, Mangey Ram (Eds.)
ISBN 978-3-11-060450-4, e-ISBN (PDF) 978-3-11-061905-8
e-ISBN (EPUB) 978-3-11-060763-5

Volume 3
Computational Intelligence: Theoretical Advances and Advanced Applications
Dinesh C. S. Bisht, Mangey Ram (Eds.)
ISBN 978-3-11-065524-7, e-ISBN (PDF) 978-3-11-067135-3
e-ISBN (EPUB) 978-3-11-066833-9

Volume 2
Supply Chain Sustainability: Modeling and Innovative Research Frameworks
Sachin Kumar Mangla, Mangey Ram (Eds.)
ISBN 978-3-11-062556-1, e-ISBN (PDF) 978-3-11-062859-3,
e-ISBN (EPUB) 978-3-11-062568-4

Volume 1
Soft Computing: Techniques in Engineering Sciences
Mangey Ram, Suraj B. Singh (Eds.)
ISBN 978-3-11-062560-8, e-ISBN (PDF) 978-3-11-062861-6,
e-ISBN (EPUB) 978-3-11-062571-4